Noise in Electron Devices

Technology Press Books
in Science and Engineering

EDITED BY

Louis D. Smullin

and

Hermann A. Haus

Noise in Electron Devices

The M.I.T. Press

Cambridge, Massachusetts

Preface

Noise: Any undesired sound. By extension, noise is any unwanted disturbance within a useful frequency band, such as undesired electric waves in any transmission channel or device. Such disturbances when produced by other services are called interference.

*The International Dictionary
of Physics and Electronics.*
D. Van Nostrand Company

In the summer of 1955, a special two-week course on Noise in Electron Devices was held at the Massachusetts Institute of Technology, Cambridge, Massachusetts. This was one of a series of similar intensive courses in many different fields offered every summer at the Institute. These courses are offered so that technical workers may bring themselves up to date in various specialized fields. The lectures are given by specialists drawn from industry and campus.

In planning this course, it was felt that significant advances in both the qualitative and quantitative understanding of noise in thermionic and solid-state electron devices had been made within the last few years. These advances had been so rapid that only a handful of specialists in each of the subfields were conversant with the status of their various specialties. Thus, it was felt that a useful purpose would be served by presenting and comparing the various treatments of noise.

The history of the study of noise in electron devices is almost as old as the devices themselves. The first quantitative discussion of shot noise was due to Schottky in 1918. The next big steps were made in the 1930's by North, Rack, and Spenke, in which the space-charge reduction of noise in thermionic devices was explained and the effects of finite transit time pointed out. However, the motivation for the development of very low noise amplifiers was not great. Radio communication in the broadcast and high-frequency bands had little need for supersensitive receivers, since the limiting noise was generated outside the receiver by atmospherics and static. Only as radio communications pushed up in frequency toward the VHF range and beyond, where atmospherics were unimportant, did low-noise receiver design become urgent. Similarly, until the advent of the modern age of servomechanisms and instrumentation, there was little need for low-noise amplifiers at low frequencies, except in the research laboratory. The Second World War with its explosive expansion of the frequency spectrum caused by developments in radio, radar, and automatic weapon control and direction ushered in a much increased interest in noise phenomena and in the design of less noisy devices.

As even a casual reading of the ensuing pages will indicate, it is not yet possible to describe all noise phenomena by one elegant equation. Some of the differences encountered in the various theories are due to the differences between the devices. The physical structures of traveling-wave tubes, triodes, and transistors are so different that the mathematical description of the methods of current control in these devices have little similarity. This, plus the fact that the physical noise processes are not identical, makes it impossible to present a single, unified treatment of noise. Further, shot noise in electron streams appears to be a different phenomenon from flicker (f^{-1}) noise, etc. In spite of many differences like these, it will become evident on reading the following pages that many mathematical techniques and physical concepts are used in common in discussing noise in the various devices.

To the user, nearly all kinds of noise are equally annoying, since they interfere with the detection of desired signals. Yet not all noises are of equal fundamental importance. Consider, for example, microphonics and induced hum. In principle, these can be reduced to arbitrarily small levels by proper design. On the other hand, Johnson noise, shot noise, and $1/f$ noise appear to have fundamental lower limits. Only the latter types of noise are discussed here. In addition to their fundamental nature, these noise types share a very important feature. They are all random, stationary functions of time. The mathematical tools for treating such functions are well developed and are directly applicable to the analysis of noise.

The electric devices in which noise is of practical interest include nearly all components: resistors (metallic and composition), capacitors, iron-core inductors, wire connectors, solid-state (crystal) diodes, amplifiers (thermionic and solid state), oscillators, electromechanical transducers, electro-optical transducers, and electro-acoustic transducers. According to the fashions or problems of the day, something on noise in any or all of these devices is almost always in the literature. In the strictest sense, all these are "electronic" devices. However, we shall select from these a very small subgroup for discussion, in accordance with the more conventional usage of the words "electron devices."

The noise that occurs in a sensitive amplifier or transducer serves to mask weak signals and to limit the ultimate usable sensitivity of the device. On the other hand, noise may also occur in large-signal devices, amplifiers, or oscillators. In these devices, the noise output appears as a modulation of the desired signals and is usually proportional to the desired signal strength. It is undesirable, since it may hide small desired modulations or introduce cross talk in multichannel systems. Of these two classes, only the former, small-signal amplifier noise, is discussed here. Relatively little of a fundamental or quantitative nature is known about oscillator and large-signal noise.

Within the class of small-signal amplifiers and transducers, we have chosen to discuss only small-signal electronic and solid-state amplifiers of electric signals: the familiar vacuum tube (at all frequencies), transistors, and the thermionic and solid-state diodes (which are, of course, not amplifiers but are essential building blocks of corresponding amplifier types). Thus, such technically important devices as iconoscopes, orthicons, image amplifiers, etc., are not discussed here at all. This choice was made because of the limited time available for the summer course. In that short time it was only possible to cover a few subjects if the presentation was to consist of more than a series of casual descriptions. The particular selection was made to present as nearly as possible a unified picture. It was felt that this presentation would serve as a background for the more complex noise phenomena in tubes like the orthicon, comprising both thermionic and solid-state elements.

The topics included in this book correspond to the organization of the summer course. They cover the general problem of noise due to thermionic emission, the general circuit aspect of noise in microwave tubes, some of the detailed engineering solutions to the problems encountered in the design of low-noise traveling-wave tubes and space-charge control tubes, semiconductor noise, noise behavior of semiconductor diodes and transistors, and principles of low-noise transistor circuit design.

The noise sources in thermionic and solid-state devices may be divided into two basic classes:

1. *Shot noise, Johnson noise, partition noise,* etc., with mechanisms that may be said to be basically understood, and whose difficulties are wholly mathematical in nature.

2. *Flicker noise* whose true physical nature(s) is (are) not yet fully
   understood. Active research is being carried on to correlate
   the effect of tube-parameter variations, material variations,
   etc., on the amount and kind of noise generated, so that it may
   ultimately be possible to relate the noise to a few fundamental
   physical processes.

The remarks about the first group, shot noise, etc., must not be
interpreted as meaning that no major work in this area remains to be
done. These noise processes are "understood" in the sense that we
know their basic sources. They are the random emission of electrons
from a cathode and, in a solid, the random creation and destruction
of carriers. The noise may be subsequently modified by coulomb
interaction among the electrons. At least in thermionic devices no
other basic physical processes appear to be involved. Still, there are
many important problems that are only vaguely understood because
of the extreme complexity of the available mathematical approaches.
As a result, quantitative theoretical solutions can only be found after
the problem has been grossly idealized. We must then fall back upon
experiment to determine whether the approximations made were
legitimate and to disclose new modes of behavior that the mathematics
had not yet uncovered.

An important tool that may well have a strong influence on future
research is the large-scale digital computer. The pioneering work of
Tien, in the digital computation of the high-frequency reduction of
shot noise by space charge, is probably only an example of what will
be done in this area as engineers become more familiar with the
capabilities of large-scale computers. Among the problems that may
be solved by digital computation are those of the nonlinear behavior
in the region of the potential minimum, the effect of nonlaminar
flow in electron beams, and the effect of finite-beam cross section on
noise, etc.

At the present state of development we are able to predict the best
noise figure that conventional thermionic amplifiers can have, once
their internal noise sources are specified; and, indeed, we are able to
calculate on paper, at least in principle, just how this optimization
should be carried out. Thus, it might appear that in this area there
is little further room for experimentation. Actually, of course, much
work remains to be done in the design of circuits that will attain the
minimum noise figure over the widest band of frequencies and over
the widest range of operating parameters. However, even more
importantly, a great deal remains to be done that must by nature be

experimental in the investigation of the fundamental noise processes themselves. Thus, we do not yet know exactly what the ultimate noise figure of a beam-type device using a thermionic cathode must be. Several years ago, using the assumption that the current-modulation noise and velocity-modulation noise at the cathode were uncorrelated, it was predicted that the ultimate noise figure of a traveling-wave tube would be about 6 db. Since then tubes have been produced with noise figures of less than 4 db. It is still not known whether this reduction of noise is due to the existence of correlation, or whether the intensity of the noise sources is just much less than is predicted. Until new solid theoretical bases are established for the computation of an ultimate figure, it hardly behooves one to rest content with noise figures even as low as 4 db. Therefore, active experimentation probably will continue on the design of beam-acceleration circuits, external circuits, and cathode materials, in the hope that another decibel or two may be pared from the already rather good performance obtained.

The transistor amplifier has come a very long way in the few years of its existence. At low frequencies, transistors already have a noise performance comparable to the best thermionic amplifiers. At the higher frequencies, beyond a few megacycles, their ability to amplify is still limited, so that it is not yet possible to discuss their ultimate noise performance in this range. So much of the success of transistor design is bound up with the quality of the materials used and the exact geometric configuration of the junctions or rectifying contacts that no serious prediction of ultimate noise performance can be made at this early date.

A volume on noise in electron devices with this publication date should at least mention maser amplifiers. The ammonia and the solid-state masers give promise of being essentially noiseless amplifiers. In fact, recent measurements* have indicated that the noise contributed by the ammonia beam corresponds to an effective temperature of only a few tens of degrees absolute. The only measurable noise appeared to be that due to the Johnson noise of the circuit, which was not supercooled. Solid-state maser amplifiers have shown equally low-noise behavior, with the promise of considerably greater bandwidth than the ammonia maser.† Thus, within the not too distant future, it may be possible to construct amplifying devices

---

* Private communication, J. C. Helmer and M. W. Muller, Varian Associates, Inc.

† Private communication, J. W. Meyer and A. L. McWhorter, M.I.T., Lincoln Laboratory.

having effective noise temperatures only a few degrees Kelvin. At present it appears that these devices will have to operate at temperatures of liquid helium, so that for many applications they will undoubtedly not supplant the more conventional amplifiers based on thermionically emitting cathodes or on the flow of electrons and holes. It will probably turn out that each of these classes of devices will find its own particular niche wherein it combines the optimum in economy, performance, and cost for the particular application. Similarly, the parametric amplifier has already demonstrated its low-noise performance over most of the UHF band and even at microwave frequencies. The most important types are those using electron beams, or variable-capacitance, solid-state diodes. The detailed analysis of such devices has been carried out along the basic principles described here in conjunction with electron-beam and junction-diode devices. Thus, although the gain mechanism of these devices (parametric amplification) is not presented, the noise mechanisms are precisely the same as those discussed in connection with electron beams and junction diodes.

It is hoped that the material presented will bring the student to the point of understanding the noise phenomena in the devices actually discussed, and that the application of the basic analytical tools to other devices will not be too diffiult a task. A word of warning must be inserted here: in all science and technology there is art. Thus, a familiarity with the theoretical behavior of noise on electron beams does not insure that one can build quiet traveling-wave tubes. There is a tremendous area of "know-how" and of neighboring scientific disciplines that must be brought to bear on the problem of building low-noise devices. The preparation of this book will have been worth while if its study will permit one to *analyze* the noise performance of some given devices, or to translate the requirements for low-noise performance into detailed specifications for the various components of the device to be designed, and finally if it stimulates a few individuals to study any of the many unsolved scientific problems that still face us in this field.

This book is a product of many individual authors. Therefore, the treatment is by no means uniform in emphasis. The editors have tried, insofar as it is possible to use a unified notation throughout the volume. However, in many cases, usage in one subfield has already established a nomenclature or terminology which is different from that in an adjoining subfield. Symbols are defined as they are introduced. It is hoped that this will cause the reader no inconvenience, and, if it does, we, the editors, can only apologize. Mks units are used in all theoretical derivations, but in many practical examples constants are evaluated in terms of centimeters and gausses, and are so indicated. Proper bibliographic credit is always a difficult matter. Most of the references quoted here are from the United States and British literature. This is partly an indication of the various authors' reading habits, and no claim to completeness is made. It is a pleasure to acknowledge the patient and exacting editorial work of Miss Constance D. Boyd of the M.I.T. Technology Press and the careful secretarial work of Miss Joan Dordoni in the preparation of the manuscript.

December 1958   Louis D. Smullin
                                                             Hermann A. Haus

Contributors

Louis D. Smullin   *Associate Professor of Electrical Engineering*
Research Laboratory of Electronics
Massachusetts Institute of Technology

Hermann A. Haus   *Associate Professor of Electrical Engineering*
Research Laboratory of Electronics
Massachusetts Institute of Technology

Thomas E. Talpey   *Member of the Technical Staff*
Bell Telephone Laboratories, Inc.
Murray Hill, New Jersey

C. F. Quate   *Member of the Technical Staff*
Bell Telephone Laboratories, Inc.
Murray Hill, New Jersey

W. H. Fonger   *Member of the Technical Staff*
RCA Laboratories
Princeton, New Jersey

Aldert van der Ziel   *Professor of Electrical Engineering*
University of Minnesota

Rolf W. Peter   *Director, Physical and Chemical*
Research Laboratory
David Sarnoff Research Center
Princeton, New Jersey

Contents

CHAPTER 1

Shot Noise from Thermionic Cathodes

C. F. Quate

Introduction

In the majority of RF amplifiers that make use of electron beams, the random emission of electrons at the cathode surface is the primary source of noise. In these amplifiers the cathode may be operated either in the "temperature-limited" region, wherein all emitted electrons are drawn to the anode, or in the "space-charge-limited" region, wherein only a small fraction of the emitted electrons are drawn to the anode. In the temperature-limited type the noise properties of the beam consist of a noise current which is pure shot noise and a noise velocity which is calculated as the deviation from the mean value of the emission velocities, described by the Maxwellian distribution. In space-charge-limited operation the velocity and current noises are modified by the action of the potential minimum region. A large part of the following discussion will be devoted to this phenomenon.

We will treat only the problem of a one-dimensional diode consisting of two parallel planes of infinite extent—one plane being the cathode which emits the electrons and the second plane being the anode which is maintained at a positive potential with respect to the cathode. The d-c, or steady-state, conditions for space-charge-limited flow for this one-dimensional diode will first be considered by following the work of Fry and others.[1-4] This work, together with a discussion of several fluctuation parameters associated with the Maxwellian distribution function, will serve as an introduction to the noise calculations. For simplicity, we will first discuss the noise in space-charge-limited flow by using an approximate "single-velocity" model in order to give some insight into the low-frequency region where the presence of the potential minimum must result in noise current that is less than that of shot noise. After this we will consider the actual multivelocity flow problem with the aid of North's[5] solution, which again is limited to the low-frequency case of short-transit angles from cathode to anode.

At higher frequencies, where the transit angles are no longer small, we make use of the equations for one-dimensional flow as given by Llewellyn and Peterson.[6] These equations are limited to problems of electron flow with only a single velocity at one given point in the beam at a given instant of time. In the region near the potential minimum, the spread in transit time between electrons of different initial velocities may be an appreciable part of an RF cycle and the single-velocity approximations are severely strained. Because of the multivelocity character of the beam of this critical region, no rigorous analysis of the noise has yet been obtained. The single-velocity equations are valid, however, if we accept as the input boundary a plane beyond the potential minimum where the electrons have reached an average velocity several times the value of their initial velocity owing to thermal energy. The problem then becomes one of determining the noise fluctuations in velocity and current at this input plane beyond which the single-velocity equations can be used. Some approximations for the input conditions have been obtained by Robinson,[7] Watkins,[8] and Whinnery.[9] We will discuss part of this work and show that it leads to answers that are reasonably close to those obtained from experiments. It is believed, however, that the rigorous solution must be obtained with the aid of a computer. Tien has worked out one problem with this approach, using the "Monte Carlo" method, and a summary of his results will be presented as the conclusion to this chapter.

1. Steady-State Conditions for Space-Charge-Limited Flow

Single-Velocity Approximation

The simplest model that can be visualized for space-charge-limited flow in a plane-parallel diode is that used by Child. He neglected the initial velocities of the electrons and used the "single-velocity" model to obtain the three-halves power law of current versus voltage. We shall write down the steps leading to Child's law, since the more rigorous treatment, which includes the multivelocity flow, will proceed in a similar fashion. It will also be evident that in many cases the equation is a sufficiently good approximation for the potential distribution in a diode.

We solve the one-dimensional Poisson equation by specifying that both the potential and the potential gradient are zero at the input which is located at the cathode surface.

$$\frac{d^2V}{dx^2} = -\frac{\rho}{\epsilon_0} = \frac{J_0}{\epsilon_0 v} \tag{1.1}$$

where J_0 is the current density, ρ is the charge density given by $-J/v$, ϵ_0 is the dielectric constant of free space, and v is the velocity of the electron at point x. The energy equation tells us that

$$v^2 = -2\frac{e}{m}V \tag{1.2}$$

where e, a negative number, represents the charge of the electron and m denotes the mass.

If we multiply Eq. 1.1 by $2(dV/dx)$ and use Eq. 1.2, we obtain, after one integration,

$$\frac{dV}{dx} = \left(\frac{4J_0}{\epsilon_0\sqrt{\dfrac{-2e}{m}}}\right)^{\frac{1}{2}} V^{\frac{1}{4}} \tag{1.3}$$

This can be written after a second integration:

$$J_0 = \frac{4\epsilon_0}{9}\sqrt{\frac{-2e}{m}}\frac{V^{\frac{3}{2}}}{x^2} = 2.33 \times 10^{-6}\frac{V^{\frac{3}{2}}}{x^2} \tag{1.4}$$

Equation 1.4 is the familiar expression for Child's law, which describes the variation of current in a model of a diode in which the electrons have only a single velocity at a given point, and in which they leave the cathode surface with zero initial velocity.

Multivelocity Flow

In an actual diode the thermal energy of the electrons at the cathode surface imparts initial velocities to the electrons that are distributed according to Maxwell's law, and, as a consequence of this multivelocity type of flow, there must be a negative potential gradient at the cathode. If it were otherwise, all the emitted electrons would reach the anode, and the beam current would not be controlled by the anode voltage. The potential profile in the diode must therefore be somewhat as pictured in Fig. 1.1. There is a negative gradient of potential between the cathode and the minimum which is located at x_m and has a potential V_m.

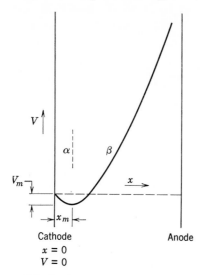

Fig. 1.1. Potential distribution in parallel-plane diode with finite emission velocity.

We designate the initial velocity of emission at the cathode, which is normal to the cathode surface, as v_s, and the velocity at an arbitrary point x by v. The value of v is a function of emission velocity v_s. We shall denote by v_m the velocity at the cathode surface of that group of electrons which just reach the potential minimum. The number of electrons emitted per unit area and unit time in the velocity range from v_s to $v_s + dv_s$ may be called $n(v_s)\, dv_s$. We see that $n(v_s)$ is the distribution function giving the number of electrons in each velocity class. As one simple application of this distribution function we may calculate the total number of electrons per unit area per unit time N from the relation

$$N = \int_0^\infty n(v_s)\, dv_s \qquad (1.5)$$

For our diode problem we must evaluate the charge density in order to apply Poisson's equation. At point x where the electrons that left the cathode with velocity v_s travel at a velocity v we have

$$d\rho = \frac{e\, n(v_s)\, dv_s}{v} \qquad (1.6)$$

For the total charge density we must integrate Eq. 1.6 between the proper limits. In the region between the minimum and the anode, which will be called the β region, we will find only those electrons that had an initial velocity great enough to overcome the retarding field of the potential minimum. Thus we must integrate Eq. 1.6 from v_m through ∞ to obtain in the β region

$$\rho_\beta = e \int_{v_m}^{\infty} \frac{n(v_s)\, dv_s}{v} \tag{1.7}$$

In the region between the cathode and potential minimum, called the α region, at point x the electrons moving away from the cathode contain all velocity groups v_s from v_x to ∞, where v_x is the velocity at the cathode of an electron that just reaches point x ($0 < x < x_m$). The group that has been returned and is now moving toward the cathode had initial velocities from v_x to v_m. We can then write

$$\rho_\alpha = e \int_{v_x}^{\infty} \frac{n(v_s)\, dv_s}{v} + e \int_{v_x}^{v_m} \frac{n(v_s)\, dv_s}{v}$$

which in turn can be written

$$\rho_\alpha = 2e \int_{v_x}^{v_m} \frac{n(v_s)\, dv_s}{v} + e \int_{v_m}^{\infty} \frac{n(v_s)}{v}\, dv_s \tag{1.8}$$

We now use Poisson's equation, and with Eq. 1.7 we can write for the potential in the β region

$$\frac{d^2 V}{dx^2} = -\frac{e}{\epsilon_0} \int_{v_m}^{\infty} \frac{n(v_s)\, dv_s}{v} \tag{1.9}$$

If we multiply both sides of Eq. 1.9 by $2(dV/dx)$, we have

$$2 \frac{dV}{dx} \frac{d^2 V}{dx^2} = -\frac{2e}{\epsilon_0} \int_{v_m}^{\infty} \frac{1}{v} \frac{dV}{dx} n(v_s)\, dv_s$$

Since $V = -mv^2/2e + \text{constant}$, where the constant represents the electron energy at the cathode surface, we can write

$$\frac{d}{dx} \left(\frac{dV}{dx} \right)^2 = \frac{m}{\epsilon_0} \int_{v_m}^{\infty} \frac{1}{v} \frac{d(v^2)}{dx} n(v_s)\, dv_s$$

or, by integrating with respect to x from x_m to x,

$$\left(\frac{dV}{dx} \right)_\beta^2 = \frac{2m}{\epsilon_0} \int_{v_m}^{\infty} n(v_s)(v - v_{ms})\, dv_s \tag{1.10}$$

where v_{ms} is the velocity at the potential minimum of an electron which left the cathode with a velocity v_s.

Similarly for the α region we can integrate from x to x_m to obtain from Eq. 1.8 (noting that the limits of integration are variable)

$$\left(\frac{dV}{dx}\right)_\alpha^2 = \frac{2m}{\epsilon_0}\left[\int_{v_m}^\infty n(v_s)(v - v_{ms})\, dv_s + 2\int_{v_x}^{v_m} v\, n(v_s)\, dv_s\right] \quad (1.11)$$

Properties of the Maxwellian Distribution

It is now necessary to discuss the distribution function $n(v_s)$. This is simply Maxwell's distribution law and can be written

$$n(v_s) = \frac{mN}{kT_c}\, v_s \exp\left(-\frac{mv_s^2}{2kT_c}\right) \quad (1.12)$$

Here k is Boltzmann's constant, and T_c is the cathode temperature in degrees Kelvin.

We wish to compute with this particular distribution of electrons the average velocity $\bar{v}$ at point x in the α region for all electrons moving in the forward direction. It is a quantity that will recur somewhat later, and it can be written

$$\bar{v} = \frac{\displaystyle\int_{v_x}^\infty vv_s \exp\left(-\frac{mv_s^2}{2kT_c}\right) dv_s}{\displaystyle\int_{v_x}^\infty v_s \exp\left(-\frac{mv_s^2}{2kT_c}\right) dv_s} \quad (1.13)$$

The velocity at point x, which is at a potential V, is related to the initial velocity v_s by

$$v^2 = v_s^2 - 2\frac{e}{m}V \quad (1.14)$$

or

$$v\, dv = v_s\, dv_s \quad (1.15)$$

and Eq. 1.13 becomes

$$\bar{v} = \frac{\displaystyle\int_0^\infty v^2 \exp\left[-\frac{m}{2kT_c}\left(v^2 + 2\frac{e}{m}V\right)\right] dv}{\displaystyle\int_0^\infty v \exp\left[-\frac{m}{2kT_c}\left(v^2 + 2\frac{e}{m}V\right)\right] dv} \quad (1.16)$$

which is evaluated with the aid of the integrals

$$\int_0^\infty x e^{-x^2} \, dx = \frac{1}{2} \tag{1.17}$$

and

$$\int_0^\infty x^2 e^{-x^2} \, dx = \frac{\sqrt{\pi}}{4} \tag{1.18}$$

to give

$$\bar{v} = \left(\frac{\pi k T_c}{2m} \right)^{\frac{1}{2}} \tag{1.19}$$

We wish to stress the significance of Eq. 1.19. It tells us that the average velocity of the electrons moving away from the cathode remains constant as we move from the cathode to the potential minimum. This is a consequence of the Maxwellian distribution, together with the fact that the field at the surface of the cathode is negative. This second condition is necessary in order that we may always integrate from 0 to ∞ in the numerator of Eq. 1.16.

It is not difficult to see why the average, Eq. 1.19, should remain constant: Although each electron loses velocity as it moves from the cathode, the more slowly moving electrons are continually sorted and returned to the cathode. Later we shall use this property to show that fluctuations in the magnitude of the potential minimum may smooth out the fluctuations in emitted current but will not change the velocity fluctuations.

The distribution function may equally well be written in terms of the average velocity $\bar{v}$, and in this form it will be useful later in a physical interpretation of some of our noise equations. Thus with Eq. 1.19 we can write

$$n(v_s) = \frac{\pi}{2} \frac{N}{\bar{v}^2} \exp\left(-\frac{\pi}{4} \frac{v_s^2}{\bar{v}^2} \right) \tag{1.20}$$

We can easily see from Eqs. 1.20 or 1.12 that the normalizing factor N is the total number of electrons emitted per unit area per unit time and is given by Eq. 1.5.

Potential Distribution for the Multivelocity Diode

Now, to return to the problem of solving Poisson's equation for the potential distribution, we can combine Eqs. 1.10, 1.11, and 1.12. First, it will prove simpler to change the variable from v_s to v with the aid of Eqs. 1.14 and 1.15.   This gives

$$\left(\frac{dV}{dx}\right)^2 = \frac{2m^2 N}{\epsilon_0 k T_c}\left\{\int_0^\infty v^2 \exp\left[-\frac{m}{2kT_c}\left(v^2 + 2\frac{e}{m}V\right)\right]dv\right.$$

$$-\int_{\sqrt{2\frac{e}{m}(V_m-V)}}^\infty vv_m \exp\left[-\frac{m}{2kT_c}\left(v^2 + 2\frac{e}{m}V\right)\right]dv$$

$$\left.\pm\int_0^{\sqrt{2\frac{e}{m}(V_m-V)}} v^2 \exp\left[-\frac{m}{2kT_c}\left(v^2 + 2\frac{e}{m}V\right)\right]dv\right\} \quad (1.21)$$

The upper sign is to be used in the α region, and the lower sign in the β region.   We note that V_m is a negative number.

We wish to convert to normalized parameters which serve to measure both voltage and distance from the potential minimum. We therefore define

$$\eta = -\frac{e(V - V_m)}{kT_c} = 11{,}605\,\frac{V - V_m}{T_c} \quad (1.22)$$

and

$$\xi = \frac{(2m\pi)^{\frac{1}{4}}}{(kT_c)^{\frac{3}{4}}}\left|\frac{eJ_0}{\epsilon_0}\right|^{\frac{1}{2}}(x - x_m) = 9.19 \times 10^5\,\frac{J_0^{\frac{1}{2}}}{T_c^{\frac{3}{4}}}(x - x_m) \quad (1.23)$$

where J_0 = current density beyond the potential minimum.

We define the error function by the relation

$$\text{erf}\,(x) = \frac{2}{\sqrt{\pi}}\int_0^x e^{-x^2}\,dx \quad (1.24)$$

and note that, for $x \gg 1$, erf $(x) = 1$.

Equation 1.21 can now be evaluated with Eqs. 1.22 and 1.23 to be

$$\left(\frac{d\eta}{d\xi}\right)^2 = e^\eta - 1 \pm e^\eta\,\text{erf}\,\sqrt{\eta} \pm \frac{2}{\sqrt{\pi}}\sqrt{\eta} \quad (1.25)$$

Here again the upper signs apply to the α region between 0 and x_m, and the lower signs apply to the β region beyond x_m.

Equation 1.25 cannot be integrated explicitly, but it has been tabulated,[1-4] with the latest and most complete tables being given in reference 10.   In Fig. 1.2 there is shown a plot of η vs. ξ.   We should note that at the cathode ξ_c approaches a limit of -2.55 for large values of η_c.   In order to determine what is meant by large values of

η_c, let us evaluate the current which is given by

$$J_0 = - e \int_{v_m}^{\infty} n(v_s) \, dv_s \tag{1.26}$$

or

$$J_0 = J \exp \left(- \frac{e V_m}{k T_c} \right) \tag{1.27}$$

where $J = $ current density at cathode. Therefore

$$V_m = - \frac{k T_c}{e} \ln \frac{J_0}{J} = \frac{T_c}{11,605} \ln \frac{J_0}{J} \tag{1.28}$$

and

$$\eta_c = \frac{e V_m}{k T_c} = - \ln \frac{J_0}{J} \tag{1.29}$$

Approximate Solutions of the Multivelocity Flow

In Eq. 1.29 we find that large values of η_c correspond to small values of J_0/J. This then states that the limiting value of $\xi_c = -2.55$ is determined by the requirement that the current density passing through the potential minimum is a small fraction of the emitted current density.

Thus, since η_c is very large for small values of J_0/J, we can replace erf (x) by 1, and Eq. 1.25 has the following approximate form:

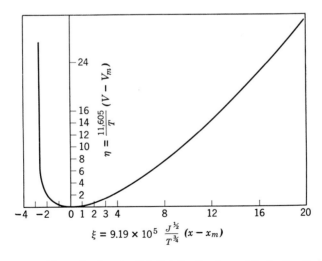

Fig. 1.2. Normalized potential distribution in multivelocity diode.

In the α region at the cathode surface,

$$\left(\frac{d\eta_c}{d\xi_c}\right)^2 = e^{\eta_c} - 1 + e^{\eta_c} \operatorname{erf} \sqrt{\eta_c} - \frac{2}{\sqrt{\pi}} \sqrt{\eta_c} \approx 2e^{\eta_c} \qquad (1.30)$$

In the β region at the anode (if we assume that V_a is large) η_a will be large, and Eq. 1.25 becomes

$$\left(\frac{d\eta_a}{d\xi_a}\right)^2 = e^{\eta_a} - 1 - e^{\eta_a} \operatorname{erf} \sqrt{\eta_a} + \frac{2}{\sqrt{\pi}} \sqrt{\eta_a} \qquad (1.31)$$

$$\approx \frac{2}{\sqrt{\pi}} \sqrt{\eta_a} - 1$$

For positive ξ, and for values of η greater than 8, the solution of Eq. 1.25 can be expressed in the form

$$\xi = 1.255\eta^{3/4} + 1.688\eta^{1/4} - 0.51 - 0.1677\eta^{-1/4} + \cdots \qquad (1.32)$$

If we use the first term of Eq. 1.32, we find from Eqs. 1.22 and 1.23 that

$$J_0 = 2.33 \times 10^{-6} \frac{(V - V_m)^{3/2}}{(x - x_m)^2} \qquad (1.33)$$

Equation 1.33 is the familiar Child's law, Eq. 1.4, for the diode which is derived on a single-velocity basis. However, the potential difference and the diode spacing are now measured from the potential minimum rather than from the cathode.

If we use the first two terms of Eq. 1.32 and note that we must square ξ to obtain the current density, we obtain

$$J_0 = 2.33 \times 10^{-6} \frac{(V - V_m)^{3/2}}{(x - x_m)^2} \left(1 + \frac{2.66}{\sqrt{\eta}}\right) \qquad (1.34)$$

or

$$J_0 = 2.33 \times 10^{-6} \frac{(V - V_m)^{3/2}}{(x - x_m)^2} \left[1 + \frac{0.025 T_c^{1/2}}{(V - V_m)^{1/2}}\right] \qquad (1.35)$$

Langmuir discusses this approximation in reference 4, page 244, and points out that it is an accurate representation of the actual η vs. ξ curve (Fig. 1.2) down to values of $\eta = 1$ with an error in ξ of about 2%. Therefore, in most practical cases it is sufficient to use Eq. 1.35 together with the limiting value of $-\xi_c = 2.55$.

It may give us more insight into Eq. 1.34 and will be helpful in a later discussion of noise if we re-express this equation in terms of the average velocity rather than in terms of cathode temperature.

We can write from Eq. 1.19

$$\frac{kT_c}{e} = \frac{2}{\pi}\frac{m}{e}\bar{v}^2$$

and Eq. 1.35 becomes

$$J_0 = 2.33 \times 10^{-6}\frac{(V - V_m)^{3/2}}{(x - x_m)^2}\left[1 + \frac{2.66\left|\frac{2m}{\pi e}\right|^{1/2}\bar{v}}{(V - V_m)^{1/2}}\right] \tag{1.36}$$

Thus we see from Eq. 1.36 that a finite velocity of emission causes an increase of the current by an amount expressed by the second term on the right.

Dependence of Anode Current on Changes in Emission Current

In later sections, as we discuss noise, we will be interested in the change in anode current caused by fluctuations in the cathode current. With the foregoing equations we can[11] easily derive an expression for the variation of anode current under the following restrictions. We limit ourselves to small values of J_0/J so that Eq. 1.30 is valid, and to large values of anode voltage so that Eq. 1.31 holds. From Eq. 1.23 evaluated at the anode, $x = x_a$,

$$\xi_a = K J_0^{1/2}(x_a - x_m)$$

$$K = \text{constant} = \frac{9.19 \times 10^5}{T_c^{3/4}} \tag{1.37}$$

At the cathode, $x = 0$, we have

$$\xi_c = -KJ_0^{1/2}x_m \tag{1.38}$$

These two equations can be combined to give

$$\xi_a - \xi_c = Kx_aJ_0^{1/2} \tag{1.39}$$

and

$$J_0 = \frac{(\xi_a - \xi_c)^2}{(Kx_a)^2} \tag{1.40}$$

If we differentiate Eq. 1.40 with respect to the emitted current density J, we obtain

$$\frac{\partial J_0}{\partial J} = \frac{2(\xi_a - \xi_c)}{(Kx_a)^2}\left(\frac{\partial \xi_a}{\partial J} - \frac{\partial \xi_c}{\partial J}\right)$$

and, from Eq. 1.22,

$$\frac{\partial \xi_a}{\partial J} = \frac{\partial \xi_a}{\partial \eta_a} \frac{\partial \eta_a}{\partial J} = \frac{\partial \xi_a}{\partial \eta_a} \frac{\partial V_m}{\partial J} \frac{e}{kT_c}$$

and

$$\frac{\partial \xi_c}{\partial J} = \frac{\partial \xi_c}{\partial \eta_c} \frac{\partial \eta_c}{\partial J} = \frac{\partial \xi_c}{\partial \eta_c} \frac{\partial V_m}{\partial J} \frac{e}{kT_c}$$

Therefore, we have

$$\frac{\partial J_0}{\partial J} = \frac{2eJ_0}{kT_c(\xi_a - \xi_c)} \left(\frac{\partial \xi_a}{\partial \eta_a} - \frac{\partial \xi_c}{\partial \eta_c} \right) \frac{\partial V_m}{\partial J} \tag{1.41}$$

Now we see, from Eq. 1.30, that

$$\frac{\partial \xi_c}{\partial \eta_c} = \frac{1}{2} \exp \left(-\frac{1}{2} \eta_c \right)$$

and, from Eq. 1.31,

$$\frac{\partial \xi_a}{\partial \eta_a} = \frac{1}{\left(\frac{2}{\sqrt{\pi}} \sqrt{\eta_a} - 1 \right)^{\frac{1}{2}}}$$

Since η_c is large under our assumptions, we see that

$$\frac{\partial \xi_c}{\partial \eta_c} \ll \frac{\partial \xi_a}{\partial \eta_a}$$

and Eq. 1.41 can be written

$$\frac{\partial J_0}{\partial J} = \frac{2eJ_0}{kT_c(\xi_a - \xi_c)} \frac{\partial \xi_a}{\partial \eta_a} \frac{\partial V_m}{\partial J} \tag{1.42}$$

We can eliminate $\partial V_m / \partial J$ through the use of Eq. 1.27:

$$\frac{\partial J_0}{\partial J} = \exp \left(-\frac{eV_m}{kT_c} \right) - \frac{Je}{kT_c} \exp \left(-\frac{eV_m}{kT_c} \right) \frac{\partial V_m}{\partial J}$$

or

$$\frac{\partial V_m}{\partial J} = \frac{kT_c}{eJ_0} \left(\frac{J_0}{J} - \frac{\partial J_0}{\partial J} \right) \tag{1.43}$$

Equation 1.42 becomes

$$\frac{\partial J_0}{\partial J} \left(1 + \frac{2}{\xi_a - \xi_c} \frac{\partial \xi_a}{\partial \eta_a} \right) = \frac{2J_0}{J(\xi_a - \xi_c)} \frac{\partial \xi_a}{\partial \eta_a} \tag{1.44}$$

The term $\partial\xi_a/\partial\eta_a$ is not convenient to use, and we can easily express this in terms of g, which is defined by

$$g = \frac{\partial J_0}{\partial V_a} = \frac{2(\xi_a - \xi_c)}{(Kx_a)^2}\left[\frac{\partial\xi_a}{\partial\eta_a} - \left(\frac{\partial\xi_a}{\partial\eta_a} - \frac{\partial\xi_c}{\partial\eta_c}\right)\frac{\partial V_m}{\partial V_a}\right]\frac{|e|}{kT_c}$$

$$\approx \frac{2J_0}{\xi_a - \xi_c}\left(\frac{\partial\xi_a}{\partial\eta_a}\right)\left(1 - \frac{\partial V_m}{\partial V_a}\right)\frac{|e|}{kT_c} \qquad (1.45)$$

From Eq. 1.27 we know that

$$g = \frac{eJ_0}{kT_c}\frac{\partial V_m}{\partial V_a} \qquad (1.46)$$

and Eq. 1.45 becomes

$$g = \frac{2J_0}{\xi_a - \xi_c}\left(\frac{\partial\xi_a}{\partial\eta_a}\right)\left(1 - \frac{kT_c}{e}\frac{g}{J_0}\right)\frac{|e|}{kT_c}$$

Since $\partial V_m/\partial V_a \ll 1$, we find for the above, with the aid of Eq. 1.46,

$$g\left[1 + \frac{2}{\xi_a - \xi_c}\left(\frac{\partial\xi_a}{\partial\eta_a}\right)\right] = \frac{2J_0}{\xi_a - \xi_c}\left(\frac{\partial\xi_a}{\partial\eta_a}\right)\frac{|e|}{kT_c} \qquad (1.47)$$

Equation 1.47 can be combined with Eq. 1.44 to give

$$\frac{\partial J_0}{\partial J} = \frac{kT_c}{|e|J}g \qquad (1.48)$$

Numerical Example for a Typical Diode

To conclude this section we will consider a numerical example to determine the order of magnitude of some of the quantities that we have been discussing. Consider a cathode operating at $1100°\,\mathrm{K}$ with a cathode emission of 2 amperes per cm^2 and an anode-current density of 0.1 ampere per cm^2. Thus $J/J_0 = 20$, and we can use the asymptotic value of $-\xi_c = 2.55$.

We find with Eq. 1.23 that

$$x_m = \frac{2.55}{9.19 \times 10^5}\frac{T_c^{\frac34}}{J_0^{\frac12}} = 0.0017 \text{ cm}$$

From Eq. 1.28,

$$V_m = \frac{1100}{11{,}605}\ln 20 = 0.29 \text{ volt}$$

Now for a diode operating at 10 volts we have for the last factor of Eq. 1.35, which is a correction to Child's law,

$$1 + \frac{0.025 T_c^{1/2}}{(V - 0.28)^{1/2}} = 1.26$$

whereas, if $V = 100$ volts, the correction factor becomes 1.08.

Thus for the steady-state conditions the corrections to the single-velocity (Child's law) expression are of importance only for low-voltage diodes. However, we shall find this multivelocity treatment of great importance and indeed the starting point for the succeeding calculations on noise in the presence of the potential minimum.

2. Noise Properties of an Electron Stream at the Surface of the Cathode

We will discuss in this section the noise current and noise velocity in a temperature-limited diode. The term temperature-limited is used to describe the condition of operation wherein all the emitted electrons are drawn to the anode. The resulting relations will be valid for the input conditions at the surface of the cathode for space-charge-limited flow which will be studied later. The diode will be assumed to have a small cathode–anode distance so that transit time effects can be neglected.

Shot Noise

The noise current in a temperature-limited diode, called shot noise, was first described by Schottky.[12,13] Here we wish to give two arguments which have been used by Pierce[14,15] that may or may not make the shot-noise formula more plausible.

The first example considers a short diode consisting of two emitting surfaces opposing each other. Both surfaces are assumed to be identical, and the entire diode is held at the same temperature. If V is the voltage of cathode 2 with respect to cathode 1, we then have, from Eq. 1.27, that the current flowing to cathode 2, which is held at $-|V|$ with respect to cathode 1, is given by

$$I_0 = I \exp\left(-\frac{eV}{kT}\right) \tag{1.49}$$

(note: e is a negative quantity) where I is the total emitted current of cathode 1. The diode conductance is defined

$$g \equiv \frac{\partial I_0}{\partial V} = -\frac{eI}{kT_c} \exp\left(-\frac{eV}{kT_c}\right) \tag{1.50}$$

We now short-circuit the diode, and, since there are no external sources of power, the electric-energy flow in the lead connecting cathode 1 to cathode 2 must be the Johnson noise attributable to the resistance of the diode.

Johnson noise is given in the equation

$$\overline{i^2} = 4kTg\,\Delta f \tag{1.51}$$

where g is the conductance of the noise element, k is Boltzmann's constant, T is the temperature in degrees Kelvin, and Δf is the bandwidth of the measuring system in cycles per second. From Eq. 1.50, when $V = 0$,

$$g = \frac{|e|I}{kT_c}$$

so that we may write for the short-circuited diode, using Eq. 1.51, and the fact that the diode is at equilibrium temperature T_c,

$$\overline{i^2} = 4kT_c\frac{|e|I}{kT_c}\,\Delta f$$

$$\overline{i^2} = 4|e|I\,\Delta f$$

If $V = 0$, a current flows from cathode 2 to 1 equal to that from cathode 1 to 2, so that the noise associated with a current I_0 flowing from cathode 1 is given by

$$\overline{i^2} = 2|e|I\,\Delta f \tag{1.52}$$

This is recognized as the shot-noise formula.

The second argument, which may make Eq. 1.52 appear reasonable, begins by considering a periodic procession of electrons, one electron every T sec. These electrons (carrying a charge e) form a series of current impulses, and the spectrum of current flow is made up of a number of harmonics spaced by

$$\Delta f = \frac{1}{T} \tag{1.53}$$

The peak amplitude of each harmonic is given by

$$i_x = \frac{2|e|}{T} \tag{1.54}$$

We assume there to be many such periodic processions of electrons, each producing its own set of harmonics, and this large number of electrons produces a net harmonic component at a given frequency of peak amplitude i_m. To this we add one more periodic procession of electrons carrying a current with a peak harmonic amplitude i_x and phase θ with respect to i_m.

The square of the total current I is given by

$$\left|I^2\right| = \tfrac{1}{2}(i_m + i_x e^{j\theta})(i_m + i_x e^{-j\theta}) \tag{1.55}$$

$$= \tfrac{1}{2}(i_m{}^2 + i_x{}^2 + 2i_m i_x \cos \theta)$$

If θ is chosen at random, the result of a large number of additions of i_x will just as likely have negative values of θ as positive values, and the third term of Eq. 1.55 does not contribute to the total current. Therefore, by adding a current i_x at a random phase, we increase $\left|I\right|^2$ by an amount $i_x{}^2$.

Thus, for p sets of electrons per sec, the mean-square value $\overline{i^2}$ for each harmonic will be, from Eq. 1.54,

$$\overline{i^2} = \frac{p}{2}\left(\frac{2e}{T}\right)^2 = 2p\,\frac{e^2}{T^2} \tag{1.56}$$

The average current is given by

$$I_0 = p\,\frac{|e|}{T} \tag{1.57}$$

and therefore, from Eqs. 1.53, 1.56, and 1.57, we obtain

$$\overline{i^2} = 2|e|\,I_0\,\Delta f \tag{1.58}$$

which is again the shot-noise formula.

Noise-Velocity Parameters of the Temperature-Limited Diode

Let us now turn our attention to a second noise parameter which will appear frequently in our discussions; namely, the mean square deviation of the velocity, $\overline{\delta v^2}$.

Just as in Section 1, where we calculated the average velocity from the Maxwellian distribution, we can now calculate the square of the deviation from this average. Let

$$\delta v^2 \equiv (v - \bar{v})^2 \tag{1.59}$$

and thus

$$\overline{\delta v^2} = \frac{\displaystyle\int_0^\infty (v - \bar{v})^2 v \exp\left(-\frac{mv^2}{2kT_c}\right)dv}{\displaystyle\int_0^\infty v \exp\left(-\frac{mv^2}{2kT_c}\right)dv} \tag{1.60}$$

or

$$\overline{\delta v^2} = \frac{1}{\displaystyle\int_0^\infty v \exp\left(-\frac{mv^2}{2kT_c}\right) dv} \left[\int_0^\infty v^2 v \exp\left(-\frac{mv^2}{2kT_c}\right) dv \right.$$

$$\left. - 2\bar{v} \int_0^\infty vv \exp\left(-\frac{mv^2}{2kT_c}\right) dv + \bar{v}^2 \int_0^\infty v \exp\left(-\frac{mv^2}{2kT_c}\right) dv \right] \quad (1.61)$$

With Eq. 1.13 this can be written

$$\overline{\delta v^2} = \overline{v^2} - \bar{v}^2 \quad (1.62)$$

From Eq. 1.19 we have

$$\bar{v}^2 = \frac{\pi k T_c}{2m} \quad (1.19)$$

and further

$$\overline{v^2} = \frac{\displaystyle\int_0^\infty v^3 \exp\left(-\frac{mv^2}{2kT_c}\right) dv}{\displaystyle\int_0^\infty v \exp\left(-\frac{mv^2}{2kT_c}\right) dv}$$

$$\overline{v^2} = \frac{2kT_c}{m} \quad (1.63)$$

Thus we have

$$\overline{\delta v^2} = \frac{kT_c}{2m}(4 - \pi) \quad (1.64)$$

Now we must consider how this is related to the mean-square fluctuation of velocity which is created by shot noise in the emission current. Suppose we have a current I_s which forms part of a current I and leaves the cathode with a velocity v_s. Then the average velocity $\bar{v}$ will be

$$\bar{v} = \frac{\Sigma I_s v_s}{I} \quad (1.65)$$

We can find the fluctuation in $\bar{v}$ due to the fluctuation $\Delta I_s = i_s$ in the current I_s as

$$\Delta \bar{v}_s = \left(-\frac{1}{I^2}\frac{dI}{dI_s}\Sigma I_s v_s + \frac{v_s}{I}\right) i_s \quad (1.66)$$

Since a change in I_s by an amount ΔI_s also represents a change in I

by ΔI_s, we have

$$\frac{dI}{dI_s} = 1 \tag{1.67}$$

and therefore

$$\Delta \bar{v}_s = \left(-\frac{1}{I^2} \bar{v} I + \frac{v_s}{I}\right) i_s$$

or

$$\Delta \bar{v}_s = (v_s - \bar{v}) \frac{i_s}{I} \tag{1.68}$$

Since each velocity class is emitted independently, we may assume that the fluctuation i_s is a shot-noise current, $(2|e| I_s \Delta f)^{1/2}$, independent of the fluctuation in any other stream, and the total mean-square fluctuation in velocity will be given by

$$\overline{(\Delta \bar{v}_s)^2} = \sum_s (v_s - \bar{v})^2 \frac{\overline{i_s^2}}{I^2}$$

$$= \sum_s (v_s - \bar{v})^2 \frac{2|e| I_s \Delta f}{I^2}$$

$$= \frac{2|e|}{I^2} \Delta f \sum_s (v_s - \bar{v})^2 I_s \tag{1.69}$$

but

$$I_s = K v_s \exp\left(-\frac{mv_s^2}{2kT_c}\right) \Delta v_s \tag{1.70}$$

where Δv_s is the width of the velocity class and

$$I = K \sum_s v_s \exp\left(-\frac{mv_s^2}{2kT_c}\right) \Delta v_s \tag{1.71}$$

Using Eqs. 1.70 and 1.71 in Eq. 1.69, and replacing the summation by integration, we obtain

$$\overline{\Delta \bar{v}_s^2} = \frac{2|e| \Delta f}{I} \frac{\int_0^\infty (v_s - \bar{v})^2 v_s \exp\left(-\frac{mv_s^2}{2kT_c}\right) dv_s}{\int_0^\infty v_s \exp\left(-\frac{mv_s^2}{2kT_c}\right) dv_s} \tag{1.72}$$

and, from Eq. 1.72

$$\overline{\Delta \bar{v}_s{}^2} = \frac{2|e| \, \Delta f}{I} \, \overline{(v^2 - \bar{v}^2)} = \frac{2|e| \, \Delta f}{I} \, \overline{\delta v^2} \tag{1.73}$$

which then gives, for the mean-square fluctuation within the frequency band Δf,[16]

$$\overline{\Delta \bar{v}_s{}^2} = \frac{|e| \, \Delta f k T_c}{mI} \, (4 - \pi) \tag{1.74}$$

In the next section we will treat the problem of noise in a space-charge-limited diode and find that fluctuations in the potential minimum will act to smooth the shot-noise current of emission. However, from Eq. 1.19 we see that the average velocity is not changed by fluctuations in the potential minimum, and therefore we shall assume that the fluctuation velocity given by Eq. 1.74 does not change with changes in the potential minimum.

The above quantity is of sufficient importance to introduce a separate symbol for it. We shall denote it hence simply by $\overline{v^2}$. Although v has been used before to denote the velocity of single electron groups, we believe that there will be no occasion for confusion.

3. Space-Charge Reduction of Noise Using the Single-Velocity Approximation

In the next section we will present a more general treatment of the low-frequency noise in a space-charge-limited diode and show that this noise is reduced much below shot noise. Since this treatment is complicated by the multivelocity character of the beam, we will consider first some approximate derivations of this reduction factor which are based on simplified models. The model will be assumed to have an electron stream which has only a single velocity at a given plane, x, and the mean-square fluctuation in this velocity will be assumed to be equal to the mean-square fluctuation in velocity which was calculated for the multivelocity beam in Section 2. The space-charge reduction of noise based on a single-velocity theory was first demonstrated by Rack[16] to give results that were in good agreement with the multivelocity treatment. Rack's work was based on the Llewellyn–Peterson equations, and we will take this up somewhat later. First, however, we can illustrate rather easily from the equations of Section 1 that a given noise fluctuation at the cathode associated with shot noise must result in a noise current at the anode which is reduced below shot noise by a "smoothing factor" Γ^2.

Simple Expression for Smoothing Factor Γ^2

In Eq. 1.36 we have an approximate expression which relates the anode current to the finite velocity of emission at the cathode which can be written (D. O. North pointed out this method)

$$I_0 = I_{00} \left[1 + \frac{2.66\bar{v}}{\left[\frac{\pi}{2} \left| \frac{e}{m} \right| (V_a - V_m) \right]^{1/2}} \right] \qquad (1.36a)$$

where I_{00} is the diode current found from Eq. 1.33. From the numerical calculation in Section 1 (p. 14), we concluded that this is a good approximation to the average current density for voltages above 20 volts. If there is a fluctuation $\Delta \bar{v}_s$ of the average emission velocity, caused by excess current in the velocity group v_s, the resulting change in anode current ΔI is given from Eq. 1.36a as

$$\Delta I = \frac{I_{00} 2.66}{\left[\frac{\pi}{2} \left| \frac{e}{m} \right| (V_a - V_m) \right]^{1/2}} \Delta \bar{v}_s \qquad (1.75)$$

and the mean square of the fluctuation current is

$$\overline{(\Delta I)^2} = \frac{(2.66 I_{00})^2}{\frac{\pi}{2} \left| \frac{e}{m} \right| (V_a - V_m)} \overline{(\Delta \bar{v}_s^2)} \qquad (1.76)$$

If we now associate the fluctuation of velocity $\overline{\Delta \bar{v}_s^2}$ at the cathode with shot noise, the mean-square fluctuation in velocity is, from Eq. 1.74,

$$\overline{\Delta \bar{v}_s^2} \equiv \overline{v^2} = \frac{|e| \Delta f k T_c}{m I_0} (4 - \pi) \qquad (1.74)$$

Setting $I_{00} \cong I_0$, the mean-square fluctuation in anode current is, from Eq. 1.76,

$$\overline{(\Delta I)^2} = \frac{(2.66)^2}{\pi} (4 - \pi) \frac{2|e| I_0 \Delta f}{(|e|/k T_c)(V_a - V_m)}$$

$$\cong 2|e| I_0 \Delta f \frac{9}{4} \left(\frac{4 - \pi}{\eta_a} \right) \qquad (1.77)$$

The factor which relates the reduced noise current to shot noise is denoted by Γ^2, and from Eq. 1.77 we see that

$$\Gamma^2 = \frac{9}{4} \left(\frac{4 - \pi}{\eta_a} \right) \qquad (1.78)$$

which can be written from Eq. 1.22, assuming $T_c = 1100°\,\mathrm{K}$

$$\Gamma^2 \approx \left(1 - \frac{\pi}{4}\right)\frac{1}{V_a - V_m} \tag{1.79}$$

$$\approx \frac{1}{5V_a} \quad (\text{with } V_a \text{ in volts})$$

Thus we find Γ^2 considerably less than unity. The relation of Eq. 1.78 is valid only for large values of η_a since it is based on Eq. 1.36a. Also, it neglects the effect of the electrons which were returned to the cathode because of low initial velocities. We will see from the next section, which treats the problem in a more rigorous fashion, that the effect of these returning electrons is small and the limiting expression for large values of η_a is just that given by Eq. 1.78.

Smoothing Factor Γ^2 Derived from Another Simplified Model

In reference 14, Pierce discusses the noise in space-charge-limited single-velocity flow along the following lines: In an actual diode, the relation between the anode current I_0 and the emitted current I is given (from Section 1) as

$$I_0 = I \exp\left(-\frac{eV_m}{kT_c}\right) \tag{1.27}$$

If we now assume V_m to be held constant, the electrons returning to the cathode are independent of those going on to the anode, and hence the anode current must contain pure shot noise, given by

$$\overline{i^2} = 2|e|I_0\,\Delta f \tag{1.80}$$

Consider an *a-c open-circuited diode*. In order to keep the fluctuation current zero, V_m must change in such a way as to create a current equal and opposite to that given in Eq. 1.80. Therefore at x_m there must be a fluctuating voltage

$$\overline{V_m{}^2} = 2|e|I_0\,\Delta f R_m{}^2 \tag{1.81}$$

where

$$R_m = \frac{dV_m}{dI_0}$$

and, from Eq. 1.27,

$$R_m = \frac{kT_c}{|e|I_0} \tag{1.82}$$

Therefore Eq. 1.81 becomes

$$\overline{V_m{}^2} = \tfrac{1}{2}4kT_cR_m\,\Delta f \tag{1.83}$$

If there were no change in $V_a - V_m$, the fluctuation in anode voltage would also be given by Eq. 1.83. However, if we compare this to the equivalent result, Eq. 1.77 multiplied by $1/g^2$, we find that Eq. 1.83 gives a value that is much too small. We must therefore look for fluctuations in the space between the potential minimum and the anode. These variations in voltage between the potential minimum and the anode are related to the fluctuations in average velocity of the electrons in the region.

We will neglect the effect of the thermal velocities of emission, and this in turn implies that we are neglecting the effect of the electrons that return to the cathode. The assumption can be verified through comparison with the results of the next section. With these assumptions the steady-state conditions are given by Child's law

$$I_0 = 2.33 \times 10^{-6}\,\frac{V_a{}^{3/2}}{x^2} \tag{1.4}$$

From Eq. 1.4 we obtain for the conductance of the diode

$$g \equiv \frac{\partial I_0}{\partial V_a} = \frac{3}{2}\frac{I_0}{V_a}$$

and the diode resistance is

$$R \equiv \frac{1}{g} = \frac{2}{3}\frac{V_a}{I_0} \tag{1.84}$$

Consider now an electron which crossed the potential minimum at $x = 0$ at a time $t = 0$. The charge between the electron and the potential minimum is given by $-I_0t$. Since the field at the minimum is zero, the potential gradient at x is given from Gauss's theorem as

$$\frac{\partial V}{\partial x} = \frac{I_0 t}{\epsilon_0} \tag{1.85}$$

and the acceleration is

$$\ddot{x} = \left|\frac{e}{m}\right|\frac{I_0 t}{\epsilon_0} \tag{1.86}$$

which gives

$$\dot{x} = \left|\frac{e}{m}\right|\frac{I_0}{2\epsilon_0}t^2 + \dot{x}_0 \tag{1.87}$$

and

$$x = \left| \frac{e}{m} \right| \frac{I_0}{\epsilon_0} t^3 + \dot{x}_0 t = \left| \frac{e}{m} \right| \frac{I_0}{6\epsilon_0} t^3 + \dot{x}_0 t \tag{1.88}$$

where $\dot{x}_0$ is the velocity at $t = 0$, $x = 0$.

Now the voltage between the potential minimum and x (where henceforth in this computation x is the position of the anode) is given by

$$\dot{x}^2 - \dot{x}_0{}^2 = -2 \frac{e}{m} V_a \tag{1.89}$$

or

$$V_a = \frac{1}{2 \left| \frac{e}{m} \right|} \left(\frac{\frac{e}{m} I_0}{2\epsilon_0} \right)^2 t^4 + \frac{I_0}{2\epsilon_0} \dot{x}_0 t^2 \tag{1.90}$$

If at constant x we now vary $\dot{x}_0$ by a small amount, we find, from Eq. 1.88,

$$\frac{dt}{d\dot{x}_0} = \frac{-t}{\left(\frac{|e| I_0}{m\epsilon_0} t^2 + \dot{x}_0 \right)} \tag{1.91}$$

and, from 1.90,

$$dV_a = \frac{I_0 t}{\epsilon_0} \left(\frac{|e| I_0}{2\epsilon_0 m} t^2 + \dot{x}_0 \right) dt + \frac{I_0}{2\epsilon_0} t^2 \, d\dot{x}_0 \tag{1.92}$$

and, using Eq. 1.91,

$$dV_a = - \frac{I_0}{2\epsilon_0} t^2 \, d\dot{x}_0 \tag{1.93}$$

We will now evaluate t. Since the major portion of the thermal velocities at the potential minimum are small compared with the velocities in the rest of the region, we can take the value of t for $\dot{x}_0 = 0$. Thus, from Eqs. 1.87 and 1.88, we can write

$$t^2 = \left| \frac{e I_0}{2m\epsilon_0} \right|^{-1} \left(\frac{2|e| V_a}{m} \right)^{1/2} \tag{1.94}$$

and, from Eqs. 1.93 and 1.94,

$$dV_a = -2^{-1/2} \left| \frac{e}{m} \right|^{-1/2} V_a{}^{1/2} \, d\dot{x}_0 \tag{1.95}$$

If $\overline{(d\dot{x}_0)^2}$ is the mean-square fluctuation in velocity, the mean-square fluctuation in voltage $\overline{V_a{}^2}$ will be, from Eq. 1.95,

$$\overline{V_a{}^2} = 2\left|\frac{m}{e}\right| V_a \overline{d\dot{x}_0{}^2} \tag{1.96}$$

or, with Eq. 1.75, we can write

$$\overline{V_a{}^2} = 3\left|\frac{m}{e}\right| I_0 R \overline{d\dot{x}_0{}^2} \tag{1.97}$$

From Section 2, Eq. 1.74, we have

$$\overline{v^2} = \overline{d\dot{x}_0{}^2} = \frac{|e|\,\Delta f k T_c}{I_0 m}(4 - \pi)$$

and Eq. 1.97,

$$\overline{V_a{}^2} = 3(4 - \pi)kT_c R \,\Delta f$$

$$\overline{V_a{}^2} = (0.644)4kT_c R \,\Delta f \tag{1.98}$$

This is the fluctuation in voltage between the anode and potential minimum for an open-circuited diode. It is also the fluctuation in anode–cathode voltage for an a-c open-circuited diode under the assumption that we can neglect the effect of those electrons that return to the cathode. We see here the well-known result that the noise from the space-charge-limited diode is two thirds of that from a thermal resistor with resistance equal to the diode resistance.

Smoothing Factor from the Llewellyn–Peterson Equations

We will now present the approach used by Rack.[16] It will be assumed that the reader is familiar with the Llewellyn–Peterson equations[6] which can be written in the form†

$$V_b - V_a = A^*I + B^*J_a + C^*v_a \tag{1.99}$$

where $V_b - V_a$ is the alternating voltage between two planes in a diode and J_a and v_a are the a-c convection-current density and a-c velocity at plane a. I is the total a-c current density in the diode.

If we consider the case where we have no fluctuations in the diode, i.e., $J_a = 0$ and $v_a = 0$, Eq. 1.99 reduces to

$$V_b - V_a = A^*I \tag{1.100}$$

from which we identify A^* with the a-c impedance of the diode.

† See also Chapter 3 but note changes in notation.

Therefore we associate the last two terms with the voltage produced in the diode by fluctuations in the electron stream, and we write for the a-c open-circuited diode

$$V_b - V_a = B^*J_a + C^*v_a \tag{1.101}$$

The coefficients B^* and C^* are given in reference 6 as

$$B^* = j\frac{1}{\epsilon_0}\frac{T^2}{\theta^3}u_a(2P - \beta Q) \tag{1.102}$$

$$C^* = -\frac{2m}{e}(u_a + u_b)\frac{P}{\theta^2} \tag{1.103}$$

In these equations

$$u_a = \text{average velocity at plane } a$$
$$u_b = \text{average velocity at plane } b$$
$$\theta = \text{transit angle from plane } a \text{ to plane } b$$
$$T = \text{transit time from plane } a \text{ to plane } b$$

and

$$P = 1 - e^{-j\theta} - j\theta e^{-j\theta} \tag{1.104}$$

and

$$Q = 1 - e^{-j\theta}$$

respectively. For our noise problems we will consider the a plane to be just slightly beyond the potential minimum, so that we encounter no electrons returning to the cathode, but close enough to the minimum so that the d-c acceleration may be taken equal to zero. Under these conditions u_a will be considered to be zero, and, from Eq. 1.102, B^* is zero. This tells us that fluctuations in current density at the potential minimum produce no significant effect on fluctuations at the anode. We are left with the equation

$$V_b - V_a = C^*v_a \tag{1.105}$$

And from Eq. 1.104 we have, for $\theta \to 0$,

$$\frac{P}{-\theta^2} = \frac{1}{2} \tag{1.106}$$

If b represents the anode plane, Eq. 1.106 is applicable to the short-transit-angle diode problem.   Equation 1.105 becomes†

$$V_b - V_a = \frac{mu_b}{e} v_a \tag{1.107}$$

$$= - \left(2 \frac{m}{|e|} V_{ob} \right)^{\frac{1}{2}} v_a$$

and the mean-square fluctuation in anode voltage is given by

$$\overline{V^2} = 2 \frac{m}{|e|} V_{ob} \overline{v_a}^2 \tag{1.108}$$

This will be recognized as identical to Eq. 1.96, and therefore we can write again

$$\overline{V^2} = \tfrac{3}{4}(4 - \pi)4kT_cR\,\Delta f \tag{1.109}$$

In this section we have derived several consistent expressions for the noise-smoothing factor Γ^2, using models that are limited to single-velocity flow.   We must now develop the problem of multivelocity flow, and we will find that the above approximations are valid for small values of J_0/J and relatively large values of V_a ($V_a > 20$ volts). These are familiar approximations and apply for most operating conditions in the low-frequency region.

4. Space-Charge-Limited Noise for Diodes with Short Transit Angles

The problem of reduced shot noise with multivelocity flow that is encountered in space-charge-limited diodes when the transit time from cathode to anode is short compared to an RF cycle will be treated by following the discussion presented by Thompson, North, and Harris in reference 5.   As pointed out in Section 1, the mechanism of space-charge-limited flow is such that part of the emitted electrons are turned back to the cathode by the negative gradient prior to the potential minimum.   Since this gradient is established by the space charge of the electrons, it is not difficult to understand that an instantaneous increase of emitted electrons over and above the average number would result in a lowering of the potential minimum, and hence a larger number of electrons will be returned to the cathode.   It is this gating action of the potential minimum that reduces the noise below the level of pure shot noise.

When additional electrons are emitted which have velocities suffi-

---

† Note that in the Llewellyn–Peterson notation V_a corresponds to the input plane, and, in this context, to the potential minimum, not the anode!

cient to pass the potential minimum, the number of electrons passing through to the anode is momentarily increased, thus lowering the value of V_m owing to the added space charge. The more negative potential minimum turns back some electrons which would have otherwise passed to the anode. Similarly, if the number of emitted high-velocity electrons is less, V_m becomes less negative, and an additional number of the lower-velocity electrons are allowed to pass to the anode. It is this critical relation between the fluctuations in emission velocities and the corresponding fluctuations in potential minimum that we wish to study.

Formulation of Smoothing Factor

The analysis will proceed along the following lines: Assume that steady-state conditions exist within the diode. We now inject into this diode a small current which contains electrons with velocities of emission from v_s to $v_s + \Delta v_s$. The fluctuation current is given by the shot-noise formula $(2eI_s \, \Delta f)^{1/2}$, with I_s being the current carried by the velocity group v_s to $v_s + \Delta v_s$. From this we shall determine the resulting fluctuation in the potential minimum. This in turn will allow us to calculate the corresponding fluctuations in anode current. An integration over all the velocity classes will then give the total fluctuation in the anode current.

We shall consider only fluctuations of long enough duration so that they act as a succession of equilibrium states. We assume the cathode to be at zero potential and define V_s by

$$-2\frac{e}{m} V_s = v_s{}^2 \tag{1.110}$$

where v_s is the velocity of emission of a small steady increment of current, $i_s(v_s)$. We further define the parameter

$$\lambda = \frac{-e(V_s + V_m)}{kT} \tag{1.111}$$

where V_m is a negative number. Thus we can use λ rather than v_s to designate the emission velocity of electrons which comprise i_s. If we choose v_s such that the element of current in this velocity class crosses the potential minimum to the anode, λ is positive. For smaller values of v_s, such that the current is turned back, λ is negative and must lie in the range $-\eta_c \leq \lambda \leq 0$, where η_c is defined in Eq. 1.29. We see that the value $\lambda = -\eta_c$ corresponds to a value of v_s equal to zero. Also, from the η vs. ξ plot of Fig. 1.2, the point at which the electrons stop and return to the cathode is given by $\eta = -\lambda$.

We now consider the noise fluctuations in the anode current. For every value of $i_s(\lambda)$ that we inject into the diode, we will find the new equilibrium current $\hat{I}$. Thus, if I_0 represents the steady-state anode current which flows before the admission of $i_s(\lambda)$, we are able to determine the net increase in I_0. The ratio of this net increase in anode current to the increment of current $i_s(\lambda)$ will be a function only of λ. This follows from the previous argument which indicated that the change in anode current due to an incremental change in emission current would be a function of the velocity of emission and hence λ. The ratio of the net increase in anode current to i_s will be denoted by $\gamma(\lambda)$. It represents the factor by which a change in emission is converted into a change in plate current.

The fluctuation Δi_s in the incremental current is a true shot fluctuation and can be written

$$\overline{\Delta i_s{}^2} = 2|e|\,\Delta I_s\,\Delta f \tag{1.112}$$

where ΔI_s is the emission current containing electrons with emission velocities between λ and $\lambda + \Delta\lambda$. From Eq. 1.27 we can write

$$\Delta I_s = I e^{-\lambda}\,\Delta\lambda \tag{1.113}$$

or

$$\overline{\Delta i_s{}^2} = 2|e|I\,\Delta f(e^{-\lambda}\,\Delta\lambda) \tag{1.114}$$

Now we have stated that the fluctuations in anode current will be changed by the factor $\gamma(\lambda)$ from those in the emitted current. We can then write the fluctuations in an incremental element of anode current as

$$\overline{\Delta i^2} = 2|e|I\,\Delta f[\gamma^2(\lambda)e^{-\lambda}\,\Delta\lambda] \tag{1.115}$$

Since the fluctuations of Eq. 1.112 associated with one velocity class are independent of other velocity classes, we can obtain the total fluctuations in anode current by integrating Eq. 1.115 to give

$$\overline{i^2} = 2|e|I\,\Delta f \int_{-\eta_c}^{\infty} \gamma^2(\lambda)e^{-\lambda}\,d\lambda \tag{1.116}$$

which we can write as

$$\overline{i^2} = 2|e|I\,\Delta f\,\Gamma^2 \tag{1.117}$$

Γ^2 will be recognized as the space-charge reduction factor and can be written as

$$\Gamma^2 = \Gamma_\alpha{}^2 + \Gamma_\beta{}^2 \tag{1.118}$$

where

$$\Gamma_\alpha{}^2 = \int_{-\eta_o}^{0} \gamma^2(\lambda)e^{-\lambda}\,d\lambda \tag{1.119}$$

and

$$\Gamma_\beta{}^2 = \int_0^\infty \gamma^2(\lambda)e^{-\lambda}\,d\lambda \tag{1.120}$$

where α refers to the group of electrons that do not have initial velocities at the cathode sufficient to overcome the potential minimum and hence are returned to the cathode. The subscript β denotes that group of electrons which pass the potential minimum and hence reach the anode.

Evaluation of Smoothing Factor

The factor $\gamma(\lambda)$ for the α and β groups can be expressed as follows. For values of λ pertaining to the α group the total anode current after injection of i_s is $\hat{I}$, whereas before injection it was I_0. We can then write

$$\gamma(\lambda) = \frac{\hat{I} - I_0}{i_s(\lambda)} \quad (\alpha \text{ group}) \tag{1.121}$$

For the β group the total anode current after injection is $\hat{I} + i_s$, for i_s now passes on to the anode, and therefore

$$\gamma(\lambda) = 1 + \frac{\hat{I} - I_0}{i_s(\lambda)} \quad (\beta \text{ group}) \tag{1.122}$$

Let us now calculate $\hat{I}$ when we inject a small additional current i_s from the cathode. For the β group we can write, in place of Eq. 1.9,

$$\frac{d^2V}{dx^2}\bigg|_\beta = -\frac{e}{\epsilon_0}\int_{v_m}^\infty \frac{n(v_s)\,dv_s}{v} - \frac{i_s}{\epsilon_0 v} \tag{1.123}$$

Following the procedure of Section 1, we can make a first integration of Eq. 1.123 by multiplying both sides by $2(dV/dx)$ and, since

$$v = \sqrt{2\left|\frac{e}{m}\right|}\sqrt{V + V_s}$$

we obtain, in place of Eq. 1.10,

$$\left[\frac{dV}{dx}\right]_\beta^2 = \frac{-2m}{e}\int_{v_m}^\infty n(v_s)(v - v_m)\,dv_s$$

$$- \frac{i_s 2\sqrt{2}}{\epsilon_0\sqrt{|e/m|}}(\sqrt{V + V_s} - \sqrt{V_m + V_s}) \tag{1.124}$$

If we now change from x and V to ξ and η as given in Eqs. 1.22 and

1.23, we obtain, in place of Eq. 1.124,

$$\left(\frac{d\eta}{d\xi}\right)^2_\beta = \frac{i_s}{\hat{I}_0} \frac{2}{\sqrt{\pi}} (\sqrt{\eta + \lambda} - \sqrt{\lambda}) + e^\eta - 1 - e^\eta \operatorname{erf} \sqrt{\eta}$$

$$+ \frac{2}{\sqrt{\pi}} \sqrt{\eta} \quad (1.125)$$

which we can write as

$$\left(\frac{d\eta}{d\xi}\right)^2_\beta = \frac{i_s}{\hat{I}} F(\eta, \lambda) + \Phi_\beta(\eta) \quad (1.126)$$

where $F(\eta, \lambda) = \dfrac{2}{\sqrt{\pi}} (\sqrt{\eta + \lambda} - \sqrt{\lambda})$

and $\Phi_\beta(\eta)$ follows from Eq. 1.125.

Now, since we have assumed $i_s \ll \hat{I}$, we shall neglect all the higher powers of i_s/I_0 and write

$$d\xi = \left[1 - \frac{1}{2} \frac{i_s}{\hat{I}} \frac{F(\eta, \lambda)}{\Phi_\beta(\eta)}\right] \frac{d\eta}{\Phi_\beta(\eta)^{1/2}} \quad (1.127)$$

If we denote the value of ξ and η at the anode by the subscript a, we can write

$$\hat{\xi}_a = \int_0^{\hat{\eta}_a} \frac{d\eta}{\Phi_\beta(\eta)^{1/2}} - \frac{1}{2} \frac{i_s}{\hat{I}} \int_0^{\hat{\eta}_a} \frac{F(\eta, \lambda)}{\Phi_\beta(\eta)^{3/2}} d\eta \quad (1.128)$$

where $\hat{\xi}$ and $\hat{\eta}$ represent the perturbed values, whereas ξ represents the unperturbed value and is given from Eq. 1.128 with $i_s = 0$:

$$\xi_a = \int_0^{\eta_a} \frac{d\eta}{\Phi_\beta(\eta)^{1/2}} \quad (1.129)$$

Now since the quantities $(\hat{\eta} - \eta)$ and $(\hat{\xi} - \xi)$ are also first-order infinitesimals, we can write for Eq. 1.128

$$\hat{\xi}_a - \xi_a = \frac{\hat{\eta}_a - \eta_a}{\Phi_\beta(\eta_a)^{1/2}} - \frac{1}{2} \frac{i_s}{\hat{I}} \int_0^{\eta_a} \frac{F(\eta, \lambda)}{\Phi_\beta(\eta)^{3/2}} d\eta \quad (1.130)$$

A similar treatment for the α group gives

$$\hat{\xi}_c - \xi_c = \frac{\hat{\eta}_c - \eta_c}{\Phi_\beta(\eta_c)^{1/2}} + \frac{1}{2} \frac{i_s}{\hat{I}} \int_0^{\eta_c} \frac{F(\eta, \lambda)}{\Phi_\alpha(\eta)^{1/2}} d\eta \quad (1.131)$$

where the subscript c denotes the cathode surface. Now we will treat the problem wherein the anode potential is held constant as in an a-c short-circuited diode, and, from Eqs. 1.22, 1.23, and 1.29, we have

$$\hat{\eta}_a - \eta_a = \hat{\eta}_c - \eta_c = \frac{e}{kT}(V_m - \hat{V}_m) = \ln\frac{I_0}{\hat{I}} \cong -\frac{\hat{I} - I_0}{I_0} \quad (1.132)$$

and

$$(\hat{\xi}_a - \hat{\xi}_c) - (\xi_a - \xi_c) = (\xi_a - \xi_c)\left(\sqrt{\frac{\hat{I}}{I_0}} - 1\right)$$

$$= \frac{1}{2}(\xi_a - \xi_c)\frac{\hat{I} - I_0}{I_0} \quad (1.133)$$

If we subtract Eq. 1.131 from Eq. 1.130 and use Eqs. 1.132 and 1.133, we obtain

$$\frac{\hat{I} - I_0}{i_s} = -\frac{1}{2D}\left[\int_0^{\eta_c}\frac{F(\eta, \lambda)}{\Phi_\alpha(\eta)^{3/2}}d\eta + \int_0^{\eta_a}\frac{F(\eta, \lambda)}{\Phi_\beta(\eta)^{3/2}}d\eta\right] \quad (1.134)$$

where

$$D = \tfrac{1}{2}(\xi_a - \xi_c) + \Phi_\beta(\eta_a)^{-1/2} + \Phi_\alpha(\eta_c)^{-1/2} \quad (1.135)$$

Therefore, from Eq. 1.122, we write for the β region

$$\gamma(\lambda) = 1 - \frac{1}{2D}\left[\int_0^{\eta_c}\frac{F(\eta, \lambda)}{\Phi_\alpha(\eta)^{3/2}}d\eta + \int_0^{\eta_a}\frac{F(\eta, \lambda)}{\Phi_\beta(\eta)^{3/2}}d\eta\right] \quad (1.136)$$

where

$$\lambda \geq 0$$

It is now necessary to resort to numerical integration in order to evaluate Eq. 1.136. This is done for the case of complete space-charge-limited flow, which means that we can assume $I_0/I \ll 1$. This in turn means that we may use the values $\eta_c = \infty$, $\xi_c = -2.55$ and $\Phi_\alpha(\eta_c) = \infty$. Thus, in Eq. 1.136, $\gamma(\lambda)$ becomes a function of η_a and λ. The resulting values of $\gamma(\lambda)$ were obtained by North and others in the range $5 \leq \eta_a \leq 100$ and $0.05 \leq \lambda \leq 5$. Near $\lambda = 0$, $\gamma(\lambda)$ has a logarithmic discontinuity and must be evaluated separately. In Fig. 1.3 there is shown a plot of $\gamma(\lambda)$ vs. λ as taken from reference 5 for $\eta_a = 30$. The value of $\Gamma_\beta{}^2$ for $\eta_a = 30$ is now obtained from Fig. 1.3 with the aid of Eq. 1.119. This method then allows a plot of $\Gamma_\beta{}^2$ vs. η_a, which is given in Fig. 1.4.

Also shown in Fig. 1.4 is a plot of $\Gamma_\alpha{}^2$ vs. η_a, which is obtained in a manner similar to the preceding calculation for $\Gamma_\beta{}^2$. It is significant to note that $\Gamma_\alpha{}^2$ is always much smaller than $\Gamma_\beta{}^2$. This is to be expected, since the major portion of the electrons in the α group are returned to the cathode before they approach very near the potential minimum. Therefore the action of the α electrons is confined to a very short region.   We can thus neglect their contribution to the noise and set $\Gamma_\alpha{}^2 \approx 0$.   This assumption, together with single-velocity equations, allowed us to evaluate $\Gamma_\beta{}^2$ in the approximate but explicit forms of Section 3.

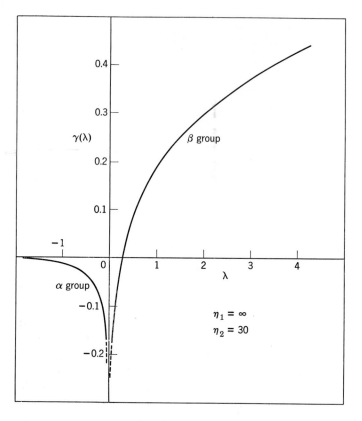

Fig. 1.3.   Shot-effect reduction factor γ as a function of the velocity parameter λ.

Smoothing Factor in Terms of the Diode Transconductance

Before concluding this section, we will discuss Eq. 1.117 in a different form in terms of the diode conductance g.   Thus Eq. 1.51 can be written

$$\overline{i^2} = \theta \cdot 4kTg\,\Delta f \tag{1.137}$$

The diode conductance g as defined by

$$g = \frac{\partial I_0}{\partial V_a}$$

was derived in Section 1, Eqs. 1.45 through 1.47. In Eq. 1.47 we set the multiplier approximately to one and replace $\xi_a - \xi_c$ by Eq. 1.135, where

$$\phi_a^{-\frac{1}{2}} \cong 0$$

We obtain

$$g = \frac{2I_0}{2D - \phi_\beta(\eta_a)^{-\frac{1}{2}}} \frac{\partial \xi_a}{\partial \eta_a} \frac{|e|}{kT_c}$$

Further, using Eqs. 1.31, 1.125, and 1.126, we can replace $\partial \xi_a/\partial \eta_a$ by $\phi_\beta(\eta_a)^{-\frac{1}{2}}$ and obtain

$$g = \frac{2I_0}{2D\phi_\beta(\eta_a)^{\frac{1}{2}} - 1} \frac{|e|}{kT_c}$$

At large η_a we can neglect the one in the denominator, and finally obtain

$$\frac{|e|}{kT_c} \frac{I_0}{g} = D\Phi_\beta(\eta_a)^{\frac{1}{2}} \tag{1.138}$$

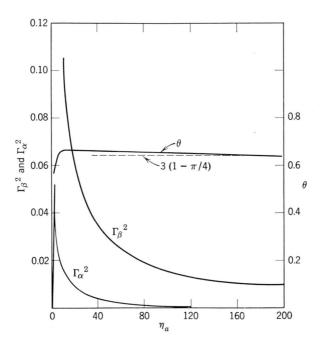

Fig. 1.4. Shot-noise reduction factor for complete space-charge-limited flow $(I_s/I \gg 1)$.

With Eqs. 1.117 and 1.137 we can write for θ

$$\theta = \tfrac{1}{2}\Gamma^2 \frac{|e|}{kT_c} \frac{I_0}{g} = \tfrac{1}{2}\Gamma^2 \, D\Phi_\beta(\eta_a)^{\frac{1}{2}} \tag{1.139}$$

This value of θ is also plotted in Fig. 1.4, and we see that for $\eta_a \to \infty$ we have a limiting value for θ given by

$$\theta \sim 3\left(1 - \frac{\pi}{4}\right) = 0.644 \tag{1.140}$$

This familiar result agrees with that found by the approximate methods of Section 3. Again it can be stated that a space-charge-limited diode generates a mean-square noise power equal to the thermal noise in a resistor whose resistance is equal to two thirds of the a-c resistance of a diode.

5. Discussion of Emission Noise at High Frequencies When the Transit Time Is Not Small

There is no adequate treatment of electron streams with Maxwellian distribution of velocities for high frequencies, and therefore the analysis of the noise problem must be based on the single-velocity approximation. Rack[16] has extended the analysis of Section 4 to include finite transit angles. Peterson[17] has used the Llewellyn–Peterson equations for calculating the noise in a high-frequency tetrode. Pierce[18] used this method in calculating the noise in the stream of a traveling-wave tube. In applying this method, he assumed that at the input plane near the potential minimum the average velocity was zero, and hence any fluctuation in current density at this point produced no effect at a later point in the stream. The only source of noise was the fluctuation in velocity which was taken to be equal to the mean-square deviation as calculated for the multivelocity stream. This analysis is subject among other things to the defect that the average velocity at the potential minimum is not zero but finite as calculated in Section 2, and hence the velocity fluctuation is not the only source of noise. There is a second source of noise in the current fluctuation at the potential minimum. The question that remains to be answered relates to the magnitude of these current fluctuations. Is it equal to shot noise or reduced shot noise, and is it correlated or uncorrelated with the noise velocity?

α Plane Beyond Potential Minimum Considered as Input

The first attempt to answer these questions was made by Robinson.[7] The single-velocity equations break down because the spread in transit angles between the potential minimum and the anode is not small. In fact, MacDonald[19] has given an expression for the anode transit time $\tau(E)$ of an electron with an initial energy E at the minimum to be

$$\tau(E) = \tau(0) \left(1 - 0.85 \left| \frac{E}{eV_a} \right|^{\frac{1}{4}} \right) \tag{1.141}$$

$\tau(0)$ = transit time of electron with zero initial energy

Thus for a typical microwave tube $\tau(0)$ might be 4 cycles and $V_a = 1000$ volts. If we use $E = kT$, the average energy of the minimum, we arrive at a value of $\omega[\tau(0) - \tau(E)] = 2\pi/3$ radians, which is not small. Robinson points out that this spread takes place largely in a region very close to the minimum, and hence, if we consider our input plane to be a given distance beyond the potential minimum, the single-velocity theory should be valid at the high frequencies. That such a plane exists can be seen from Eq. 1.141, for, if we choose 4 electrons with initial energies of 0, kT, $2kT$, and $3kT$, their transit times in the above example would be $\tau(0)$, $0.915\tau(0)$, $0.898\tau(0)$, and $0.888\tau(0)$. Thus we see that the spread in transit angle between the third and fourth electrons of 0.08π is certainly small. If we choose our input plane, called the α plane, at a point beyond the minimum where the potential is $\alpha(kT/e)$, we see that the slowest electron will have an energy of αkT at this plane. If α is sufficiently large, the spread in transit angle beyond this plane will be small.

At the α plane Robinson assumes the current fluctuations to be equal to shot noise, and the velocity fluctuations to be given by the mean-square velocity deviation for the multivelocity stream at the α plane, which is calculated in a manner similar to that used in Section 2.

Estimate of Smoothing Factor at High Frequencies for Noise Current

We see that the approach just considered would be very good if we knew the true value of the current and velocity fluctuations at the α plane. Robinson assumed pure shot noise with no correlation between velocity and current. Following Watkins,[8] we may argue that the current fluctuations at the α plane should be somewhat less than true shot noise.† Consider the diode to be divided into two diodes in series: the first between the cathode and the potential minimum and

† Compare Sec. 2, page 229, Chapter 5.

the second between the potential minimum and the α plane. The first diode prior to the minimum is a retarding field diode of area A, which has, for small transit angles, an a-c admittance given by

$$Y = \left(\frac{|e| J_0}{k T_c} + j \frac{\omega \epsilon}{d} \right) A \tag{1.142}$$

where J_0 is the current density through the diode, and d is the distance between the cathode and the potential minimum. If we consider the diode to be a-c short-circuited, the current fluctuation through the diode will be shot noise, as argued in Section 3, and we can write

$$i_{sc} = \left| 2 e I_0 \, \Delta f \right|^{1/2} \tag{1.143}$$

If we assume a linear system, Eqs. 1.142 and 1.143 can be used with Thévenin's theorem to give the open-circuit noise voltage as

$$V_{oc} = \frac{i_{sc}}{Y} = \frac{\left| 2 e I_0 \, \Delta f \right|^{1/2}}{\left(\dfrac{|e| J_0}{k T_c} + j \dfrac{\omega \epsilon_0}{d} \right) A} \tag{1.144}$$

We now consider this diode to be in series with the second diode, and, as a result of the large transit angle beyond the potential minimum, the latter diode can be considered to be open-circuited for alternating current. With this condition, the total alternating current must be zero, or

$$i + j \omega \epsilon_0 A E = 0 \tag{1.145}$$

where i is the alternating convection current.

$$i = j \omega \epsilon_0 A \frac{V_{oc}}{d} = \frac{j \omega \epsilon_0}{d} \frac{\left| 2 e I_0 \, \Delta f \right|^{1/2}}{|e| J_0 / k T_c + j \omega \epsilon_0 / d} \tag{1.146}$$

or, for the mean-square current fluctuation, we can write

$$\overline{i^2} = \left| 2 e I_0 \, \Delta f \right| \Gamma^2 \tag{1.147}$$

where

$$\Gamma^2 = \frac{1}{1 + (e J_0 d / \omega \epsilon_0 k T_c)^2} \tag{1.148}$$

We can write d in Eq. 1.148 in terms of the parameter ξ_c defined by Eq. 1.23. Using a new parameter a,

$$a = \frac{\omega}{2 \pi^{1/4} (m / 2 k T_c)^{1/4} \left| \dfrac{e J_0}{m \epsilon_0} \right|^{1/2}} = \frac{\omega}{\omega_{pm}} \tag{1.149}$$

where ω_{pm} may be defined as a plasma frequency at the potential minimum. Equation 1.148 can be written

$$\Gamma^2 = \frac{1}{1 + \xi_c{}^2/4\pi a^2} \tag{1.150}$$

Now, recalling that, for complete space charge $J_0/J \ll 1$, we have as the limiting value of $\xi_c = 2.55$. With this value Eq. 1.150 is written

$$\Gamma^2 = \frac{1}{1 + 0.52/a^2} \tag{1.151}$$

For a numerical example we will use the following values:

$$T = 1020° \text{ K}, \qquad f = 3000 \text{ Mc}, \qquad J_0 = 0.1 \text{ ampere/cm}.$$

in which case $a = 0.665$ and $\Gamma^2 = 0.46$. The mean-square fluctuation in current is reduced by a factor of 0.46 below shot noise.

Watkins points out that the assumption of small transit angles from cathode to potential minimum is not justified as it is found to be about 1.4 radians in the above example. Whinnery[9] has discussed some aspects of the noise at the potential minimum at high frequencies by using the physical picture of Thompson, North, and Harris as in Section 4. He discusses the effects of using different values of η_c in numerically evaluating the integrals of Section 4. He considers the perturbation of the potential minimum as one injects an excess of charge at the cathode and finds that the potential minimum "overcompensates." It oscillates back and forth at a frequency corresponding to the plasma frequency calculated at the potential minimum. He further points out that this plasma frequency for typical tubes occurs in the microwave region from about 2000 to 4000 Mc.

Treatment of Multivelocity Problem with the Use of a Computer

In each of the approaches that has just been presented it was necessary to make rather severe assumptions; so far it has proved difficult to assess their validity. Therefore, these solutions have limitations, for they do not give a complete answer to the multivelocity flow problem near the potential minimum. In Watkins' analysis a short transit time from cathode to potential minimum is assumed, an assumption which is not generally true. In Whinnery's work the electrons that return to the cathode are all assumed to return at the point of the potential minimum, $x = x_m$, whereas in reality the point of return is distributed between the cathode and x_m. In an analysis of this sort one must make use of a linear theory. Since the d-c velocity is small

in the region of interest, the a-c velocities can be of comparable amplitude, which violates the assumptions necessary for the linear theory.

Because of these difficulties Tien and Moshman[20] attacked the
problem by using numerical integration to trace individual electrons
through the potential minimum of a typical diode. We can see in
principle what is required for this task. The d-c conditions of the
diode can be established from the equations of Section 1. Then one
electron injected at the cathode can be traced step by step through
the minimum to the anode. By injecting a sufficient number of
electrons and computing their cumulative effect, it should be possible
to find the noise current and noise velocity at the α plane just beyond
the potential minimum.

The number of electrons that are injected at the cathode is necessarily limited when a computer is used, and the question immediately
arises as to how these electrons should be initiated so that the emission
noise is properly simulated. It is necessary to know the number of
electrons injected, their time of injection, and the velocity at which
they are injected. These initial properties were determined by Tien
and Moshman by the "Monte Carlo method" of statistics, which is
characterized by the use of random numbers.

We will first discuss briefly random numbers so as to illustrate how
they were used in obtaining the initial conditions of the injected
electrons. Consider the simple problem of finding the area under the
curve of Fig. 1.5. We would begin by dividing the interval into n
equal spaces (Δx) and sampling the heights of the curve at each inter-

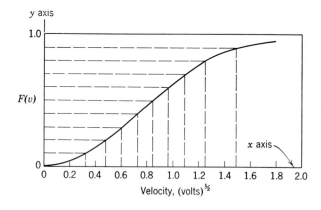

Fig 1.5 Plot showing transformation from uniform probability distribution to
xe^{-ax^2}.

val. If we desired to increase the accuracy of the computations, we would subdivide the interval and use $2n$ steps. Thus we are limited to a discrete number of steps. With the use of random numbers we would not use the uniform interval but choose n random numbers which were distributed with uniform probability in the desired interval. If we chose 100 intervals, for example, the random numbers might be 1.1, 2.9, 3.1, 3.8, etc. However, with random numbers we can increase the number that we use without regard to the discrete steps as before, and we can equally well use 99 numbers or 101 or 138, and they would still have the same distribution. As the number of samples increases, the accuracy increases.

Random numbers with a uniform distribution are readily obtainable from tables, or, if a computing machine is involved, it may be faster to generate the numbers. However, there are many instances when one wishes to use a distribution other than uniform, and this is obtained by a simple transformation. Let us consider how we might transform a set of uniformly distributed numbers to a distribution which is expressed by a probability density function

$$xe^{-ax^2}$$

This is illustrated in Fig. 1.5, where we see the evenly spaced points along the vertical axis and the transformed points along the horizontal axis. It can be seen that the points along the horizontal axis are obtained from the intersection of the equispaced vertical intervals with the curve R, where

$$R = \int_0^x xe^{-ax^2}\, dx = 1 - e^{-ax}$$

This is the integral of the desired probability density function. Since the first factor is constant, the function $R = e^{-ax^2}$ may equally well be used to obtain the desired distribution.

Now let us present a physical picture of the problem which Tien and Moshman studied. The d-c potential profile shown in Fig. 1.6 is computed from the equations of Section 1. The computer must memorize every electron in transit since interaction is considered. Therefore, for an assumed cathode-current density we must limit the area of the diode under consideration. If it is too large, we cannot handle the required number of electrons in the computer. On the other hand, if the area chosen is too small, the problem loses its random character. Therefore, an area of $(\pi/4)x^2{}_m$ was chosen since it was felt that all electrons emitted within this area had an equally important effect at the potential minimum. Bear in mind that this is a one-

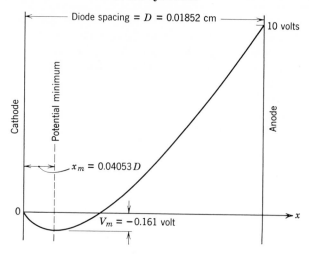

Fig. 1.6.   D-c potential distribution used in Tien–Moshman computations.

dimensional analysis, and so it contains none of the effects of sideways displacement and velocities, etc.

The velocity distribution used is as shown in Figure 1.7. For the calculation the time is quantized into intervals of 2×10^{-12} sec, and during this interval an average of 8.152 electrons are emitted in the area $(\pi/4)x_m{}^2$.

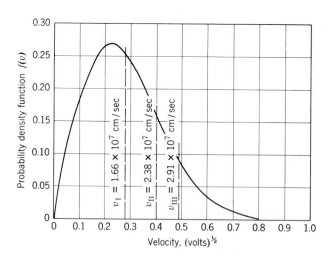

Fig. 1.7.   Maxwellian velocity distributions.

In the time interval from t_r to $t_r + \Delta t$, we must know the number of electrons emitted, the time of emission, and the velocity of emission. These quantities are found with the use of random numbers. For the number of electrons emitted, a Poisson distribution normalized about the average of 8.152 was used. The function

$$f(s) = \frac{e^{-8.1525} \times 8.1525^s}{s!}$$

is used to transform the random numbers, generated with a uniform distribution, to a set with a Poisson distribution.

Another set of random numbers with a uniform distribution is generated and used for the time of emission, since the emission would normally occur at a uniform rate throughout the velocity distribution.

For the velocity of emission a set of uniform numbers is transformed according to the Maxwellian distribution

$$f(v) = \frac{mv}{kT_c} \exp\left(-\frac{mv^2}{2kT_c}\right)$$

and

$$F(v) = \int_0^v f(v)\, dv = \left[1 - \exp\left(-\frac{mv^2}{2kT_c}\right)\right]$$

which gives

$$R_i = 1 - \exp\left(-\frac{mv^2}{2kT_c}\right) \quad \text{or} \quad R_i = \exp\left(-\frac{mv^2}{2kT_c}\right)$$

The velocity of emission corresponding to the uniformly distributed random number R_i is given by

$$v_i = \left(\frac{2kT_c}{m}\right)^{1/2} (-\log R_i)^{1/2}$$

This establishes the initial conditions, and we turn the crank until the diode is filled with 363 electrons. The process is then repeated 2000 times to obtain the final data on the noise current and velocity at the α plane, which is taken to be at $x = 1.2x_m$.

If we transform the current as a function of time to a plot of current as a function of frequency with the use of the autocorrelation function, we obtain a plot of Γ^2 versus frequency, as in Fig. 1.8. This curve answers the question which we initially posed, for we now have a picture of the noise current that is appropriate for the α plane. The peak occurs just above the frequency of oscillation of Whinnery's compensating current, and the dip occurs somewhat below this frequency. It might be stated that at the lower frequencies there is almost complete compensation of the initial disturbing pulse, and at the higher frequencies there is insufficient compensation. It seems that there occurs one frequency where the compensation is nearly complete. We should state that the effect of transverse velocities has not been evaluated and might well camouflage this effect.

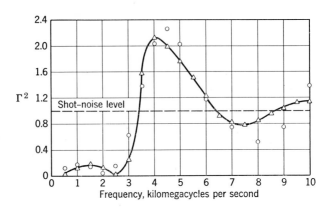

Fig. 1.8. Computed space-charge reduction factor Γ^2 as a function of frequency.

We also need to know the noise velocity at the α plane. The results of the computation can be summarized by stating that the computed noise velocity corresponds very closely with the expression of Eq. 1.74. Furthermore, within the limits of the accuracy of the computation, there is apparently no correlation between velocity and current.

This treatment is for an a-c short-circuited diode. At the frequency corresponding to the "dip" of the Γ^2 curve the transit angle is slightly less than 2π. Since physical diodes, or electron guns, are more nearly open-circuited, there is some concern as to whether the "dip" in noise current can be realized. Siegman and Bloom[21] have discussed† some linear models that enlarge upon the work of Whinnery and Watkins for the open-circuited diode and find no evidence of the minimum in noise current. On either side of this frequency region the agreement is fairly good.

Another limitation of this analysis must be kept in mind. At the location of the $x = 1.2x_m$ plane there is still a large spread in electron velocities. Thus in the region immediately beyond the plane, the single-velocity description may not be adequate. Work by Siegman, Watkins, and Hsieh[22] indicates that the multivelocity character of the

† See Chapter 5, Section 2, page 229.

beam is important for some distance beyond $1.2x_m$. They use a linearized theory to predict that correlation is produced between the current and velocity fluctuations as the beam passes from the potential minimum to the α plane.

In summarizing this discussion we see that the noise from a thermionic cathode is fairly well understood at low frequencies. In a high-frequency diode, however, the description of noise requires a knowledge of noise current, noise velocity, and their correlation at the α plane. The α plane is defined as that point beyond the potential minimum where the spread in velocities between electrons is small. Beyond the α plane one can use, with these noise parameters as input conditions, the single-velocity theory, and this will be fully treated in later sections. According to the work of Tien and Moshman at the plane $x = 1.2x_m$, the noise current is given by Fig. 1.8, the velocity has the value given in Eq. 1.74, and their correlation is zero. The later work of Siegman, Watkins, and Hsieh predicts that some correlation is introduced in the multivelocity beam as it drifts between the $1.2x_m$ plane and the α plane. It will be evident from later chapters how these parameters enter into the over-all noise figure for a high-frequency amplifier. It is sufficient here to point out that some degree of correlation between the noise velocity and current may lead to a noise figure for an amplifier which is less than that from an uncorrelated beam.†

REFERENCES

1. P. S. Epstein, "Theory of Space Charge Effects," *Verhandl. deut. physik. Ges.*, **21**, 85 (1919).
2. T. C. Fry, "The Thermionic Current between Parallel Plane Electrodes: Velocities Distributed according to Maxwell's Law," *Phys. Rev.*, **17**, 441 (1921).
3. I. Langmuir, "The Effect of Space Charge and Initial Velocities on the Potential Distribution and Thermionic Current between Parallel Plane Electrodes," *Phys. Rev.*, **21**, 419 (1923).
4. I. Langmuir and K. T. Compton, "Electrical Discharges in Gases," *Revs. Mod. Phys.*, **3**, 191 (1931).
5. B. J. Thompson, D. O. North, and W. A. Harris, "Fluctuations in Space-Charge-Limited Currents at Moderately High Frequencies," *RCA Rev.*, part I, **IV**, 269 (1940); part II, **IV**, 441 (Apr. and July 1940).

† *Editor's Note:* Recent measurements by Prof. S. Saito are in essential agreement with the predictions of Siegman, Watkins, and Hsieh. The measurements indicated a finite correlation between the velocity and current fluctuations under space-charge-limited operation. Under temperature-limited condition, the correlation disappeared.[23]

6. F. B. Llewellyn and L. C. Peterson, "Vacuum Tube Networks," *Proc. IRE*, **32**, 144 (1944).
7. F. N. H. Robinson, "Space Charge Smoothing of Microwave Shot Noise in Electron Beams," *Phil. Mag. London*, **43**, 51 (1952).
8. D. A. Watkins, "Noise at the Potential Minimum in the High Frequency Diode," *J. Appl. Phys.*, **26**, 622 (1955).
9. J. R. Whinnery, "Noise Phenomena in the Region of the Potential Minimum." *Trans. IRE* (PGED), **ED-1**, 221 (1954).
10. P. H. J. A. Kleijnen, "Extension of Langmuir's (ξ, η) Tables for a Plane Diode with Maxwellian Distribution of Electrons," *Philips Research Repts.*, **1**, 81 (1946).
11. A. Van der Ziel, *Noise*, chap. 14, Prentice-Hall, New York (1954).
12. W. Schottky, "Spontaneous Current Fluctuations in Various Conductors," *Ann. Physik*, **57**, 541 (1918).
13. S. O. Rice, "Mathematical Analysis of Random Noise," part I, *Bell System Tech. J.*, **23**, 282 (1944).
14. J. R. Pierce, "Noise in Resistances and Electron Streams," *Bell System Tech. J.*, **27**, 158 (1948).
15. J. R. Pierce, "General Sources of Noise in Vacuum Tubes," *Trans. IRE* (PGED), **ED-1**, 135 (1954).
16. A. J. Rack, "Effect of Space Charge and Transit Time on the Shot Noise in Diodes," *Bell System Tech. J.*, **17**, 592 (1938).
17. L. C. Peterson, "Space Charge and Transit Time Effects on Signal and Noise in Microwave Tetrodes," *Proc. IRE*, **35**, 1264 (1947).
18. J. R. Pierce, *Traveling-Wave Tubes*, chapter X, D. Van Nostrand Co., New York (1950).
19. D. K. C. MacDonald, "Variation of Space Charge Reduction Factor and Signal in Electron Streams under Certain Potential Distributions," *Phil. Mag. London*, **41**, 863 (1950).
20. P. K. Tien and J. Moshman, "Monte Carlo Calculation of Noise near the Potential Minimum of a High-Frequency Diode," *J. Appl. Phys.*, **27**, 1067 (1956).
21. A. E. Siegman and S. Bloom, "An Equivalent Circuit for Microwave Noise at the Potential Minimum," *Trans. IRE* (PGED), **4**, 295 (1957).
22. A. E. Siegman, D. A. Watkins, and Hsung-Cheng Hsieh, "Density-Function Calculations of Noise Propagation on an Accelerated Multi-Velocity Electron Beam," *J. Appl. Phys.*, **28**, 1138 (1957).
23. S. Saito, "New Method of Measuring the Noise Parameters of the Electron Beam, Especially the Correlation between Its Velocity and Current Fluctuations," *MIT Research Lab. Electronics Tech. Rept. 333* (1957); a shorter version has been submitted to the *Proc. IRE*.

CHAPTER 2

Low-Frequency Noise in Vacuum Tubes (Flicker Effect)

A. van der Ziel

1. Introduction

Flicker effect is the large amount of noise over and above shot noise found in vacuum tubes at low frequencies. It was discovered by J. B. Johnson in 1925 in saturated tubes.[1] In 1926 W. Schottky published a theory of the effect,[2] and in 1937 he showed that flicker effect should be more strongly suppressed by space charge than in shot effect.[3] One of the characteristics of the effect is its peculiar spectrum, which is usually of the form $f^{-\alpha}$ with α close to unity. Exceptions do occur, however; in those cases the spectrum is often of the type $1/(1 + \omega^2\tau^2)$. Sometimes (as, for example, in anomalous flicker effect) τ is quite small (10^{-5} sec) so that the spectrum is flat over a wide frequency range. In other cases τ is so large that the spectrum is of the form f^{-2} for all frequencies at which it has been measured (tungsten cathodes).

A study of flicker noise is important for the following reasons:

(*a*) The sensitivity of audio, subaudio, and d-c amplifiers is limited by the flicker noise generated in the first tube.

(*b*) If the flicker noise were better understood, flicker-noise measurements might be used as a tool in the study of the emission process.

In this chapter the theoretical and experimental data will be presented with little emphasis upon the understanding of the $1/f$ spectrum as such, but with a strong emphasis upon the mechanism(s) causing the noise. In my opinion it is premature to look for an explanation of the $1/f$ spectrum at a time when the exact mechanism of the noise has not yet been established.

This chapter is divided into six parts. The method of noise characterization is discussed in the first section. The second section discusses the theories proposed by Schottky and their application to diodes, triodes, and pentodes. In addition a discussion is given of the flicker noise in secondary-emission tubes and in phototubes. The third section discusses the influence of tube dimensioning and operating conditions upon flicker noise. The influences of the base nickel and of the coating upon flicker noise are discussed in the fourth section. The fifth section gives an account of noise in tungsten cathodes (including anomalous flicker effect), in thoriated tungsten cathodes, and in L cathodes. The last section discusses the implication of certain recent experiments and deals with some recent theories of flicker noise.

Noise Characterization

We now turn to the characterization of noise in two-terminal networks (diodes, coating) and in active four-terminal networks (triodes, pentodes, etc.).

In a two-terminal network the noise in a small frequency interval Δf can be characterized by the equivalent emf $\sqrt{\overline{e^2}}$ in series with the device, or by the equivalent current generator $\sqrt{\overline{i^2}}$ in parallel with the device. These quantities are often expressed per unit bandwidth, but it is more convenient to introduce instead the equivalent noise resistance R_n, the equivalent saturated diode current I_{eq}, and the noise ratio n of the device, since these quantities are independent of bandwidth Δf.

The noise resistance R_n is defined by

$$\overline{e^2} = 4kTR_n\,\Delta f \tag{2.1}$$

the equivalent saturated diode current I_{eq} is defined by

$$\overline{i^2} = 2eI_{eq}\,\Delta f \tag{2.2}$$

The definitions (2.1 and 2.2) follow directly from the well-known formulas for Johnson and shot noise. R_n for example is the resistance at the equilibrium temperature T which would exhibit across its terminals the open-circuit noise voltage $\overline{e^2}$.

An actual device with a resistance R may produce a higher noise voltage across its terminals than an ideal resistor of the same resistance and temperature. It is often convenient to define a noise ratio n that accounts for the deviation.

$$\overline{e^2} = n \times 4kTR\,\Delta f, \qquad \overline{i^2} = n \times 4kTG\,\Delta f \tag{2.3}$$

In the first definition of Eq. 2.3, the impedance Z of the device is represented by a resistance R and a reactance X in series; in the second definition, the admittance Y of the device is represented by a conductance G and a susceptance B in parallel. The following relations are easily proved:

$$n = \frac{R_n}{R}, \qquad n = \frac{e}{2kT}\frac{I_{eq}}{G} \simeq 20\,\frac{I_{eq}}{G} \tag{2.4}$$

$$I_{eq} = \frac{2kT}{e}\,R_n\left|Y\right|^2 \simeq \tfrac{1}{20}R_n\left|Y\right|^2 \tag{2.5}$$

Knowing the quantities R_n, I_{eq}, and (or) n is often an important step in determining the exact cause of the noise.

In an active two-terminal-pair network, two sources are needed to characterize the noise behavior. In an equivalent π representation, the most general circuit has three noise-current generators, i_{ab}, i_{ac}, and i_{bc}, connected between the three nodal points a, b, and c. This can be transformed into an equivalent circuit consisting of a current generator $(i_{ab} + i_{ac})$ connected between a and b, and a current generator $(i_{ac} + i_{bc})$ connected between b and c, Fig. 2.1. In an equivalent T representation the most general circuit consists of three noise emf's in series with the three leads; this can be reduced to an equivalent

Fig. 2.1. Equivalent noise-current-generator configuration for two-terminal-pair networks.

circuit with only two noise emf's in a similar manner.   We thus have the following possibilities:

1. Two current generators i_1 and i_2 connected across input and output, respectively.
2. Two noise emf's e_1 and e_2 connected in series with input and output, respectively.
3. One noise emf e and one current generator i; this case has four possible configurations, the most important ones are:

(a) Noise emf in series with *input*, noise current generator in parallel to *output*.
(b) Noise emf in series with *input*, noise current generator in parallel to *input*.

The two noise sources are in general partially correlated.   For a proper determination of the noise behavior of the network, we have to determine four quantities.   For example, in case 1 we must know $\overline{i_1^2}$, $\overline{i_2^2}$, and the real and imaginary parts of $\overline{i_1^*i_2}$.   In some simple cases one of the two noise sources may be negligible (reducing the four quantities to only one), or the two noise sources may be independent (reducing the four quantities to two).   The former case is found in low-frequency tubes, because no induced grid noise occurs at low frequencies (Chapter 4).

The case of flicker noise is thus particularly simple in that only one quantity is needed; the noise can be represented either by a noise current generator $\sqrt{\overline{i^2}}$ in parallel to the output or by an equivalent noise emf $\sqrt{\overline{e^2}}$ in series with the input.   As with the two-terminal network, we may introduce the equivalent noise resistance R_n with the help of Eq. 2.1 or the equivalent saturated diode current with the help of Eq. 2.2.   The two quantities are of course related by the equation

$$4kTR_n\,\Delta f \cdot g_m^2 = 2eI_{eq}\,\Delta f, \qquad I_{eq} = \left(\frac{2kT}{e}\right)R_ng_m^2 \qquad (2.6)$$

where g_m is the transconductance of the tube.   It is important to measure both quantities, however, since it is not known in advance which of the two quantities offers the simplest interpretation of the observed phenomena.   For measuring methods we refer to a paper by Nielsen and van der Ziel.[4]

2. Theory of Flicker Noise

Schottky's Theory of Flicker Effect

Schottky ascribed the effect of flicker noise to foreign atoms arriving at and departing from the surface at random.[2] He assumed that these atoms spent an average time τ_0 on the surface, and calculated the noise spectrum in the following way.

A single foreign atom, arriving at a point of the cathode and producing an electric moment p causes a change θ in potential: The effect will extend nearly uniformly over a circle of area σ around the point† so that $\theta = p/\sigma\epsilon_0$. If J_0 is the average cathode-current density of the *temperature-limited* cathode, then a change θ in surface potential will change the current density J_0 of that area by a factor $\exp(e\theta/kT_c) \simeq (1 + e\theta/kT_c)$, where T_c is the cathode temperature. The atom thus produces a current pulse with the amplitude

$$i_p = \frac{\sigma J_0 e\theta}{kT_c} = FJ_0, \qquad F = \frac{pe}{kT_c\epsilon_0} \qquad (2.7)$$

of average duration τ_0. Let N foreign atoms sit on the surface; the fluctuating current is then, if $\bar{N}$ denotes the average value of N,

$$i(t) = (N - \bar{N})i_p \qquad (2.8)$$

Assuming a Poisson distribution for N, we have $\overline{(N - \bar{N})^2} = \bar{N}$, and hence we obtain according to Chapter 6, if averages are taken over an ensemble,

$$\overline{(N - \bar{N})_t(N - \bar{N})_{t+\tau}} = \bar{N}e^{-\tau/\tau_0} \qquad (2.9)$$

Making a Fourier analysis of $i(t)$ and applying the Wiener–Khintchine theorem yield for the mean-square value of the Fourier coefficient i of $i(t)$

$$\overline{i^2} = \frac{4\bar{N}i_p{}^2 \,\Delta f\tau_0}{1 + \omega^2\tau_0{}^2} = \frac{4\bar{N}F^2J_0{}^2 \,\Delta f\tau_0}{1 + \omega^2\tau_0{}^2} \qquad (2.10)$$

For saturated (temperature-limited) tubes, $\overline{i^2}$ should thus be proportional to the square of the current density; this agreed reasonably well with Johnson's experiments.[1]

The frequency dependence also agreed reasonably well. For a

---

† The atom is assumed to be absorbed as an ion; if the ion has a charge e and its image charge is located at a distance d from the ion, then $p = ed$. To assume that the effect will extend nearly uniformly over a circle of area σ is of course an approximation, which is probably justified since σ does not appear in the final result given in Eq. 2.7.

tungsten filament the spectrum varied as f^{-2}; this follows from Eq. 2.10 if $\omega\tau_0 \gg 1$. This implies a large value of τ_0, which in itself is not unreasonable. For an oxide-coated cathode, it appeared that the spectrum could be approximated by Eq. 2.10, since Johnson's experiments indicated the possibility of a leveling off of the spectrum at low frequencies. We now know that usually this leveling off does not occur and that the spectrum is of the form $f^{-\alpha}$ with α close to unity. We may explain this by assuming not a single lifetime τ_0, but a suitably chosen wide distribution of lifetimes instead.† We refer here to the literature.

Space-Charge Suppression of Shot and Flicker Noise

In the theory of shot effect, the emission of an electron is considered to be an independent event occurring at random. In the above theory, however, flicker effect is ascribed to fluctuations in the emission current *as a whole*. Schottky showed in 1937[3] that this leads to a difference in space-charge suppression of flicker effect and of shot effect. Since his theory contains an error, we give it here in amended form.[7,8] At the same time we give the equation for shot and flicker noise in a diode for the three parts of its characteristic: *saturated region*

† If $\phi_{ii}(\omega)$ is the self-power density spectrum of a noise mechanism with a time constant τ_0 and $g(\tau_0)\,\Delta\tau_0$ is the probability of finding a time constant between τ_0 and $(\tau_0 + \Delta\tau_0)$, then

$$\phi_{ii}(\omega) = \int_0^\infty \phi_{ii}(\omega, \tau_0)\, g(\tau_0)\, d\tau_0$$

Van der Ziel[5] assumed that $\phi_{ii}(\omega, \tau_0) = [\text{const } \tau_0/(1 + \omega^2\tau_0^2)]$, in which case the distribution function

$$g(\tau_0) = 1/[\tau_0 \ln (\tau_2/\tau_1)] \quad \text{for } \tau_1 < \tau_0 < \tau_2$$

$$g(\tau_0) = 0 \qquad\qquad \text{for all other } \tau_0$$

gives

$$\phi_{ii}(\omega) = \frac{\text{const } [\tan^{-1} (\omega\tau_2) - \tan^{-1} (\omega\tau_1)]}{\omega}$$

which varies as $1/\omega$ in a wide frequency range if τ_1 and τ_2 are far enough apart. If the value of τ_0 is determined by an activation energy E_0, so that[5]

$$\tau_0 = \tau_{00} \exp (eE_0/kT)$$

Then the distribution in τ_0 corresponds to the following simple distribution $g(E_0)\,\Delta E_0$ in E_0:

$$g(E_0) = \text{const} \quad \text{for } E_1 < E_0 < E_2, \qquad g(E_0) = 0 \quad \text{otherwise}$$

Burgess proved in a recent paper that the above formula for $\phi_{ii}(\omega)$ is correct under certain conditions.[6]

(anode current I_a = emission current I_s), *space-charge-limited region* (potential minimum between cathode and anode), and *exponential region* (no potential minimum, anode voltage <0).

Let ΔI_s be the average current emitted with an initial energy between V_s and $(V_s + \Delta V_s)$ volts, and let the emission current ΔI_s in the energy interval show a small fluctuation $\delta(\Delta I_s)$. This will give rise to an anode-current fluctuation $\delta(\Delta I_a)$ that is linearly related to $\delta(\Delta I_s)$, so that

$$\delta(\Delta I_a) = \gamma(V_s)\delta(\Delta I_s), \qquad \Delta i_a = \gamma(V_s)\,\Delta i_s \qquad (2.11)$$

where Δi_a and Δi_s are the Fourier coefficients of $\delta(\Delta I_s)$ and $\delta(\Delta I_a)$. The coefficient $\gamma(V_s) = 1$ for all electrons in the saturated region; in the exponential region, $\gamma(V_s) = 1$ for electrons arriving at the anode, and $\gamma(V_s) = 0$ for electrons returning to the cathode. The expression $\gamma(V_s)$ is much more complicated for the space-charge-limited region. Equation 2.11 holds for both shot noise and flicker noise.

For shot noise the fluctuations in the individual energy intervals are independent, and

$$\overline{\Delta i_s{}^2} = 2e\,\Delta I_s\,\Delta f \qquad (2.12)$$

so that

$$\overline{i_a{}^2} = \Sigma\,\overline{\Delta i_a{}^2} = \Sigma\gamma^2(V_s)\,\overline{\Delta i_s{}^2} \qquad (2.13)$$

Thus, in the exponential and saturated regions:

$$\overline{i_a{}^2} = 2eI_a\,\Delta f \qquad (2.13a)$$

In the space-charge-limited region (compare Chapter 1),

$$\overline{i_a{}^2} = \theta 4kT_c g\,\Delta f \qquad (2.13b)$$

where k is Boltzmann's constant, g the diode conductance, and $\theta \simeq 3(1 - \pi/4) = 0.644$. Since $g = (eI_a/kT_c)$ in the exponential region, Eq. 2.13b also holds for the exponential region but with $\theta = \frac{1}{2}$. Another way of writing the equation is

$$\overline{i_a{}^2} = 2eI_a\Gamma^2\,\Delta f \qquad (2.14)$$

where Γ^2 is the space-charge suppression factor. $\Gamma^2 = 1$ in the exponential and saturated regions, and

$$\Gamma^2 = 2\theta\,\frac{kT_c}{e}\,\frac{g}{I_a} \qquad (2.14a)$$

in the space-charge-limited region.

For flicker noise the fluctuations in the individual energy intervals are completely correlated, so that

$$\overline{i_a{}^2} = \overline{(\Sigma \, \Delta i_a)^2} = \overline{[\Sigma \gamma(V_s) \, \Delta i_s]^2} = \overline{(i_{sf} \, \partial I_a/\partial I_s)^2}$$
$$= \overline{i_{sf}{}^2}(\partial I_a/\partial I_s)^2 \quad (2.15)$$

where i_{sf} is the Fourier coefficient of the emission-current fluctuation caused by flicker noise: $\partial I_a/\partial I_s \simeq kT_c g/eI_s$ in the space-charge-limited region, $\partial I_a/\partial I_s = I_a/I_s = kT_c g/eI_s$ in the exponential region, and $\partial I_a/\partial I_s = 1$ in the saturated region.

We now rewrite the formula for shot noise as follows:

$$\overline{i_a{}^2} = \chi \overline{i_s{}^2} \, (\partial I_a/\partial I_s), \qquad \overline{i_s{}^2} = 2eI_s \, \Delta f \quad (2.16)$$

Here $\chi = 1$ for the exponential and the saturated region, and $\chi = 2\theta \simeq 1.29$ for the space-charge-limited region.

Since $\partial I_a/\partial I_s < 1$ in the space-charge-limited region, flicker noise, being dependent on $(\partial I_a/\partial I_s)^2$, is more strongly suppressed by space charge than shot noise. This comes from the different way in which the contributions of the individual energy intervals had to be added. The validity of this conclusion hinges on the basic assumption that flicker noise is a fluctuation in the emission current.

It is common practice to represent the noise of a nonsaturated diode, high-gain triode, or pentode either by its equivalent-saturated diode current I_{eq} or by its equivalent noise resistance R_n, according to the definitions:

$$\overline{i_a{}^2} = 2eI_{eq} \, \Delta f = 4kT R_n \, \Delta f g^2 \quad (2.17)\dagger$$

Adding the contributions from shot noise and flicker noise, we have, from Eq. 2.15,

$$I_{eq} = \left(\frac{2\theta kT_c}{e}\right) g + g^2 \left(\frac{kT_c}{eI_s}\right)^2 \left(\frac{\overline{i_{sf}{}^2}}{2e \, \Delta f}\right) = I_{eq,s} + I_{eq,f} \quad (2.17a)$$

$$R_n = \left(\frac{\theta T_c}{T}\right)\frac{1}{g} + \left(\frac{kT_c}{eI_s}\right)^2 \frac{\overline{i_{sf}{}^2}}{4kT \, \Delta f} = R_{ns} + R_{nf} \quad (2.17b)$$

The first term is due to shot noise, and the second term is due to flicker noise; we denote these terms by the subscripts s and f, respectively.

The shot-noise resistance R_{ns} decreases with increasing plate current I_a, while $I_{eq,s}$ increases. (Measurements on diodes at frequencies where flicker noise is unimportant show more noise than predicted. This excess noise comes from electrons reflected at the anode.)[9]

† If Eq. 2.17 represents noise in a triode, g must be replaced by the transconductance g_m.

The flicker-noise resistance R_{nf} is independent of the tube current I_a, whereas $I_{eq,f}$ increases with increasing I_a. g is proportional to $I_a{}^p$, where p is found to be a slow function of the current: p is equal to 1 in the exponential part of the characteristic and gradually decreases with increasing current to the limiting value $p = \frac{1}{3}$, required by Child's law, at high currents. Therefore $I_{eq,f}$ should be proportional to $I_a{}^{2p}$, whereas $I_{eq,s}$ is proportional to $I_a{}^p$. We shall come back to this later.

Measurements at low frequencies at which flicker noise predominates show that R_{nf} is roughly independent of I_a for small I_a (but often increases with increasing plate current for large I_a). This might seem to verify the basic assumption of flicker-noise theory, according to which the noise is due to a fluctuation in emission. Later we shall see that the agreement is probably accidental.

Flicker Noise in Triodes[8]

We distinguish between two cases (see Fig. 2.2):

(a) The inhomogeneity in the potential distribution extends only to the region around the grid wires; the cathode current density is uniform. This condition should exist far from cut-off.

(b) The inhomogeneity in the potential distribution extends to the cathode, and the cathode-current density is nonuniform. In extreme cases only parts of the cathode area contribute to the anode current (island effect).

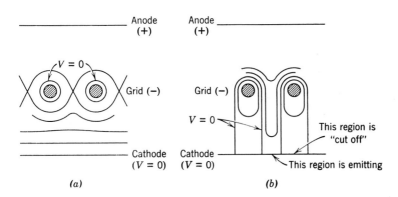

Fig. 2.2. Potential distribution in cathode-grid region (a) far above cut-off; (b) large negative bias. (a) No island effect. (b) Island effect.

In the first case we can apply the standard approximation method of replacing the grid structure by an equivalent electrode at an equivalent potential V_e. If V_g is the grid potential, V_a the anode potential, and μ_a the amplification factor of the tube, then

$$V_e = \frac{V_g + V_a/\mu_a}{1 + 1/\mu_a + 1/\mu_c} \tag{2.18}\dagger$$

The transconductance g_m of the actual triode is slightly smaller than the conductance g_e of the equivalent diode formed by the cathode and the equivalent electrode; it is easily seen that

$$g_m = g_e(\partial V_e/\partial V_g) = \sigma g_e \tag{2.19}$$

According to Eq. 2.18, σ is given as

$$\sigma = \frac{1}{1 + 1/\mu_a + 1/\mu_c} \tag{2.18a}$$

If anode and grid are connected, $V_a = V_g$, and

$$\sigma = \frac{1 + 1/\mu_a}{1 + 1/\mu_a + 1/\mu_c} \tag{2.18b}$$

The equivalent saturated diode currents of the triode and of the equivalent diode are equal. We can use the diode formulas for the equivalent diode if we replace g by g_e;‡ to obtain the correspond-

---

† The proof of Eq. 2.18 goes as follows: Let V_g, V_a, and V_c be the grid, anode, and cathode potentials, respectively; then the potential at any point in the tube must be a linear function of V_g, V_a, and V_c. Hence V_e must also be a linear function of V_g, V_a, and V_c, so that it may be written

$$V_e = aV_g + bV_a + cV_c$$

If $V_g = V_a = V_c$, then $V = V_g$ everywhere, and hence $V_e = V_g$, so that

$$a + b + c = 1$$

Substituting $b/a = 1/\mu_a$ and $c/a = 1/\mu_c$, and putting $V_c = 0$ yields Eq. 2.18. The calculation of μ_a and μ_c is a potential theory problem.[10]

μ_c has the following meaning: If the anode and cathode are interchanged, the new tube thus obtained has an amplification factor $\mu_c = \mu'$ if space charge is neglected. In general, we write $\mu_c = c_0\mu'$. The quantity c_0 is often taken to be $\frac{3}{4}$; it takes account of the space charge between cathode and grid. It may be shown that, in this approximation, $\mu_c/\mu_a \simeq (\frac{3}{4})d_{cg}/d_{ag}$, where d_{cg} and d_{ag} are the cathode-grid and anode-grid distances, respectively.[10]

‡ This rule should be applied with some caution; it holds for I_{eq} but not for R_n (compare Eqs. 2.20a and 2.20b). The reason is that I_{eq} and R_n are related by Eq. 2.6.

ing formula for I_{eq} in our triode we replace g_e by g_m/σ. Assuming Schottky's theory of flicker noise to be correct, we have, from Eq. 2.17a,

$$I_{eq} = \left(\frac{2\theta k T_c}{\sigma e}\right) g_m + g_m{}^2 \left(\frac{kT_c}{\sigma e I_s}\right)^2 \left(\frac{\overline{i_{sf}{}^2}}{2e\,\Delta f}\right) \qquad (2.20a)$$

and, instead of Eq. 2.17b, because of Eq. 2.6:

$$R_n = \left(\frac{\theta T_c}{T\sigma}\right)\frac{1}{g_m} + \left(\frac{kT_c}{\sigma e I_s}\right)^2 \frac{\overline{i_{sf}{}^2}}{4kT\,\Delta f} \qquad (2.20b)$$

For good tubes, $\sigma = 0.8$ to 0.9, but for tubes with wide grid spacing, σ may be smaller. The influence of the grid structure is thus relatively small, and, even if Schottky's theory is incorrect, we would not expect a large effect. In particular we would expect only a very small difference between the flicker noises of a tube in diode or in triode connection. Experiments at the University of Minnesota agree reasonably well with the latter prediction.

If the inhomogeneous potential distribution extends to the cathode, the case is quite different. Whereas in many tubes R_n is constant for a range of currents around 0.5 to 1.0 ma, its value usually increases with decreasing current at low current levels, often as $I_a{}^{-1}$. In addition, the triode is often much noisier than the same tube connected as a diode at low-current levels. We shall come back to this problem later.†

Flicker Noise in Pentodes[8]

If the noise of a pentode is known for the triode connection, it is easy to predict the noise in pentode connection. Let the noise in the triode connection be represented by a noise-current generator $\sqrt{\overline{i_c{}^2}}$ consisting of a shot-noise part and a flicker-noise part,

$$\overline{i_c{}^2} = 4kTR_{nt}\,\Delta f \cdot g_{mt}{}^2 = \overline{i_{cs}{}^2} + \overline{i_{cf}{}^2} \qquad (2.21)$$

where R_{nt} is the noise resistance and g_{mt} the transconductance of the tube in the triode connection. Let I_c be the cathode current, I_a the anode current, and I_2 the screen-grid current. It is usually assumed

---

† The shot-noise measurements agree reasonably well with Eq. 2.20b. Shot noise and flicker noise behave differently as far as their current dependence is concerned. In the case of shot noise, R_n also varies as $I_a{}^{-1}$ at low current levels, but for a different reason; it comes about because g_m varies as $I_a{}^{-1}$ for small I_a. But, whereas the flicker-noise resistance goes through a shallow minimum for plate currents around 0.5 to 1.0 ma, this is not true of shot noise; its value usually decreases with increasing plate current.

that the distribution of electrons between screen grid and anode is a random process such that each electron has the same probability (I_a/I_c) of arriving at the anode.† In that case the transconductance g_m of the pentode is $g_m = g_{mt}(I_a/I_c)$ and the cathode noise current i_c gives rise to an anode noise current i_{ac}:

$$\overline{i_{ac}^2} = \overline{i_c^2}\left(\frac{I_a}{I_c}\right)^2 = \overline{i_c^2}\left(\frac{g_m}{g_{mt}}\right)^2 \tag{2.22}$$

The random distribution of electrons between screen grid and anode also gives rise to an additional partition noise current i_p flowing from screen grid to anode such that (Chapter 4)

$$\overline{i_p^2} = 2e\left(\frac{I_a I_2}{I_c}\right)\Delta f \tag{2.23}$$

No extra flicker-noise term should be expected here if the above assumption really holds and if flicker is a fluctuation of emission as a whole. Putting

$$\overline{i_a^2} = \overline{i_{ac}^2} + \overline{i_p^2} = 4kTR_n\,\Delta f \cdot g_m^2 \tag{2.24}$$

we find, if R_{np} is the partition noise resistance,

$$R_n = R_{nt} + R_{np}, \qquad R_{np} = \left(\frac{e}{2kT}\right)\left(\frac{I_a I_2}{I_c}\right)\frac{1}{g_m^2} \tag{2.24a}$$

At frequencies at which flicker noise predominates, we would expect no difference in noise resistance between the pentode and triode connections. This agrees reasonably well with experiments conducted at the University of Minnesota. In addition, many earlier experiments at higher frequencies have verified Eq. 2.24a for shot and partition noise. More accurate experiments by Tomlinson showed that flicker noise in the pentode connection is slightly larger than in the triode connection;[11] recent work at the University of Minnesota seems to agree with his results.

Flicker Noise in Secondary-Emission Tubes

Experiments at the University of Minnesota seem to indicate that a flicker noise with a $1/f$ spectrum is associated with the secondary-emission process. Measurements were carried out on Philips EFP60

† Experiments at the University of Minnesota seem to indicate that this assumption is incorrect for flicker noise.

tubes,† and it was found that the noise resistance of the tube in pentode connection was definitely smaller than when the tube was used as a secondary-emission tube. The relative magnitude of this secondary-emission flicker noise, which is not very strong in some tubes, increases with increasing primary current, and also increases with increasing primary energy. The effect can be interpreted as a fluctuation in the secondary emission factor δ.

If I_p is the primary current and g_{mp} the primary transconductance, then the anode current I_a and the transconductance g_m are

$$I_a = I_p \delta, \qquad g_m = g_{mp} \delta \qquad (2.25)$$

If fluctuations ΔI_p and $\Delta \delta$ occur in I_p and δ,

$$\Delta I_a = \delta \, \Delta I_p + I_p \, \Delta \delta \qquad (2.25a)$$

Taking mean-square values and dividing by the square of the transconductance, we obtain, for the equivalent noise emf ΔE of the tube,

$$\overline{\Delta E^2} = \frac{\overline{\Delta I_a^2}}{g_m^2} = \delta^2 \frac{\overline{\Delta I_p^2}}{g_m^2} + \left(\frac{I_p}{g_m}\right)^2 \overline{\Delta \delta^2} = \overline{\Delta E_p^2} + \left(\frac{I_p}{g_{mp}}\right)^2 \frac{\overline{\Delta \delta^2}}{\delta^2} \qquad (2.26)$$

where ΔE_p is the equivalent noise emf due to fluctuations in the primary current.

We know from triode and pentode measurements that for moderate primary current ($\simeq 1$ ma) the flicker-noise resistance is not very strongly dependent upon current and that (I_p/g_{mp}) increases considerably with increasing current. Fluctuations in δ would thus explain why the relative magnitude of the secondary emission noise increases with primary current. The observed dependence of $\overline{\Delta E^2}$ on primary energy means that $(\overline{\Delta \delta^2}/\delta^2)$ increases with increasing primary energy.

No flicker noise was observed in photomultipliers; these tubes have a noise spectrum that is flat down to 1 cycle and that can be interpreted as shot noise. Since the secondary-emission flicker noise varies as I_p^2, and the photo-emission flicker noise also varies as I_p^2 (see next section), we would not expect observable flicker noise because of the low primary-current levels involved.

† The EFP60 contains a cathode, control grid, and screen grid, as any pentode. By electron-optical means the electron beams coming out of the screen grid are deflected towards the secondary-emission electrode(s) or dynode(s), and the slow secondary electrons are collected by the anode, which has a higher potential than the dynode. Using the tube in "pentode connection" means that the dynode and the anode are connected together so that the secondary emission is made ineffective.

Flicker Noise in Photo-Emission [12]

Let I_p be the photo-emission current and ϕ be the work function of the surface. Fluctuation in work function should show up as fluctuations ΔI_p in I_p. Since the photons arrive at random, and the release of an electron by a photon is a random process of probability p, we would also expect noise on that account. We write, if n is the number of photons arriving per second,

$$I_p = enp = en_1 \qquad (2.27)$$

where $p(\phi)$ is a function of ϕ and $n_1 = np$ is the number of photoelectrons emitted per second. Fluctuations Δn_1 may occur in n_1 owing to three causes:

(a) Fluctuations $\Delta n = (n - \bar{n})$; they give a contribution $e^2 p^2 \overline{\Delta n^2}$ to $\overline{\Delta I_p{}^2}$.†

(b) Spontaneous fluctuations in n_1; they give a contribution $e^2 np(1 - p)$ to $\overline{\Delta I_p{}^2}$, according to the theory of partition noise.

(c) Induced fluctuations in n_1, caused by fluctuations $\Delta\phi$ in ϕ; they give a contribution $e^2 n^2 (dp/d\phi)^2 \overline{\Delta\phi^2}$ to $\overline{\Delta I_p{}^2}$.

Adding these three contributions quadratically and assuming that n has a Poisson distribution, so that $\overline{\Delta n^2} = n$, we obtain

$$\overline{\Delta I_p{}^2} = e^2 p^2 n + e^2 np(1 - p) + e^2 n^2 \left(\frac{dp}{d\phi}\right)^2 \overline{\Delta\phi^2}$$

$$= eI_p + I_p{}^2 \left(\frac{1}{p}\frac{dp}{d\phi}\right)^2 \overline{\Delta\phi^2} \quad (2.28)$$

The first term describes the shot noise of the photo-emission, and the second term describes the flicker noise. We see that the second term is proportional to $I_p{}^2$, whereas the first term is proportional to I_p. The flicker noise should thus become important only at the higher current levels. Work at the University of Utrecht[13] seems to indicate the presence of this effect.

3. Influence of Tube Dimensions and of Operating Conditions

Influence of Cathode Area

Consider n identical triodes of unit cathode area, each having a current J, transconductance g'_m, noise resistance R'_n, and equivalent

---

† ΔI_p is the fluctuation in current for a unit time interval (1 sec); the noise spectrum is related to $\overline{\Delta I_p{}^2}$, but not identical with it.

saturated diode current I'_{eq}. If they are connected in parallel, we have

$$I_a = nJ, \quad g_m = ng'_m, \quad I_{eq} = nI'_{eq}, \quad R_n = \frac{R'_n}{n}, \quad A = n \quad (2.29)$$

where A is the total cathode area. Hence, *at constant-current density J*, the noise resistance R_n is inversely proportional to the cathode area A, and the equivalent saturated diode current I_{eq} is proportional to the cathode area.

To find out how these quantities depend on the cathode area A at *constant current I_a*, we assume the following laws:

$$R'_n = CJ^\alpha, \qquad I'_{eq} = DJ^\beta \qquad (2.30)$$

and obtain

$$R_n = \frac{R'_n}{n} = CI_a^\alpha A^{-(1+\alpha)}, \qquad I_{eq} = nI'_{eq} = DI_a^\beta A^{1-\beta} \quad (2.30a)$$

If we substitute $g_m = \text{const } I_a^p$ as mentioned in Section 2, page 53, we obtain from Eq. 2.17,

$$\alpha = \beta - 2p \qquad (2.31)$$

Experimentally it turns out that β is often not strongly dependent on current for flicker noise. Since p changes from a value of $\frac{1}{3}$ at high currents to a value of $+1$ at low currents, α will change considerably with current. At current densities below 0.05 ma per cm^2, we often find $\alpha \simeq -1$ (since $p = 1$, this means $\beta \simeq 1$). With increasing current density, α increases, goes through zero at a current density of 0.5 to 1.0 ma per cm^2, and becomes slightly positive at higher current densities.

R_n decreases with increasing A as long as the cathode current density J is sufficiently high so that $\alpha > -1$. If A increases and I_a is kept constant, J becomes smaller and smaller, and R_n decreases less and less with increasing $A (\alpha \to -1$ for $J \to 0)$. Finally, if $\alpha \simeq -1$, R_n is independent of A. We may thus improve R_n at constant I_a by going to larger cathode areas; a very small cathode area should give a large noise resistance.

Influence of Inhomogeneity in Cathode-Current Density

The grid structure and the grid bias may cause a considerable inhomogeneity in the cathode-current density J. To investigate this, we split the cathode area into elements ΔA and assume that an area ΔA

gives contributions

$$\Delta I_{\text{eq}} = DJ^\beta \Delta A, \qquad \Delta I_a = J \Delta A \qquad (2.32)$$

to I_{eq} and I_a. We then have

$$I_a = \Sigma J \Delta A$$

$$I_{\text{eq}} = D\Sigma J^\beta \Delta A = DI_a^\beta A^{1-\beta} \left(\frac{\Sigma J^\beta \Delta A}{I_a^\beta A^{1-\beta}} \right) \qquad (2.32a)$$

For a homogeneous cathode-current density $\Sigma J^\beta \Delta A = I_a^\beta A^{1-\beta}$, and we obtain Eq. 2.30a. If $\beta > 1$, and I_a is kept constant, the value of $\Sigma J^\beta \Delta A$ *increases* with increasing inhomogeneity in I_s. For $\beta = 1$, the value of I_{eq} is *independent* of the inhomogeneity in J. For $\beta < 1$, the value of $\Sigma J^\beta \Delta A$ *decreases* with increasing inhomogeneity in J.

This has been verified experimentally in many tubes by increasing the negative grid bias and simultaneously increasing the plate voltage such that the plate current remains constant. The more negative the grid is made, the stronger the inhomogeneity in cathode-current density J will be. Most tubes have $\beta > 1$ for flicker noise, though some tubes have β close to unity; usually $\beta < 1$ for shot noise at current densities above a few tenths of a milliampere per square centimeter. Consequently I_{eq} should increase with increasingly negative grid bias for flicker noise, except in these cases where $\beta \simeq 1$ for which I_{eq} should be independent of grid bias. For shot noise, however, I_{eq} should decrease with increasingly negative grid bias. This agrees roughly with experiments. The influence of negative grid bias upon I_{eq} at constant plate current I_a is especially strong at small plate current because island effect shows up strongest under that condition.

The noise resistance R_n is directly related to I_{eq}

$$R_n = \frac{e}{2kT} \frac{I_{\text{eq}}}{g_m^2} \qquad (2.33)$$

An increase in negative grid bias at constant plate current I_a usually causes a considerable decrease in g_m, especially at small I_a. Since $\beta > 1$ in most tubes, I_{eq} usually increases if V_g is made more negative; the noise resistance for constant I_a may thus increase strongly if V_g is made more negative. For that reason V_g should be made as little negative as possible, about -1.5 volts for all tubes with indirectly heated cathodes. If R_n is measured as a function of I_a at $V_g = -1.5$ volts, it is found that for most tubes R_n has a shallow minimum around plate-current densities of 0.5 to 1 ma per cm^2 at frequencies at which flicker noise predominates, whereas shot noise usually has a minimum noise resistance at the normal operating current densities. Consequently, normal tubes should be used at current densities of 0.5 to 1 ma at low frequencies; at higher frequencies for which shot noise becomes more important, the most suitable anode-current density shifts to higher values; the normal operating current should be used if shot noise predominates.

In a similar way we can discuss the difference in the flicker noise of a triode under normal conditions and the same tube used in diode connection ($V_a = V_g$) at the same anode current. Assuming that the quantity D is the same in diode and in triode connection, we find that, owing to island effect, I_{eq} is largest for the triode connection if $\beta > 1$ and largest for the diode connection if both $\beta < 1$; both are the same if $\beta = 1$. This agrees roughly with experiments, but the exceptions are more frequent than in previous experiments. Tomlinson has found that I_{eq} was larger for the diode connection in the exponential part of the characteristic;[14] this result is hard to explain.

The derivation of Eqs. 2.32a and 2.33 shows that, in tubes with a strong inhomogeneity in the cathode-current density, the quantity I_{eq} is of more fundamental importance than R_n. The latter depends upon g_m, which can vary from tube to tube. To understand the noise behavior of a tube, we should thus always measure I_{eq}. In comparing two different tube types, however, it is important to measure both R_n and I_{eq}. If I_{eq} does not differ much for the two types and R_n differs more strongly, then that difference is mainly due to a difference in g_m. We find for instance that 6AK5 tubes have a much larger flicker noise than 6AG5 tubes, especially for small values of I_a and large negative grid bias. A major part of this difference is due to a difference in g_m. But, even at $V_g = -1.5$ volts and even if the difference in g_m is taken into account, the 6AK5 tube seems noisier. This is probably due to a difference in the cathodes of the two types; its exact cause is not known.

Differences in noise may be very pronounced if the tubes show island effect, in which case only parts of the cathode contribute to I_a. The effect is especially strong in remote-cut-off tubes, where the remote-cut-off characteristic is obtained by putting one or more big holes in the grid. For large currents the total cathode area A contributes to I_a, but, if the grid bias is sufficiently negative, only those parts of the cathode right under the grid holes contribute to I_a. R_n for small I_a is then large, since

1. The effective cathode area is much smaller than A for small I_a.
2. The transconductance g_m is very small for small I_a. Remote-cut-off tubes are thus not recommended for low-noise circuits.

Dependence of Flicker Noise upon the Cathode–Grid Distance

Lindemann's experiments[15] on planar diodes and triodes with oxide-coated cathodes indicate that I_{eq} is independent of the cathode–grid distance d if the anode current I_a is kept constant in the process. Since the transconductance g_m is proportional to $1/d^2$, g_m will increase strongly if d is decreased, and hence R_n will decrease strongly with decreasing d. *Therefore tubes with very small cathode–grid distance are the most suitable ones for low flicker noise.* We shall discuss the implications of this result for the explanation of flicker effect later; we first deal with its implications for tube design.

Consider a tube such that the inhomogeneity in cathode-current density is negligible. In that case we have, according to Eq. 2.32a,

$$I_{eq} = DI_a{}^\beta A^{1-\beta} \tag{2.34}$$

We now change the cathode area A, the cathode–grid distance, and the grid geometry in such a manner that, for constant equivalent grid voltage V_2, the current I_a and the transconductance g_m remain constant. Then I_{eq} does not change if $\beta \simeq 1$, and it decreases with increasing d if $\beta > 1$. Hence, according to Eq. 2.33, the noise resistance R_n does not change for $\beta = 1$, whereas R_n decreases with increasing d if $\beta > 1$. Since miniature tubes are usually made in such a way that the characteristic is roughly equal to that of a larger prototype (I_a is thus kept constant in the scaling-down process), we conclude that miniaturization and subminiaturization in itself may not adversely influence the flicker-noise resistance R_n too strongly if $\beta \cong 1$.

The above conclusion holds under the condition "other things being equal." If different nickel alloys or different sealing and pumping techniques have to be used, the noise resistance may be adversely affected; all we have said is that it is not due to the changes in A and d.

Lindemann measured the GL6299 and WE416A microwave triodes, which have very small grid-cathode spacings. The first one, which has a very small cathode area, had a noise resistance comparable to the best high g_m tubes (150 kilohms at 10 cycles), whereas the second, which has a larger cathode area and a much larger transconductance, was considerably better than the best tube measured up to now (30 kilohms at 10 cycles). It should be pointed out that these two tube types are processed with extreme care, a fact that may by itself improve their noise performance.

Influence of Cathode Temperature

If the cathode temperature is lowered, e.g., by lowering the heater voltage in steps of 0.5 volt, and the plate current is kept constant at a value above a few hundred microamperes, then the noise resistance R_n increases at first very slowly and then somewhat more rapidly until the tube becomes saturated. This increase is mostly due to an increase in I_{eq} and is not due to a change in g_m, except close to saturation. At these currents the tubes work better than at the normal cathode temperature.

If the plate current is kept constant at a much lower value, e.g., 25 μa, a new effect occurs because much lower cathode temperatures can be used without driving the tube into saturation. First R_n increases slowly, then more rapidly, goes through a maximum, and then drops to a relatively low value before saturation. At the best operating point of those tubes, the transconductance g_m has already dropped to a lower value (as can be deduced from the increase in shot-noise resistance), but nevertheless the flicker-noise resistance may be relatively low so that that mode of operation could perhaps be recommended in some cases.

The reason for this behavior is not clear, and more work has to be done to gain a better understanding of the effect.

4. Influence of Cathode Material, Poisoning

The nickel used for the cathode sleeve contains some impurities which act as reducing agents, reducing part of the BaO to Ba. An excess concentration of Ba in the oxide coating is needed to reduce the work function and thus make the cathode a good emitter. We deal here with the influence of the impurities in the cathode nickel and of the coating upon flicker noise.

Interface Effects

If the cathode nickel sleeve contains a few tenths of a per cent of Si (such a nickel is called an "active" nickel because it readily activates the cathode), the silicon reduces part of the coating and forms a high-resistance Ba_2SiO_4 layer at the interface between nickel and coating. This layer gradually forms during aging. Its resistance increases with time and produces undesirable feedback effects.

It has been found that a huge amount of flicker noise may be generated in this layer.[16,17,18] The noise emf e generated in the layer can be represented by an equivalent emf e' in series with the grid, and e' in turn can be represented by the noise resistance R_n. Since $e' \simeq e$, R_n is a good measure for the interface noise.† The experiments can

† If other sources of noise are present, each may be represcribed by a series voltage generator in the grid circuit, or by a noise resistance R_n. The over-all noise resistance is the sum of the individual R_n's.

be represented by a formula of the type

$$R_n = \frac{C I_a{}^2 R_i{}^2}{f^\alpha} \qquad (2.35)$$

At 10 cycles, R_n may be 10^6 times as large as normal flicker noise at that frequency. Here I_a is the plate current, R_i the interface resistance, and a a factor that increases from a value of 1.1 for small R_i to a value of 1.5 for large R_i. The constant C depends only very slightly upon the way in which R_i is formed and Eq. 2.35 holds under a variety of conditions. We give here several examples:

If a tube is aged, R_i increases gradually, and so does R_n; the variation of R_n with R_i is given by Eq. 2.35. If the cathode temperature is decreased, R_i increases strongly with decreasing temperature, and so does R_n; the variation of R_n and R_i is again given by Eq. 2.35. R_i may also be varied by drawing current for a long time. It takes some time before R_i has dropped to its limiting value; this limiting value decreases with increasing amount of current drawn. After the interface layer has thus become "current-activated," it takes up to several hours before the old value of R_i is restored. By measuring quickly, it is thus possible to study R_n as a function of R_i; the result is again given by Eq. 2.35.

The dependence of R_n upon R_i and upon current indicates that the interface layer resembles a depletion layer (see Chapter 6). Noise due to a fluctuating carrier concentration would vary as $R_i{}^3$, but noise generated in a depletion layer actually varies as $R_i{}^2$. The resemblance between the interface and a depletion layer was also found by Nergaard in other experiments.[19]

In other respects the interface layer resembles a rectifying junction. Rectification has been observed by other investigators;[20] noise measurements made at the University of Minnesota at high frequencies point in the same direction. We found that the interface noise spectrum changed from $f^{-1.1}$ at low frequencies to $f^{-1.6}$ at high frequencies; the change-over occurred in the frequency range determined by the time constant of the interface layer. This is similar to the change in the spectrum of the shot-noise resistance in junction diodes which is constant at low frequencies and varies as $f^{-\frac{1}{2}}$ at high frequencies owing to lifetime effects.[21] The shot-noise spectrum is due to white-noise sources operating upon the junction; $1/f$ noise sources associated with these white-noise sources will give a contribution to the noise resistance that varies as $1/f^{-1}$ at low frequencies and as $f^{-1.5}$ at high frequencies.

From the above we see that high interface resistance results in a large excess noise at low frequencies. Therefore cathode base metals which tend to form high interface resistances should be avoided in low-frequency, low-noise applications. For example, high-silicon nickel, such as Inco 225, should not be used. More passive nickels (e.g., 499 nickel or perhaps Inco 220) or active nickels that do not form a high-resistance interface (e.g., Cathaloy A30 with 0.1% Al) are recommended.

Influence of Nickel Alloy upon Flicker Noise[15]

Comparative measurements on 6SN7 tubes with different cathode-nickel sleeves showed an influence of the cathode nickel upon flicker noise that was not due to interface formation. Arranged in order of increasing flicker-noise resistance, the alloys were: Cathaloy A30 (0.1% Al), 499 (passive), Inco 220, Inco 225 (active), and Cathaloy A31 (4% W). The difference in noise resistance between the first and the last alloy was about a factor 20, thus showing the importance of choosing the proper alloy for low-noise tubes. Later experiments on European magnesium–nickel alloys showed that these are even better than Cathaloy A30. These alloys are widely used by European tube manufacturers; the results thus give a reason why European tubes often have relatively low flicker noise.

The above measurements were carried out as follows: A number of tubes with different cathode nickels were built and processed in as nearly the same manner as possible. Often the two triodes in a single bulb contained cathode sleeves made of different alloys; the two halves were thus exposed to the same atmosphere all the time, which should reduce the uncontrolled differences in R_n. The tubes were then aged for 160 hours, after which all appeared fully activated; the trans-conductances agreed with those of normal tubes, and the pulsed emission was well within the range of variation of normal tubes. The noise resistance R_n was then measured at $I_a = 400$ μa, and logarithmic averages were taken over a considerable number of tubes. (If n tubes had noise resistances $R_{n1}, \cdots, R_{nn}$, respectively, we put log $R_n =$ $\left(\sum_{i+1}^{n} \log R_{ni} \right) /n$; that is, we expressed the noise in decibels and took the average, which seems a reasonable procedure if the noise resistance follows a log-normal distribution.)

Influence of Cathode Coating and of Emission

A comparison of the noise resistances of thick and thin coatings showed that thin coatings are in general somewhat less noisy than thick coatings.

There is little difference between cataphoretically deposited coatings and fluffier sprayed coatings.   Cataphoretic coatings are not so easily activated, but after activation there is little difference.

The "activity" of an oxide coating is usually measured by its pulsed emission, or sometimes by its low-temperature emission.   Earlier experiments on triodes showed little correlation between flicker noise and pulsed emission or between flicker noise and low-temperature emission.   These experiments were not very accurate because different triodes were compared; large individual differences of unknown origin between different tubes, which always seem to be present, could easily have masked the effect.   Later, more accurate experiments were performed on 2X2 diodes.[22]   These diodes have the advantage that the cathode activity can be changed in a reversible manner by flashing (heating the cathode to a high temperature) and subsequent reactivation; by measuring on the same tube, the individual differences between different tubes are eliminated.   These studied indicated a definite correlation between flicker noise and pulsed emission; low flicker noise (low I_{eq}) and low pulsed emission seem to go together.   Since the conductance of the diode does not depend strongly upon the emission current I_s as long as I_s is considerably larger than the plate current I_a, a low value of I_{eq} also corresponds to a low noise resistance R_n (Eq. 2.5).   This is not true if I_a and I_s are comparable; g then becomes very small.

In triodes with very inactive cathodes, g_m may be very small because the tube is operating almost in the saturated region; hence, though I_{eq} may be normal or even below normal, R_n is very large.   In one particularly inactive tube, R_n decreased by a factor 10^4 upon activation, whereas I_{eq} remained relatively constant.   This shows that it is important to operate tubes not too close to saturation.

Noise during Early Life, Cathode Poisoning

If tubes are aged, it is found that the noise usually drops during the first 100 hours.   Tubes to be used in low-noise audio circuits should therefore be properly aged before use.

The reason for the effect is not very clear.   In some inactive tubes, the effect might be caused by further activation during aging, but this is certainly not the general reason.   In one 6SN7 tube that was aged at $I_a = 20$ ma at an anode voltage of 20 volts (diode connection), the flicker noise showed first a rapid and strong increase followed by a considerable decrease to a value lower than the initial one.   Such a transient increase in noise was sometimes observed in other tubes; the increase in I_{eq} was often coupled with a slight decrease in g_m, and consequently with somewhat larger increase in R_n; it took considerable time before the tube recovered.

Probably these effects are due to poisoning caused by the electron bombardment of the anode. One may think here either of gas evolution from the anode or of decomposition of cathode material on the anode by the electron bombardment. It is not unreasonable to assume that some form of cathode poisoning exists during early life and that the decrease in R_n during early life is often due to recovery from these initial poisoning effects. This may also be one of the uncontrolled parameters in the flicker-noise generation. More work on this problem is needed.

5. Noise in Other Types of Cathodes

Tubes with Tungsten Filaments, Anomalous Flicker Effect[15,23]

If a tube with tungsten cathode (e.g., 5722 diode) is operated under saturated conditions, it will show a considerable increase in I_{eq} at low frequencies. In good tubes, the spectrum is flat down to about 30 cycles and then starts to increase,[24] sometimes as f^{-2}, sometimes as f^{-3}. In poorer tubes the noise is flat down to about 300 cycles and then shows the same type of spectrum; those tubes can be improved considerably by "flashing" the filament. Apparently the noise reduction is due to the evaporation of foreign atoms from the surface.

The f^{-2} spectrum may be understood as a spectrum of the type [const $\times \tau_0/(1 + \omega^2\tau_0^2)$] with $\omega\tau_0$ sufficiently large. An f^{-3} noise might be caused by a double mechanism, perhaps a $1/f$ noise mechanism modulating an emission source with a long time constant τ_0, this would give $(A/f)/(1 + \omega^2\tau_0^2)$ which varies as $1/f^3$ if $\omega\tau_0 \gg 1$.†

If the tube is operated space-charge-limited, one observes "anomalous flicker effect"; the noise spectrum contains a large component of the form [const $\times \tau_0/(1 + \omega^2\tau_0^2)$] with τ_0 of the order of 10^{-5} sec. Displayed on an oscillograph this extra noise shows up as a number of random pulses, all with the same polarity.

The effect is ascribed to positive ions emitted by the cathode and trapped in the potential minimum in front of the cathode. The ion may move around the cathode many times before it is actually captured by the cathode or neutralized by capturing an electron. As long as the ion will move around, it will raise the potential minimum and thus cause a large current pulse.

---

† If, for example, a $1/f$ noise source modulates the quantity F (or, what amounts to the same thing, the dipole moment p) of Eq. 2.10, then this would result in the required product spectrum.

To test this idea, Lindemann built a diode with an extra, solid electrode between filament and plate (Fig. 2.3). If the electrode and anode were connected together, and a certain amount of current drawn, anomalous flicker effect occurred; the effect disappeared if the electrode was made negative. This is understandable, for under that condition the electrode acts as an "ion sink," preventing the ions from being trapped in the space charge. This shows that positive ions are indeed the cause of the effect.

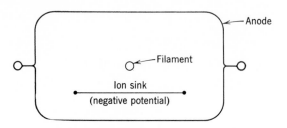

Fig. 2.3. Lindemann's experimental tube for measuring the effect of positive ions in the potential minimum region.

The effect disappeared in normal diodes as soon as the anode voltage was made negative, even though there was still a potential minimum and shot effect was considerably suppressed by space charge. This is understandable now, for the ions can be captured by the anode. It was also found that the anomalous effect disappeared before the tube showed full saturation. This is understandable, for, if the potential minimum is too close to the cathode, it is impossible to trap ions in the space charge.

A study of the individual current pulses causing the anomalous flicker effect revealed that they all showed a very rapid rise time followed by a slow exponential decay. The pulses could be approximated by the equation

$$i(t) = 0 \quad \text{for} \quad t < t_0, \qquad i(t) = i_0 \exp\left[-(t - t_0)/\tau_0\right]$$
$$\text{for} \quad t > t_0 \quad (2.36)$$

with random t_0. Making a Fourier analysis of these current pulses, one obtains indeed a noise spectrum of the type [const $\times \tau_0/(1 + \omega^2\tau_0^2)$], and the values for τ_0 obtained from the study of the pulse shape and from the noise spectrum agreed quite well. This indicates that the frequency dependence of the spectrum has to be ascribed to the shape of the individual pulses.

The size of the pulses is so large that it seems very unlikely that they are caused by single ions. The only reasonable explanation is that they are caused by relatively large "bursts" of ions which are gradually captured by the cathode; this may also explain the exponential decay.

In another experiment a triode with tungsten filament was built that did not show anomalous flicker effect. This was achieved by inserting a solid electrode (ion sink) into the tube covering half the grid and connected to the grid. By keeping the grid and the ion sink negative, no ions could be trapped in the space-charge cloud, so that no anomalous flicker effect could occur.

For tubes with tungsten filaments, flicker noise should be a surface effect, most likely caused by foreign atoms adsorbed on the surface. We would thus expect Schottky's theory of space-charge suppression to be true. Therefore, if $\overline{i_s^2}$ is the mean square of the Fourier coefficient of the emission fluctuation, and the tube is operating in the exponential part of the characteristic, we would expect (compare Eq. 2.15):

$$\overline{i_a^2} = \overline{i_s^2} \left(\frac{\partial I_a}{\partial I_s}\right)^2 \tag{2.37}$$

so that the equivalent saturated diode current I_{eq} should be much smaller in the exponential region than in the saturated region. Actually it was found that I_{eq} was not much different. Apparently the noise mechanism at low currents must be different from the noise mechanism in the saturated region, in contrast with Schottky's theory.

Noise in Tubes with Thoriated Tungsten Filaments

Measurements on RCA 8025 tubes with Th–W filaments indicated that the spectrum was quite current-dependent; at some currents the noise was small, and it varied as f^{-1}; at others it varied as f^{-3}, and the noise was several orders of magnitude larger. In the quiet operating conditions the noise resistance is very low (42,000 ohms at 10 cycles at $I_a = 3$ ma), but at the noisy operating points the noise resistance is very high. The reason for this behavior is not clear. Laboratory-built tubes with Th–W filament showed similar behavior.

Noise in Tubes with L Cathode

Measurements were conducted on tubes with L cathode. This cathode supposedly consists of a monoatomic layer of barium on oxygen on tungsten. For that reason Schottky's theory might apply, and, since the surface is almost completely covered, one might hope that the random fluctuations in work function causing the flicker noise might be small.

We found indeed that tubes with L cathodes had relatively low noise resistance. The flicker noise was of the $1/f$ type. In several tubes a noise term of the form [const $\times \tau_0/(1 + \omega^2\tau_0{}^2)$] was observed with τ_0 strongly temperature-dependent. This suggests that the value of the time constant is controlled by an "activation energy" E. In such a case we would expect (see footnote, Section 2, page 50):

$$\tau_0 = \tau_{00} \exp\,(eE/kT_0) \qquad (2.38)$$

where τ_{00} is a constant. Experimental values for τ_0 agreed with Eq. 2.38 and indicated a value for E of about 4 ev. The cause of the effect is not clear; further work will be needed to clarify the situation.

6. Present State of the Theory

The Implications of the Movable-Anode Experiments[15]

Movable-anode experiments in diodes and movable-grid experiments in triodes indicate that I_{eq} at constant plate current I_a is independent of the anode–cathode (or grid–cathode) distance d. This is in direct contrast with any theory based on the assumption that the noise is due to emission fluctuations, that is, to spontaneous changes in I_s unassociated with external fields. For, according to Eqs. 2.17a and 2.17b, R_n should be independent of d, and I_{eq} should depend upon d because it contains the factor g^2; it is well known that g at constant I_a depends strongly upon d. We must therefore conclude that flicker noise is not due to an emission fluctuation.

Neither can the noise be generated deep in the coating as a resistance noise. For in that case the noise can be represented by an emf e in series with the coating; $\overline{e^2}$ may depend upon the current flowing through the coating but should not depend upon the anode–cathode distance. Hence again R_n at constant I_a should be independent of d; I_{eq} should again depend upon d because it contains the factor g^2. This is ruled out by the experiment.

We are thus forced to conclude that the flicker noise is a true fluctuation in the plate current (I_{eq} independent of d), and that it cannot be interpreted either as a fluctuation in the emission current or as a noise emf generated deep in the coating. This also excludes Schottky's flicker-noise mechanism (Section 2, page 49) as a *direct* cause of the noise. What is left?

Further experiments by Lindemann gave the answer. The experiments were repeated at much lower plate currents ($I_a = 10$ to $50\ \mu$a), and there a new effect appeared. I_{eq} now remained constant down to a certain critical discharge d_c, and decreased for $d < d_c$. This can be associated with the existence of a potential minimum. As long as a potential minimum exists and I_a is kept constant, the depth and the position of the potential minimum and the potential distribution between potential minimum and cathode will not change. But, if d becomes smaller than d_c, no potential minimum exists, and the field strength at the cathode will *increase* with decreasing d. Apparently this causes a decrease in I_{eq}. How can the flicker noise of the tube depend upon the d-c field strength at the cathode? Only if the noise is caused by the field distribution in a thin surface layer of the coating; one may then understand how the field strength right outside the coating can influence the field distribution in the surface layer.

In a subsequent experiment Lindemann showed the existence of an appreciable potential drop in the surface layer, even at relatively low plate currents. A planar diode with movable anode was operated at $I_a = 20$ μa, and the cathode–anode distance was chosen such that no potential minimum existed between cathode and anode. The anode voltage was then kept constant, and the anode–cathode distance was gradually decreased to a very small value. If the potential of the oxide surface remained constant, I_a would not change. It was found, however, that I_a increased from 20 to 50 μa in the process, indicating a change in surface potential of about 0.1 volt. The work of Nergaard[19] on the existence of a thin surface layer (depletion layer) depleted of donors (and hence of electrons) points in the same direction; according to Nergaard, most of the voltage drop in the oxide coating is concentrated in this depletion layer. Work at Cornell University[25] has also shown the existence of a thin surface layer of decreased carrier density of a somewhat different character, whereas Jansen, Loosjes, and Compaan[26] have shown the existence of a high-resistance surface layer under pulsed conditions.

We may thus assume that a voltage drop exists in the surface layer of the coating. It is reasonable to assume that this voltage drop will fluctuate in time.† It will now be shown how this fluctuating voltage may indirectly cause a true fluctuation in the plate current for tubes with porous coatings.

The potential drop across the surface grains will cause a direct voltage and a fluctuating voltage to be applied across the surface pores. The first gives the electrons emitted by the pores sufficient energy to

---

† Van der Ziel[27] has discussed a model of such a thin surface layer which shows this effect (see Section 6, page 74). However, the noise mechanism suggested by van der Ziel is not the *direct* cause of the flicker noise for porous cathode coatings.

pass the potential minimum, the latter modulates the direct pore
current and causes a fluctuating pore current, which will be hardly
suppressed by the space charge because of the high initial energy of the
electrons coming out of the pores. Since the direct voltage and the
fluctuating voltage depend only upon the direct current flowing
through the coating, one obtains indeed a true fluctuation in *current*
that way.

Lindemann proved this with the help of a special tube with a very
dense oxide coating. The potential drop across the surface layer was
found to be of the same order of magnitude as in a tube with a porous
cathode, but the flicker noise differed from that of porous coatings, in
that the dependence of I_{eq} at constant I_a upon the anode–cathode
distance was more like what one would expect for a fluctuation in the
emission current. The porous structure of the cathode coatings may
thus explain the observed flicker noise in normal tubes.

Lindemann's experiments thus indicate that the flicker noise in tubes
with porous coatings is caused by injection of fast electrons into the
anode–cathode region in a way that is geometry-independent and
unaffected by the space-charge smoothing mechanism operating in that
region. They also point to the pores as the most likely source of this
injection mechanism. The reader should note that the pore mechanism
generates noise in an indirect manner; the experiments do not indicate
which mechanism is responsible for the fluctuating emf developed
across the pores.

We can now also understand some other results. If flicker noise is a
true fluctuation in the plate *current* I_a, then the fundamental formula
of flicker noise is the formula for I_{eq} given by Eq. 2.32a, and the
formula for R_n is derived from it. Experiments indicate that β is
often almost independent of current, whereas we know that the quantity p is strongly current-dependent. It is therefore not so surprising to find a current range in which R_n is independent of current.
It simply means that $\beta \simeq 2p$ in that current range (Eq. 2.31); it
should not be quoted to support Schottky's theory of space-charge
suppression.

The flicker noise generated thus depends upon the field distribution
in the surface layer; the same field distribution will also influence the
cathode activity and temperature, thus making flicker noise depend
upon cathode activity and upon cathode temperature in a complicated
manner.

Other Theories for Oxide-Coated Cathodes

Inuishi and his co-workers have found that the flicker noise decreased
with increasing cathode activity. He explained this effect by assum-

ing that the noise was due to fluctuations in the donor concentration. We shall show that the noise generated by this mechanism is many orders of magnitude too small, and that another effect should give several orders of magnitude more noise.[27]

Consider first a noise mechanism causing a fluctuation $\delta\phi$ in work function ϕ. According to Richardson's law, this gives an emission fluctuation

$$\delta I_s = - \left(\frac{e I_s}{k T_c} \right) \delta\phi, \qquad \overline{\delta I_s^2} = I_s^2 \left(\frac{e}{k T_c} \right)^2 \overline{\delta\phi^2} \qquad (2.39)$$

If this noise is suppressed by space charge, the anode-current fluctuation is

$$\delta I_a = \delta I_s (\partial I_a / \partial I_s) = -g_m \, \delta\phi \qquad (2.39a)$$

because of the expression for $(\partial I_a / \partial I_s)$. A fluctuation $\delta\phi$ in work function can thus be represented by an equivalent fluctuating emf $e = -\delta\phi$ in series with the input. The flicker-noise measurements on normal tubes indicate $\sqrt{\overline{e^2}} \simeq 10^{-6}$ volt, and hence an equivalent work-function fluctuation $\sqrt{\overline{\delta\phi^2}} \simeq 10^{-6}$ can explain the noise.

Now, according to the semiconductor theory of the oxide-coated cathode, the emission current I_s is proportional to the square root of the number of donors, n_d.[28] If δn_d is the fluctuation in the number of donors and $\overline{\delta n_d^2} = n_d$, as required for a Poisson distribution, we have

$$I_s = C n_d^{\frac{1}{2}}, \qquad \delta I_s = \frac{1}{2} I_s \frac{\delta n_d}{n_d}, \qquad \overline{\delta I_s^2} = \frac{1}{4} I_s^2 \frac{1}{n_d} \qquad (2.40)$$

so that indeed $\overline{\delta I_s^2}$ decreases with increasing activity (increasing n_d at constant I_s). By comparing Eqs. 2.39 and 2.40 we can estimate how large n_d should be to explain the noise. An equivalent work-function fluctuation $\sqrt{\overline{\delta\phi^2}} \simeq 10^{-6}$ can explain the noise and $(k T_c / e) \simeq$ 0.1 volt; hence a value $n_d \simeq 2.5 \times 10^9$ is needed to explain the noise.

This is many orders of magnitude too small. According to Nergaard,[19] a well-activated cathode has about 10^{18} donors per cm^3 at 1000° K. A cathode of 50 μ thickness and 1 cm^2 area thus has about 5×10^{15} donors. That should give only an equivalent noise emf of less than 10^{-9} volt. Some improvement may be obtained by assuming that only the donors in a thin surface layer contribute to the noise. It is obvious, however, that this cannot supply the factor 2×10^6 needed to obtain the required value of n_d.

A related mechanism gives much more noise. Donor levels in BaO lie at about 1.4 volts and 2.3 volts below the bottom of the conduction band. Only a very small portion of the donors is thus ionized at

1000° K; Nergaard mentions a value of 10^{15} electrons per cm^3 at 1000° K, which means that only 0.1% of the donors is ionized. The emission current I_s is proportional to the number n of electrons inside the material. Fluctuations δn in n should occur due to processes of the type

Free electron + ionized donor $\rightleftarrows$ neutral donor

Assuming $\overline{\delta n^2} = n$, we have

$$\delta I_s = I_s \frac{\delta n}{n}, \qquad \overline{\delta I_s^2} = I_s^2 \frac{\overline{\delta n^2}}{n^2} = I_s^2 \frac{1}{n} \qquad (2.41)$$

so that again $\overline{\delta I_s^2}$ would decrease with increasing activity at constant I_s. Comparing Eqs. 2.40 and 2.41, we see that 10^{10} electrons are needed to explain the noise. A density of 10^{15} electrons per cm^3 would give 5×10^{12} free electrons in a cathode coating of 1 cm^2 area and 50 μ thickness. This mechanism thus gives much more noise than Inuishi's mechanism. The right order of magnitude is obtained if only fluctuations in a thin surface layer contribute to the noise.

We know that this picture is too simple. Lindemann's experiments indicate indeed that the noise is generated in a thin surface layer. But the electron density in the surface layer is not constant, and I_s is not proportional to this density; the problem is much more complex. Moreover, we saw that the pore mechanism of Section 6, page 71, seems to be the predominant source of the flicker noise.

In the above model most of the coating resistance is concentrated in the surface layer; the fluctuations in the carrier density of that layer will thus also give rise to a noise emf of the type to be discussed in the next section.[22]

Noise Generated in the Coating[22]

Measurements on two coatings pressed together indicate a $1/f$ noise term proportional to $I^2 R^3$ for small R, where I is the current flowing through the coating and R the coating resistance. For larger values of R, depletion effects occur at one of the electrodes, causing an increase in R and a $1/f$ noise term varying as $I^2 R^2$ (see Chapter 6, page 319).

Inuishi and his co-workers[29] measured saturation effects in coatings at 950° K. They used a thick coating held between two nickel electrodes and opposed by an anode, such that emission effects and conduction effects could be measured in the same tube. Using relatively inactive coatings and measuring the equivalent saturated diode current I_{eq} as a function of plate current, they found a large *increase* in noise at the current at which the change-over from space-charge limitation to saturation occurred. They attributed these results to the space-charge suppression in the space-charge-limited region. They then measured I_{eq} for the coating noise and found also a large rise in noise at a certain current. They assumed that the conduction of the coating was pore conduction; the $1/f$ noise was thought to be suppressed by the space charge of the pores, and the large increase in noise was attributed to saturation effects in the pores.

Hannam[22] has not been able to verify this. On the contrary, he finds that I_{eq} drops in going from the space-charge-limited region to the saturated region. The measurements were carried out at low cathode temperatures, at which the emission current was of the order of a few hundred microamperes. The drop in I_{eq} can be explained as follows: If the flicker noise has to be described partly by a current generator $\sqrt{\overline{i^2}}$ between top of coating and anode and partly by an emf $\sqrt{\overline{e^2}}$ in series with the coating, then

$$2eI_{eq}\,df = \overline{i^2} + \overline{e^2}g_m{}^2 \tag{2.42}$$

The second term drops to zero if the tube becomes saturated, and the first term should vary only gradually at the onset of saturation; hence the drop in I_{eq}.

Hannam found exceptionally large drops in I_{eq} in tubes showing interface effects; this is easily understood, for in that case most of the noise is indeed generated in the coating. He found, however, that the same effect, sometimes smaller, sometimes larger, occurred in tubes that did not show any interface effects and he was thus forced to attribute the change in I_{eq} to noise generated in the coating. (The effect was observed only at low cathode temperatures ($T \simeq 650°$ K); coating noise decreases rapidly with increasing temperature so that it is probably unimportant at the normal operating temperature of the cathode).

Effects similar to those found by Inuishi have been observed by Hannam[22] in tubes with a 499 cathode nickel sleeve. Upon drawing saturation currents at a relatively low temperature for a long time, the emission current gradually dropped and settled down to a low value; other cathode nickels did not show that effect. After a steady emission current had been reached, the dependence of I_{eq} upon I_a was exactly as found by Inuishi. That effect had nothing to do with changes in the space-charge suppression factor of flicker noise but was caused by physical changes in the coating.

REFERENCES

1. J. B. Johnson, *Phys. Rev.*, **26**, 71 (1925).
2. W. Schottky, *Phys. Rev.*, **28**, 74 (1926).
3. W. Schottky, *Physica*, **4**, 175 (1937).
4. E. G. Nielsen and A. van der Ziel, *Rev. Sci. Instr.*, **25**, 899 (1954).
5. A. van der Ziel, *Physica*, **16**, 359 (1950).
6. R. E. Burgess, *Brit. J. Appl. Phys.*, **6**, 185 (1955).
7. J. G. van Wijngaarden and K. M. van Vliet, *Physica*, **18**, 683 (1952).
8. A. van der Ziel, *Noise*, Prentice-Hall, New York, (1954).
9. G. E. Duvall, *MIT Research Lab. Electronics Tech. Rept. 82*, (1948).
10. K. R. Spangenberg, *Vacuum Tubes*, McGraw-Hill Book Co., New York (1948).
11. T. B. Tomlinson, *J. Brit. IRE*, **14**, 515 (1954).
12. R. C. Schwantes, H. J. Hannam, and A. van der Ziel, *J. Appl. Phys.*, **27**, 573 (1956).
13. C. Th. J. van Alkemade, Utrecht, private communication.
14. T. B. Tomlinson, *J. Appl. Phys.*, **23**, 894 (1952).
15. W. W. Lindemann, Ph.D. thesis, University of Minnesota, (1955); W. W. Lindemann and A. van der Ziel, *J. Appl. Phys.*, **27**, 1179 (1956); **28**, 448 (1957).
16. W. W. Lindemann and A. van der Ziel, *J. Appl. Phys.*, **23**, 1410 (1952).
17. H. O. Berktay, *Wireless Engr.*, **30**, 48 (1953).
18. H. J. Hannam and A. van der Ziel, *J. Appl. Phys.*, **25**, 1336 (1954).
19. L. S. Nergaard, *RCA Rev.*, **13**, 464 (1952).
20. A. S. Eisenstein, *Advances in Electronics*, vol. I, pp. 1–64, Academic Press, New York (1948).
21. R. L. Anderson and A. van der Ziel, *Trans. IRE*, *PGED-1*, 20 (1952).
22. H. J. Hannam, Ph.D. thesis, University of Minnesota (1956), and unpublished results.
23. W. Graffunder, *Telefunkenroehre*, **5**, 41 (Apr. 1939).
24. J. G. van Wijngaarden, K. M. van Vliet, and C. J. van Leeuwen, *Physica*, **18**, 689 (1952).
25. E. O. Kane, "Thermionic Emission from Semi-conductors," *Contract Nonr* 401(08) *Tech. Rept. 3*, Cornell University (1954).
26. C. G. J. Jansen, R. Loosjes, and K. Compaan, *Philips Research Repts.* **9**, 241 (1945).
27. A. van der Ziel, *Physica*, **20**, 327 (1954).
28. G. Hermann and S. Wagener, *The Oxide-Coated Cathode*, vol. 2, Chapman & Hall, London (1951).
29. Y. Inuishi and T. C. Yang, *J. Phys. Soc. Japan*, **8**, 565 (1953); J. Nakai, Y. Inuishi, and T. C. Yang, *J. Phys. Soc. Japan*, **10**, 437 (1955).
See also: "Study of the Cause and Effect of Flicker Noise in Vacuum Tubes," *U. S. Signal Corps Contract DA36-039 SC-15355 Repts. 1-9*; *DA36-039 SC-56683 Repts. 1–8*, University of Minnesota (1951–1955).

CHAPTER 3

Signal and Noise Propagation along Electron Beams

H. A. Haus

1. Analysis of Signal Propagation along Electron Beams

Introduction

The power amplification of a conventional triode is based on the control of the current from the cathode by a potential applied to the grid. At low frequencies the control of the current can be achieved without expenditure of power if the grid current is reduced to zero by a proper bias. The electron flow can be analyzed mathematically like a time-independent stationary process. Any slow time variation of the process is represented by a simultaneous time variation of all parameters of the process. Such an analysis might be called "quasistationary." At frequencies at which the electron transit time through the tube is not negligibly small compared to a period of the applied RF grid voltage, the quasistationary analysis is inadequate. Transit-time effects cause induced grid currents. The existence of grid currents calls for a supply of RF power to the control grid. Successful attempts to minimize transit-time grid loading and other undesirable high-frequency effects have led to the modern microwave triode.

Simultaneously, new principles of amplification have been recognized and put to use in a new class of amplifiers: microwave-beam amplifiers. In these amplifiers transit-time effects are used to advantage. In this chapter we shall deal with the noise performance of one subclass of microwave-beam amplifiers, *longitudinal-beam amplifiers.* They were chosen for attention partly because the noise in these amplifiers is by now fairly well understood, partly because all low-noise microwave-beam amplifiers that have been built to date are members of this class.

The analysis of longitudinal-beam amplifiers has some features in common with the analysis of the high-frequency triode and other related tubes. These we shall call "space-charge-control" tubes, referring to their basic principle of operation. The mathematical approach on pages 83 to 94 is also used in the high-frequency analysis of noise in space-charge-control tubes.

One feature distinguishes microwave longitudinal-beam amplifiers from conventional space-charge-control tubes. In the latter tubes the applied RF fields act on the electron beam while it passes through the potential minimum in front of the cathode. The former tubes employ an electron beam formed in an electron gun that is free of applied RF fields. Examples of longitudinal-beam amplifiers are the traveling-wave tube,[1] the klystron,[2-4] the resistive-wall amplifier,[5] the space-charge-wave amplifier,[6] the double-stream amplifier,[7] the rippled-wall and rippled-stream amplifier,[8] and the backward-wave amplifier.[9] The designer of a low-noise, longitudinal-beam amplifier can take advantage of the fact that the electron beam is formed in a region free of applied RF fields. He can design structures surrounding the beam in front of the RF interaction region of the amplifier. Such structures, if properly chosen, can reduce the noise output of the amplifier without affecting its gain. The theory of such noise-reducing schemes and their limitations, will be the topic of this chapter. An expression will be derived for the minimum obtainable noise figure of a longitudinal-beam amplifier, using a beam with a given noise. A logical definition of the "noisiness" of an electron beam follows from the expression for the minimum noise figure. It will be shown that noise-reducing structures preceding the amplifier are, in principle, sufficient for attaining the minimum noise figure, and that elaborate feedback and noise-cancelation schemes within the amplifier cannot lead to a lower minimum noise figure. Finally, it is shown that a traveling-wave tube with negligible loss in its RF structure preceded by conventional noise-reducing schemes[10] may attain, in principle, the minimum noise figure.

Assumptions

An exact analysis of the propagation of signals and noise along electron beams is extremely difficult. Certain approximations have to be made before the problem becomes amenable to a mathematical treatment.

In all longitudinal-beam amplifiers, amplification is obtained through an energy transfer from the electron motion to predominantly longitudinal electric fields. If *a very large longitudinal magnetic focusing field confines the motion of the beam,* the energy transferred by the electrons comes entirely from the kinetic energy associated with the longitudinal motion. In the analysis of longitudinal-beam amplifiers, the assumption of an infinite magnetic focusing field is made quite often because it represents the physical facts adequately for many purposes, and leads to mathematical simplicity.

The electrons emitted at random from the cathode in the electron gun of a longitudinal-beam amplifier form a beam in which the velocity and the density of the electrons passing any reference cross section fluctuate statistically. Part of the noise output of an amplifier which employs the beam is caused by the currents induced in the amplifier structure by the fluctuations in the beam. Also, some of the electrons may be intercepted by the RF structure of the amplifier in a more-or-less random fashion if the beam is inadequately focused. Noise produced by this mechanism is commonly called "partition noise." In a well-designed amplifier, the interception current can be kept to less than 0.5% of the total beam current. Under such conditions *the effect of partition noise is negligible* compared to the noise induced in the RF structure. The present analysis will deal solely with the latter.

An electron beam consists of a large, but finite, number of electrons. The electrons in the beam interact with each other by virtue of their Coulomb repulsion force. Contributions to the force on any particular electron come partly from the next neighbors, partly from electrons farther away. The forces exerted upon any electron by its next neighbors fluctuate rapidly with time. These forces are usually referred to as "short-range collisions." The forces exerted upon the electron by electrons farther away behave more regularly. These forces are the "long-range collisions." Up to the present time, all signal and noise analyses of electron beams have neglected the granular nature of the charge (See reference 11 for an interesting discussion of this approximation.) *The electron beam is treated as a "fluid,"* made up of an infinite number of infinitesimally small particles with an infinitesimal charge. *The effect of the short-range collisions among the particles is neglected* in high-vacuum electron beams. An argument that this approximation is legitimate has been given by Mott-Smith[12] for beams of reasonable length and current density. Experimental results corroborate the validity of this approximation.[13]

The granular nature of the electron beam is the ultimate cause of noise. In the analysis of an electron beam as a "fluid," this effect is taken into account in terms of appropriate noise input conditions at the cathode, or at some reference plane further on in the beam.

The *small-signal theory* is used for the analysis of noise in electron beams. The assumption is made that the excitation of the beam can be treated as a small perturbation of the time-average conditions in the beam. The approximations of small-signal theory are apparently good anywhere along the electron beam except at the potential minimum in front of a cathode operated under space-charge-limited conditions. Whether or not the small-signal assumption is applicable to the region of the potential minimum at high frequencies is not clear.

Electrons emerging from the cathode have different velocities according to the Maxwellian distribution. The analysis of an electron beam as a charged "fluid" still retains this picture. Parts of the fluid with higher velocities drift through parts with lower velocities without friction. Friction is neglected as soon as the effect of short-range collisions is disregarded. The *single-velocity theory* makes the assumption that the beam can be treated as if all electrons passing a beam cross section had the same velocity. This theory has been adopted almost exclusively in microwave work. Its justification has been discussed by various authors.[14—16] The conclusion is that single-velocity theory yields results in good agreement with the more sophisticated multi-velocity theory, as long as the range of velocities possessed by the majority of the electrons is small compared to their average velocity. Such a situation prevails in a beam that has been accelerated to a few volts above the potential minimum. At the potential minimum in front of a space-charge-limited cathode this condition is obviously violated. The single-velocity theory applied to the potential-mini-mum region cannot give better than qualitative answers. This difficulty is circumvented in the analysis of noise in electron beams in this chapter. The noise-input conditions to the electron beam are stated at a cross section beyond the potential minimum, chosen so that the small-signal and single-velocity theories are applicable at, and beyond, the cross section. No specific values are assumed for the noise parameters at the reference cross section. The evaluation of the noise parameters is left to a detailed analysis of the potential minimum region which does not make the single-velocity assumption.†

Finally, there is the assumption of the *one-dimensional theory*. Only a single spatial co-ordinate, the co-ordinate along the electron beam, is retained in the analysis. The one-dimensional theory in conjunction with all the assumptions introduced above leads to a characterization of a perturbation in the beam in terms of two modulation parameters: for example, the velocity and current modulations. Strictly speaking, this assumption implies that the electron beam is of an infinite parallel-plane (or spherical) geometry, with the motion of the electrons confined to the longitudinal axis (in the direction of the radius vector). All parameters of the beam are assumed to be independent of the coordinates transverse to the beam. A more practical example which fits into the one-dimensional formalism is a freely drifting electron beam of finite cross section in which only two space-charge waves are excited[19,20] If the beam passes through transition regions in which its shape or time-average velocity is changed, cross-coupling among the different modes of the beam occurs, and, in addition to the original two waves, many other waves travel along the beam. In such cases, the one-dimensional formalism yields only approximate answers; but, the smaller the cross-coupling of modes in a transition region, the better the approximation should be. This obtains in a thin beam, and the one-dimensional theory gives good approximate answers. Recently, the noise analysis has been extended to systems propagating any number of modes.[21] The results are, however, rather complex, so that their presentation here does not seem warranted.

Before the noise in electron beams can be analyzed, the propagation of signals along an electron beam has to be understood. Section 1, page 83, and all of Section 2 are devoted to this problem.

The Basic Equations

The first two equations of Maxwell give the electric field $\mathbf{E}(\mathbf{r}, t)$ and the magnetic field $\mathbf{H}(\mathbf{r}, t)$ produced by a given current distribution $\mathbf{J}(\mathbf{r}, t)$. The vectors $\mathbf{E}$, $\mathbf{H}$, and $\mathbf{J}$, are all functions of the radius vector $\mathbf{r}$, and the time t.

$$\nabla \times \mathbf{E}(\mathbf{r}, t) = -\mu_0 \frac{\partial}{\partial t} \mathbf{H}(\mathbf{r}, t) \tag{3.1}$$

---

† In the meantime, such an analysis has been carried out by P. K. Tien[17] on a digital computer in a way that avoids the small-signal assumption. Tien's analysis is carried from the cathode to a plane just beyond the cathode. Later work by Siegman, Watkins, and Hsieh[18] carries the analysis from the potential minimum to a plane corresponding to a beam potential of several volts.

$$\nabla \times \mathbf{H}(\mathbf{r}, t) = \mathbf{J}(\mathbf{r}, t) + \epsilon_0 \frac{\partial}{\partial t} \mathbf{E}(\mathbf{r}, t) \qquad (3.2)$$

The force equation gives the relation between the acceleration of charged particles and the field. If the motion of the particles is confined by an infinite magnetic focusing field in the z direction, or, if the velocity is z-directed for other reasons, we have

$$\frac{\partial v(\mathbf{r}, t)}{\partial t} + v(\mathbf{r}, t) \frac{\partial v(\mathbf{r}, t)}{\partial z} = \frac{d}{dt} v(\mathbf{r}, t) = \frac{e}{m} E_z(\mathbf{r}, t) \qquad (3.3)$$

where e is the charge of the particle (for an electron it is a negative quantity), and m is its mass. Equation 3.3 neglects the force upon the particle produced by RF magnetic fields, an approximation legitimate at nonrelativistic velocities. The quantity $v(\mathbf{r}, t)$ is the z component of the velocity. For the sake of brevity no subscript z is used. The continuity equation is, if the current is entirely z-directed,

$$\frac{\partial}{\partial z} J(\mathbf{r}, t) = - \frac{\partial}{\partial t} \rho(\mathbf{r}, t) \qquad (3.4)$$

If the single-velocity assumption is made, the current density is given as the product of the velocity and space-charge density:

$$J(\mathbf{r}, t) = v(\mathbf{r}, t)\, \rho(\mathbf{r}, t) \qquad (3.5)$$

Under the small-signal assumption all quantities can be split into a time-average part and a time-varying part which is much smaller in amplitude than the time-average part. In evaluating the time-dependent parts, cross products of the time-varying quantities can be neglected. The resulting equations for the time-varying quantities become linear, and thus a sinusoidal excitation of frequency ω causes all time-dependent quantities to vary at the same frequency. The superposition principle can be applied under the small-signal approximation. Complex notation can be used to represent the time-varying quantities. We can write

$$\mathbf{E}(\mathbf{r}, t) = \mathbf{E}_0(\mathbf{r}) + \mathrm{Re}\,[\boldsymbol{E}(\mathbf{r})e^{j\omega t}]$$

$$\mathbf{H}(\mathbf{r}, t) = \mathbf{H}_0(\mathbf{r}) + \mathrm{Re}\,[\boldsymbol{H}(\mathbf{r})e^{j\omega t}]$$

$$J(\mathbf{r}, t) = J_0(\mathbf{r}) + \mathrm{Re}\,[J(\mathbf{r})e^{j\omega t}] \qquad (3.6)$$

$$v(\mathbf{r}, t) = u(\mathbf{r}) + \mathrm{Re}\,[v(\mathbf{r})e^{j\omega t}]$$

$$\rho(\mathbf{r}, t) = \rho_0(\mathbf{r}) + \mathrm{Re}\,[\rho(\mathbf{r})e^{j\omega t}]$$

Complex vector quantities are indicated by boldface italic type. These definitions introduced into Eqs. 3.1 through 3.5, lead to a separation between the time-average and time-varying parts. The time-dependent part of Maxwell's equation is

$$\nabla \times E(\mathbf{r}) = -j\omega\mu_0 \, H(\mathbf{r}) \tag{3.7}$$

$$\nabla \times H(\mathbf{r}) = \mathbf{a}_z \, J(\mathbf{r}) + j\omega\epsilon_0 \, E(\mathbf{r}) \tag{3.8}$$

where $\mathbf{a}_z$ is the unit vector in the z direction. The current has been assumed to be entirely z-directed. The time-average and time-dependent parts of the force equation are under the same assumptions:

$$u(\mathbf{r}) \, \frac{\partial}{\partial z} \, u(\mathbf{r}) = \frac{e}{m} \, E_{0z}(\mathbf{r}) \tag{3.9a}$$

$$j\omega \, v(\mathbf{r}) + \frac{\partial}{\partial z} \, [u(\mathbf{r}) \, v(\mathbf{r})] = \frac{e}{m} \, E_z(\mathbf{r}) \tag{3.9b}$$

The continuity equation gives

$$\frac{\partial}{\partial z} \, J_0(\mathbf{r}) = 0 \tag{3.10a}$$

$$\frac{\partial}{\partial z} \, J(\mathbf{r}) = -j\omega \, \rho(\mathbf{r}) \tag{3.10b}$$

and from Eq. 3.5 we have

$$J_0(\mathbf{r}) = u(\mathbf{r}) \, \rho_0(\mathbf{r}) \tag{3.11a}$$

$$J(\mathbf{r}) = u(\mathbf{r}) \, \rho(\mathbf{r}) + \rho_0(\mathbf{r}) \, v(\mathbf{r}) \tag{3.11b}$$

The Infinite-Parallel-Plane Beam

The study of the infinite-parallel-plane beam is worth while for two reasons. First, the analysis of this model is relatively simple. With minimum effort, a good understanding can be obtained of the physical processes of current and velocity modulation and of power transfer between the fields and the electrons. Second, with minor modifications the results of the infinite-parallel-plane beam can be applied to the dominant space-charge waves (page 98) in a beam of finite cross section. These space-charge waves play an important role in microwave amplification.

We shall now analyze the propagation of signals along such an electron beam of infinite cross section. The time-average current density and velocity are constant throughout the cross section. The positive z direction is picked as the direction of positive velocity and current (see Fig. 3.1). Thus, a convection-current density of negatively charged particles with a positive velocity is, by convention, negative.

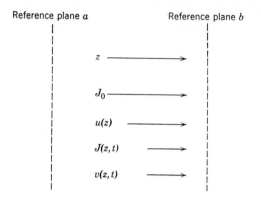

Fig. 3.1.   Direction of positive current and velocity.

The time-average velocity can be found from an integration of the time-average part of the force equation, Eq. 3.9a.   Since the electron flow is assumed to depend merely upon the z co-ordinate, we can replace the independent variable **r**, which consists of all three co-ordinates, by z.   We have, from Eq. 3.9a,

$$\frac{m}{2e} \frac{d}{dz}[u(z)^2] = E_{0z}(z) \tag{3.12}$$

The field, in turn, is determined by the space-charge distribution in the electron beam.   From Gauss's law in its one-dimensional form, we have

$$\frac{d}{dz} E_{0z}(z) = \frac{\rho_0(z)}{\epsilon_0} \tag{3.13}$$

Since the time-average current density is independent of z (according to Eq. 3.10a) and is in turn related to the time-average velocity and space-charge density by Eq. 3.11a, we can express the space-charge density in terms of J_0 and $u(z)$.   Once this is done, Eqs. 3.12 and 3.13 contain only two unknown functions, $u(z)$ and $E_{0z}(z)$, and they can be solved subject to appropriate boundary conditions.   The variety of such boundary conditions is great.   It is conceptually possible to construct an arbitrary d-c potential distribution with the aid of infinitely permeable grids, open-circuited for radio frequencies, to which arbitrary d-c potentials are applied.

Let us suppose that the time-average velocity and space-charge density have been found in the way described above. Let us further assume that the RF excitation applied to the electron beam is independent of the transverse co-ordinates x and y. This assumption concludes the set of assumptions that has become known under the name of "the infinite-parallel-plane, one-dimensional beam model." Under the stated assumption we can have wave solutions with finite E_x or E_y. These waves cannot couple to a beam confined to motion in the z direction by an infinite magnetic field. These waves are the familiar TEM waves, and they propagate unaffected by the beam. We shall ignore the TEM waves for reasons that will become more obvious later.

Next, let us consider the solution which couples to the beam by virtue of a finite E_z field. With $\partial/\partial x = \partial/\partial y = 0$, there is no z component of the curl of $\boldsymbol{H}(\mathbf{r})$. We thus have, from Eq. 3.8,

$$J(z) + j\omega\epsilon_0\, E_z(z) = 0 \tag{3.14}$$

Equation 3.14 shows that the sum of the convection- and displacement-current densities is zero. It can also be observed that there is no transverse electric field and the motion of the electrons is entirely longitudinal. Omitting, from now on, any explicit indication of the z dependence of the process, we obtain, from Eqs. 3.14 and 3.9b,

$$j\omega v + \frac{d}{dz}(uv) = -\frac{e}{m}\frac{1}{j\omega\epsilon_0}\, J \tag{3.15}$$

Equations 3.10b and 3.11b lead to the expression

$$j\omega J + u\frac{d}{dz}J = j\omega\rho_0 v \tag{3.16}$$

Equations 3.15 and 3.16 are put into a symmetrical form by introducing the new dependent variable,

$$V = \frac{m}{e}\, uv \tag{3.17}$$

(The quantity V was introduced by L. J. Chu as the *kinetic-voltage modulation*.[22] Its particular significance will become apparent later.) Further, we write

$$\frac{e}{m}\frac{\rho_0}{\epsilon_0} = \omega_p^{\,2} \tag{3.18}$$

ω_p has the dimensions of frequency and is, in general, a function of distance. It is commonly called the plasma frequency, because an electron plasma of uniform density ρ_0 oscillates at this frequency.

Noting that the d-c current density J_0 is, according to Eq. 3.11a,

$$J_0 = \rho_0 u \qquad (3.11a)$$

we may rewrite Eqs. 3.15 and 3.16 in the form

$$\left(j\frac{\omega}{u} + \frac{d}{dz}\right) V = j\frac{\omega_p{}^2}{u^2}\left(\frac{u}{\omega}\right)\frac{(m/e)u^2}{J_0} J \qquad (3.15a)$$

$$\left(j\frac{\omega}{u} + \frac{d}{dz}\right) J = j\frac{\omega}{u}\frac{J_0}{(m/e)u^2} V \qquad (3.16a)$$

The quantity $\dfrac{1}{2}\left|\dfrac{m}{e}\right| u^2$ is the time-average kinetic energy of an electron measured in electron volts, or, the d-c potential at the position of the electron, V_0, provided the value zero is assigned to the potential at the plane at which the electrons have zero velocity. The quantity ω_p/u has the dimension of inverse length and may thus be considered to be a propagation constant. We shall denote

$$\frac{\omega_p}{u} = \beta_p \qquad (3.19)$$

as the *plasma propagation constant*. In general, β_p is a function of distance. The quantity ω/u has also the dimension of inverse length

$$\frac{\omega}{u} = \beta_e \qquad (3.20)$$

It is usually called the *electronic propagation constant* and is a function of distance if u is distance-dependent. Introducing this new notation into Eqs. 3.15a and 3.16a, we find

$$\left(j\beta_e + \frac{d}{dz}\right) V = j\beta_p{}^2 \frac{2V_0}{\beta_e|J_0|} J \qquad (3.21)$$

$$\left(j\beta_e + \frac{d}{dz}\right) J = j\frac{\beta_e|J_0|}{2V_0} V \qquad (3.22)$$

Some salient features of the electron-beam problem are brought out in the notation of Eqs. 3.21 and 3.22. It should be recalled that Eq. 3.21 is essentially the force equation. On the left-hand side is a term proportional to the acceleration. The right-hand side is (essentially) the field set up by the current-density modulation. This field is proportional to the per cent current modulation, $J/|J_0|$, and to the space-charge density ρ_0 in the beam, since $\beta_p{}^2 \propto \omega_p{}^2$ and $\omega_p{}^2$ is in turn given by Eq. 3.18. Since ρ_0 gives a measure of the strength of the field available in a bunched beam, the proportionality to ρ_0 is easily understood. We shall make use of the above considerations later when discussing beams of finite cross section. Equation 3.22 expresses the rate of change of the current-density modulation as caused by a velocity modulation in the beam. (The kinetic voltage V is proportional to the velocity modulation according to Eq. 3.17). Thus, Eq. 3.22 is a special form of the continuity equation.

Physical reasoning leads us to a change of the dependent variables in Eqs. 3.21 and 3.22 which improves their appearance. An electron beam is a system of charged particles which interact through their space-charge repulsion forces. The system moves with the time-average velocity u. The mere fact that the electrons move with a finite time-average velocity implies that any perturbation applied to the electron beam at some point, say, $z = 0$, arrives time-delayed at a later point $z > 0$. If space-charge forces have time to act upon the electrons during their travel between the points $z = 0$ and z, the velocity of the electrons gets modified. This is an effect over and above the natural time delay. We acknowledge this time delay by introducing a transformation of the dependent variables in which the time delay is brought out explicitly.

$$V = Ve^{-j\theta} \qquad (3.23)$$

$$J = Qe^{-j\theta} \qquad (3.24)$$

where θ is the transit angle between the reference plane $z = 0$ and the point z.

$$\theta = \omega \int_0^z \frac{dz}{u} \qquad (3.25)$$

In terms of these new variables, Eqs. 3.21 and 3.22 become

$$\frac{dU}{dz} = j\beta_p{}^2 \frac{2V_0}{\beta_e|J_0|} Q \qquad (3.26)$$

$$\frac{dQ}{dz} = j \frac{\beta_e|J_0|}{2V_0} U \qquad (3.27)$$

Equations 3.26 and 3.27 are special forms of the equations formulated by Llewellyn.[23,24] They have the appearance of transmission-line equations with a pure imaginary impedance per unit length, $Z = -j\beta_p{}^2 2V_0/\beta_e|J_0|$, and a pure imaginary admittance per unit length, $Y = -j\beta_e|J_0|/2V_0$. U and Q play the role of voltage and current on the analog transmission line (see Fig. 3.2).   Similar equations have been obtained for a spherical flow[25] and one-dimensional flow of any general geometry.[26]   The fact that the impedance and admittance per unit length of the analog transmission line are purely imaginary implies that the transmission line is lossless.   Along such a

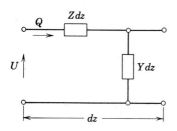

Fig. 3.2.   Analog transmission line of an electron beam.

transmission line, the power must be independent of distance.   The time-average power flow along a transmission line is given by one-half the real part of the complex product of the voltage U and current Q on the transmission line, Re $[UQ^*]/2$.   Thus, we have for the analog power

$$\tfrac{1}{2} \operatorname{Re} [U(z_1)\, Q(z_1)^*] = \tfrac{1}{2} \operatorname{Re} [U(z_2)\, Q(z_2)^*]$$

where z_1 and z_2 are the positions of two reference planes along the transmission line.   According to the definitions of Eqs. 3.23 and 3.24, we have $UQ^* = VJ^*$, and thus

$$\tfrac{1}{2} \operatorname{Re} [V(z_1)\, J(z_1)^*] = \tfrac{1}{2} \operatorname{Re} [V(z_2)\, J(z_2)^*]$$

L. J. Chu[22] defined the quantity

$$\tfrac{1}{2} \operatorname{Re} [V(z)\, J(z)^*] = \operatorname{Re} (S_k) \qquad (3.28)$$

as the *real kinetic-power density* of the electron beam.   Since we can easily check that the kinetic-power density has the dimension of power per unit area, the name seems, at least, partly justified.   Later we shall see that there are even more compelling reasons for the name. For the moment it is sufficient to note that the kinetic-power density is independent of distance in any system of infinite parallel-plane geometry, regardless of the distribution of the time-average potential. In such a system the magnetic field is zero, $\mathbf{H}(z,\,t) = 0$, and, correspondingly, there is no flow of electromagnetic power in an infinite parallel-plane beam.   In Section 2, page 95, we shall see that this fact is responsible for the conservation of kinetic power.

The "transmission-line equations" (3.26 and 3.27) lead to a differential equation of second order with two solutions. These solutions can be used to satisfy arbitrary boundary conditions. If the kinetic-voltage and current-density modulations are known at any cross section of the beam, they are known everywhere. From the mathematical point of view, the cross section at which the initial conditions are given is arbitrary. However, from a physical point of view it is not arbitrary. Since the electromagnetic power flow in an infinite parallel-plane, one-dimensional geometry is zero (ignoring the TEM waves), an excitation is not transmitted through the beam by electromagnetic radiation, but is transported along the beam by the electrons. The excitation propagates in the direction of motion of the electrons. An excitation in the region of an electron beam is thus most naturally given in terms of the boundary conditions at the entry of the beam into the region.

The Beam in a Drift Region

An electron beam that flows between two electrodes that are a finite distance apart, both at the same potential, causes a potential depression between the electrodes with a potential minimum situated halfway between them. The closer together the two electrodes are, the smaller the potential depression. The potential depression is negligibly small, and thus the time-average fields are negligible, when the spacing between the electrodes is infinitesimal. An electron beam that drifts freely with no time-average forces acting upon it can be realized by a system of electrodes, all at equal time-average potential, spaced very closely together and open-circuited for radio frequency. An electron beam between two equipotential electrodes a finite distance apart, neutralized by heavy positive ions, acts in the same way. The ions cannot follow the RF changes of the field and thus do not affect them. However, all time-average fields are eliminated, since the ions can follow them, although sluggishly, until they fill the potential minima and compensate the charge of the electrons with their own positive charge.

The two methods of realizing a drift region are artifices used to adapt the model of an infinite parallel-plane beam to represent a more physical situation: a finite longitudinal beam surrounded by a perfectly conducting cylindrical wall and confined by a large, ideally infinite, longitudinal magnetic field. In this latter case the time-average electric fields produced by space charge are entirely radial. Therefore the time-average velocity of the electrons is independent of z. The main features of propagation of signals along such a finite beam are contained in the model of the infinite parallel-plane beam in a drift

region. This accounts for the importance attributed to the problem
of the infinite parallel-plane beam in a drift region.

In the absence of time-average fields, the time-average velocity of the
electrons in the infinite parallel-plane beam cannot change. Correspondingly, the d-c charge density ρ_0 and thus the plasma frequency
ω_p are independent of distance. Equations 3.26 and 3.27 can be
solved very easily in terms of two arbitrary constants U_+ and U_-:

$$U = U_+ e^{j\beta_p z} + U_- e^{-j\beta_p z}$$

$$Q = \frac{|J_0|}{2V_0} \frac{\beta_e}{\beta_p} (U_+ e^{j\beta_p z} - U_- e^{-j\beta_p z})$$

with $\beta_p = \omega_p/u$ and $\beta_e = \omega/u$.

The kinetic voltage and the current density become

$$V = (U_+ e^{j\beta_p z} + U_- e^{-j\beta_p z}) e^{-j\beta_e z}$$

$$J = \frac{|J_0|}{2V_0} \frac{\beta_e}{\beta_p} (U_+ e^{j\beta_p z} - U_- e^{-j\beta_p z}) e^{-j\beta_e z}$$

If we single out one part of the beam of cross-sectional area σ, the time-
average current flow through this area is $I_0 = |u\sigma\rho_0|$, and the time-
varying current is $i = \sigma J$. We define the "characteristic impedance"
of the beam by

$$W = \frac{2V_0}{I_0} \frac{\beta_p}{\beta_e} \tag{3.29}$$

Note that both I_0 and V_0 are defined positive. With the aid of these
definitions we can write the solutions for the kinetic voltage and
RF current in the beam of cross section σ in the form

$$V = (U_+ e^{j\beta_p z} + U_- e^{-j\beta_p z}) e^{-j\beta_e z} \tag{3.30}$$

$$i = \frac{1}{W} (U_+ e^{j\beta_p z} - U_- e^{-j\beta_p z}) e^{-j\beta_e z} \tag{3.31}$$

According to Eqs. 3.30 and 3.31, two wave solutions exist in the beam.
Their propagation constants are, respectively, $\beta_e + \beta_p$ and $\beta_e - \beta_p$.
The wave with the propagation constant $\beta_e + \beta_p$ has a phase velocity
smaller than the time-average beam velocity u. The wave with the
propagation constant $\beta_e - \beta_p$ travels with a phase velocity larger than
the beam velocity, provided that $\beta_e > \beta_p$. The case of $\beta_e < \beta_p$ never
occurs in practice. The reason for this will be discussed in Section 2,
page 100. Thus, we shall call the wave with the propagation constant

$\beta_e - \beta_p$ simply the "fast wave," implying that the inequality $\beta_e > \beta_p$ is satisfied.

It is noteworthy that the group velocity

$$v_g = \frac{d\omega}{d\beta}$$

with

$$\beta = \beta_e \pm \beta_p$$

is equal to the beam velocity u for both the fast and the slow waves, since

$$\frac{d\beta_p}{d\omega} = 0$$

In general, a time-dependent signal (excitation) travels with a speed equal to the group velocity. Thus, an excitation of the beam at $z = 0$ reaches an observer at $z = z$ with a time delay $\tau = z/u$ that is equal to the time it takes the electrons to traverse this distance. This suggests that an excitation is transferred bodily with the electrons at a speed corresponding to the electron velocity.

A little reasoning shows that this is indeed true. In our solution, $H = 0$, and thus the electromagnetic power flow is zero. An excitation of the beam is, in essence, a deviation of the electrons and their velocities from the positions and the velocity they possess in the absence of an excitation. Since the electromagnetic power flow is zero, the only way an excitation can be communicated to another point in the beam is by a transfer of the excited electrons themselves to this point. The same reasoning also shows that signal propagation "upstream," i.e., in a direction opposite to the d-c velocity of the electrons, does not occur in our solution.

If we now turn to the evaluation of the kinetic power, we have, from Eqs. 3.28, 3.30, and 3.31,

$$\sigma_{\frac{1}{2}} \operatorname{Re}[VJ^*] = \tfrac{1}{2} \operatorname{Re}[Vi^*] = \frac{1}{2W}(|U_+|^2 - |U_-|^2) \qquad (3.32)$$

According to Eq. 3.32, the real kinetic power of the two waves is additive (orthogonality of power flow)—the fast wave carrying positive power, the slow wave carrying negative power. Since we have found that the group velocities of the two waves are both positive, we have here an indication that the sign of the kinetic power cannot give the direction of signal propagation. It has a different significance. We shall return to this question in the next section.

The kinetic power carried by a beam in a drift tube can be written in a more elegant form by defining normalized wave amplitudes a_1 and a_2 as

$$a_1 = (2W)^{-\frac{1}{2}}U_+, \qquad a_2 = -(2W)^{-\frac{1}{2}}U_- \qquad (3.33)$$

With the aid of this definition we can write the kinetic power in the form

$$\tfrac{1}{2}\,\text{Re}\,[Vi^*] = |a_1|^2 - |a_2|^2 \qquad (3.34)$$

We shall find the use of normalized wave amplitudes very convenient later.

The resemblance of Eqs. 3.30 and 3.31 to transmission-line solutions is apparent. This result is not surprising, since Eqs. 3.26 and 3.27, from which Eqs. 3.30 and 3.31 have been derived, have the form of transmission-line equations. It is important, however, to note the difference between transmission-line solutions and the solutions of a beam in a drift region. Both the voltage and current modulations are multiplied by a factor $e^{-j\beta_e z}$, which does not appear in the transmission-line solutions. Instead of one forward and one backward wave, we have a fast wave and a slow wave, both with phase velocities in the direction of the flow of the beam.

It is natural to suppose that techniques of conventional transmission-line theory can be applied to the beam problem. Equations 3.30 and 3.31 show that this is possible if proper precautions are taken. As is common practice in transmission-line analysis, we define an impedance at any cross section z by

$$Z = \frac{V}{i} = W\,\frac{1+\Gamma}{1-\Gamma} \qquad (3.35)$$

where

$$\Gamma = \frac{U_-}{U_+}\,e^{-2j\beta_p z} \qquad (3.36)$$

Equation 3.35 is formally identical with the well-known relation between the impedance and the reflection coefficient on a transmission line. The reflection coefficient is defined, as usual, as the ratio of the amplitudes of the voltage in the wave carrying negative power $(-)$ and the wave carrying positive power $(+)$. The role of the $(-)$ wave is played, in the beam problem, by the slow wave, whose kinetic power is negative, quite analogous to the negative electromagnetic power carried by the reflected wave of transmission-line theory. The role of the wavelength is played in the beam problem by the quantity $\lambda_p = 2\pi/\beta_p$, the so-called plasma wavelength. Equation 3.36 shows one important difference between the transmission-line problem and the beam problem. The angle of the reflection coefficient Γ decreases with increasing z, whereas the opposite is true for the reflection coefficient in transmission-line theory.

The bilinear relation between Z and Γ as shown in Eq. 3.35 is conveniently represented in the plane of complex Γ, the Smith chart of transmission-line theory.[27] Motion in the positive z direction along the electron beam corresponds to clockwise rotation in the Γ plane. A shift by half a plasma wavelength along the electron beam leaves Z unchanged.

The General Solutions of Llewellyn's Equations

Equations 3.26 and 3.27 can be solved for conditions other than those of a drift region. An important case is the one of an electron beam traveling between two completely permeable grids at potentials V_{0a} and V_{0b}, under the influence of its own space charge. The details of the solution are rather tedious and not within the scope of this discussion. The details are given in references 23 and 24. We list here only the results in the form of an example (Example 1). Certain changes of notation have been made to conform with our notation. In particular, the change in the sign convention for the current should be noted.

Example

1. Time-Average Solutions

Relation between potential and velocity:

$$\left| \frac{e}{m} \right| V_0 = \tfrac{1}{2} u^2$$

where $\left| e/m \right| = 1.76 \times 10^{11}$ in rationalized mks units

Definition of space-charge factor:

$$\zeta = 3 \left(1 - \frac{T_0}{T} \right)$$

where T is the transit time of the electrons between planes a and b, T_0 is the transit time between the same planes in the absence of space charge, with the potentials at the cross section unchanged.

Relation among the space-charge factor ζ, distance between reference cross sections d, transit time T, and initial and final velocities u_a and u_b:

$$d = \left(1 - \frac{\zeta}{3} \right) (u_a + u_b) \frac{T}{2}$$

Current density:

$$J_0 = \frac{\epsilon_0 m}{e} (u_a + u_b) \frac{2\zeta}{T^2}$$

Ratio of the actual current density $|J_0|$ to the maximum possible current density $|J|_{max}$:

$$\frac{|J_0|}{|J|_{max}} = \frac{9\zeta}{4}\left(1 - \frac{\zeta}{3}\right)^2$$

Maximum current density:

$$|J|_{max} = \frac{2.33 \times 10^{-6}(V_{0a}^{1/2} + V_{0b}^{1/2})^3}{d^2} \text{ amperes/unit area}$$

2. RF Solutions

$$V_b = AV_a + BJ_a$$
$$J_b = CV_a + DJ_a \tag{3.37}$$

V_a and V_b are the kinetic-voltage modulations at the reference cross sections a and b, respectively; J_a and J_b are the corresponding current-density modulations. In the equations above the assumption is made that the two grids at the cross sections a and b are RF open-circuited; the solutions obtained by Llewellyn and Peterson are more general.[23,24]

The coefficients A to D are given by:

$$A = \frac{1}{u_a}[u_a - \zeta(u_a + u_b)]e^{-j\theta} \tag{3.38}$$

$$B = \frac{-T^2}{2\epsilon_0}(u_a + u_b)(1 - \zeta)\frac{e^{-j\theta}}{j\theta} \tag{3.39}$$

$$C = \frac{2\epsilon_0\zeta}{u_aT^2}\frac{u_a + u_b}{u_b}j\theta e^{-j\theta} \tag{3.40}$$

$$D = \frac{1}{u_b}[u_b - \zeta(u_a + u_b)]e^{-j\theta} \tag{3.41}$$

2. Matrix Representation of Microwave Amplifiers

The problem of interaction between an electron beam and an electromagnetic field is solvable in closed form only under the assumption of small-signal theory. Once this assumption is made, the differential equations of the system are linear. The solutions are then linear functions of the excitation of the system on its boundaries. It is convenient to write linear relations among sets of variables in matrix form. In Section 2, page 96, we derive a basic relation of small-signal theory that will suggest a convenient matrix representation of an amplifier. Pages 104–114 are devoted to a study of the restrictions imposed on the matrices.

Kinetic-Power Theorem

The definition of the kinetic-power density was introduced by Eq. 3.28. The significance of the kinetic-power concept is studied in greater detail in this section.

It is obvious that amplification of electromagnetic energy in an electron tube occurs at the expense of the kinetic energy of the electrons. The flow of kinetic energy into a longitudinal-beam microwave amplifier minus the flow of the kinetic energy out of the tube is equal to the electromagnetic power delivered to the RF structure surrounding the beam. Unfortunately, difficulties are encountered in attempting to make use of this simple statement.

The small-signal theory linearizes the equations of the electron beam and thus facilitates a solution. But small-signal theory neglects squares and cross products of the amplitudes of the excitation. Energy and power relations involve squares and cross products of the small-signal amplitudes which are of the same order of magnitude as the terms neglected in the small-signal approximation. Thus, it might seem that a discussion of energy and power associated with an electron beam is bound to be inconsistent if it is based on small-signal assumptions. However, a closer look at the problem resolves these apparent difficulties.

An identity analogous to the Poynting theorem can be derived for the longitudinal beam of Fig. 3.3, starting from the small-signal equations, Eqs. 3.7 and 3.8. These equations hold for a beam whose electrons are confined to an entirely longitudinal motion. We take a scalar product of Eq. 3.7 with $H(r)^*$, and of the complex conjugate of Eq. 3.8 with $E(r)$. Subsequent subtraction of the two equations gives

$$-\nabla \cdot [E(r) \times H(r)^*] = E_z(r)\, J(r)^*$$
$$+ j\omega[\mu_0\, H(r) \cdot H(r)^* - \epsilon_0\, E(r) \cdot E(r)^*] \quad (3.42)$$

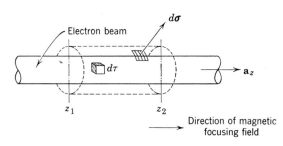

Fig. 3.3.   Volume of integration in Eq. 3.46.

Equation 3.42 looks like the conventional Poynting theorem. It differs from it in establishing an identity among the *approximate* small-signal solutions of Maxwell's equations.

Through the use of the force equation (Eq. 3.9*b*), the continuity equation (Eq. 3.10*b*), and the relation among current density, charge density, and velocity (Eq. 3.11*b*), we find

$$E_z(\mathbf{r})\, J(\mathbf{r})^* = \frac{m}{e}\left\{ j\omega\, \rho_0(\mathbf{r})\, |v(\mathbf{r})|^2 + \frac{\partial}{\partial z}\, [u(\mathbf{r})\, v(\mathbf{r})\, J(\mathbf{r})^*] \right\} \quad (3.43)$$

We define the complex kinetic-power density according to L. J. Chu (compare Eq. 3.28):

$$S_k(r) = \frac{1}{2}\frac{m}{e}\, u(\mathbf{r})\, v(\mathbf{r})\, J(\mathbf{r})^* \mathbf{a}_z = \frac{1}{2}\, V(\mathbf{r})\, J(\mathbf{r})^* \mathbf{a}_z \quad (3.44)$$

where $V(\mathbf{r})$ is the kinetic-voltage modulation defined by Eq. 3.17, and $\mathbf{a}_z$ is the unit vector in the z direction. The definition of Eq. 3.44 introduced into Eq. 3.43, and that, in turn, applied to Eq. 3.42 leads to an alternate form of the small-signal Poynting theorem. Noting that

$$\frac{1}{2}\frac{\partial}{\partial z}\, [v(\mathbf{r})\, J(\mathbf{r})^*] = \nabla \cdot S_k(\mathbf{r})$$

we find

$$-\nabla \cdot [\tfrac{1}{2}\boldsymbol{E}(\mathbf{r}) \times \boldsymbol{H}(\mathbf{r})^* + S_k(\mathbf{r})]$$
$$= \tfrac{1}{2} j\omega \left[\mu_0\, \boldsymbol{H}(\mathbf{r}) \cdot \boldsymbol{H}(\mathbf{r})^* - \epsilon_0 \boldsymbol{E}(\mathbf{r}) \cdot \boldsymbol{E}(\mathbf{r})^* + \frac{m}{e}\, \rho_0(\mathbf{r}) |v(\mathbf{r})|^2 \right] \quad (3.45)$$

S_k is a complex vector in the direction of the flow, i.e., the z direction, with the dimension of power density. We shall call it the "complex kinetic-power density." Integration of Eq. 3.45 over the volume τ enclosed by the surface σ, shown in Fig. 3.3, gives

$$-\oint [\tfrac{1}{2}\boldsymbol{E}(\mathbf{r}) \times \boldsymbol{H}(\mathbf{r})^* + S_k(\mathbf{r})] \cdot d\boldsymbol{\sigma} = \tfrac{1}{2}j\omega \int \left[\mu_0\, \boldsymbol{H}(\mathbf{r}) \cdot \boldsymbol{H}(\mathbf{r})^* \right.$$
$$\left. - \epsilon_0\, \boldsymbol{E}(\mathbf{r}) \cdot \boldsymbol{E}(\mathbf{r})^* + \frac{m}{e}\, \rho_0(\mathbf{r}) |v(\mathbf{r})|^2 \right] d\tau \quad (3.46)$$

The real part of Eq. 3.46 is

$$\text{Re} \oint [\tfrac{1}{2}\, \boldsymbol{E}(\mathbf{r}) \times \boldsymbol{H}(\mathbf{r})^* + S_k(\mathbf{r})] \cdot d\boldsymbol{\sigma} = 0 \quad (3.47)$$

Since the small-signal amplitudes of the electric and magnetic fields have been found by neglecting terms involving squares and cross products of the small-signal amplitudes, the integral Re $[\oint E(\mathbf{r}) \times$ $H(\mathbf{r})^* \cdot d\mathbf{\delta}]/2$ cannot give the electromagnetic power flow through the surface σ exactly. However, the integral gives the electromagnetic power correctly within second order of the small-signal terms. Indeed, if no time-average electron current flows to the RF structure, power can be exchanged between the beam and the RF structure only by means of the RF radiation. The radiated power involves cross products of electric and magnetic fields. Hence, contributions to the time-average power come from cross-products of the electric and magnetic field amplitudes of equal frequency. Since the electron beam is an inherently nonlinear system, an excitation of the beam at a particular frequency leads to a response at the fundamental and higher frequencies. The fundamental frequency is, clearly, by far the largest if small-signal theory is applicable. Thus, the cross-product of the electric and magnetic fields pertaining to the fundamental, a term of second order in the small-signal amplitudes, gives the largest contribution to the time-average electromagnetic power. In disregarding terms corresponding to power radiated at higher harmonic frequencies, an error is made of an order higher than second, in terms of the small-signal amplitudes. An additional error is made when the amplitudes obtained from the *approximate* small-signal analysis are used to compute the radiated power. However, this error is also of higher than second order.

From the above it is therefore clear that the integral

$$\tfrac{1}{2} Re \, [\oint E(\mathbf{r}) \times H(\mathbf{r})^* \cdot d\mathbf{\delta}]$$

represents the electromagnetic power from the beam correctly within second order of the small-signal amplitudes. By means of Eq. 3.47 *we can identify the electromagnetic power delivered by the beam in the volume* τ *by computing the net real kinetic-power flow*, Re $[\oint S_k(\mathbf{r}) \cdot d\mathbf{\delta}]$ *into the volume on a small-signal basis.* This is the content of the kinetic-power theorem first formulated by L. J. Chu.[22] (This approach should be compared with reference 28.)

The usefulness of the kinetic-power theorem stems from its generality. It is applicable to electron flows of arbitrary geometry as long as the directions of the d-c and a-c velocities of the electrons coincide;[26] in our case, this is the z direction. For example, the electron motion in a freely drifting, thin, longitudinal beam is governed by the kinetic-power theorem. If the beam is surrounded by a perfectly conducting cylinder, the sum of the electromagnetic and the kinetic powers must be independent of distance.

The Thin Beam

A beam of finite cross section differs from the infinite-parallel-plane one-dimensional model in that it allows for an arbitrary distribution of current or velocity over a cross section. The details of the distribution depend upon the nature of the excitation. To satisfy the initial conditions at some reference cross section imposed by an assumed excitation at the same cross section, many more parameters are needed than the velocity and current-modulation amplitudes found in the one-dimensional analysis of Section 1, page 83. A beam of finite cross section must propagate more than one mode. (The sole mode of the infinite-parallel-plane analysis consists of the fast and slow space-charge waves). In references 19 and 20 an analysis of an unaccelerated longitudinal electron beam of radius b in an infinite magnetic field surrounded by a drift tube of radius a is carried out in detail. It is found that the beam propagates an infinite number of so-called "space-charge modes." Each space-charge mode consists of a pair of space-charge waves having propagation constants whose arithmetic average is approximately equal to $\beta_e = \omega/u$. If the beam is thin ($\beta_e b < 1$), and the beam does not fill the drift tube ($b < a$), the lowest-order space-charge mode possesses a current and velocity modulation approximately uniform across the beam. All higher-order modes vary rapidly over the cross section. A microwave structure with an approximately uniform E_z field at the beam couples strongly to the mode with the uniform current and velocity distribution, whereas its coupling to all higher-order modes is considerably less. For this reason, the interaction of a thin beam with a microwave structure can be analyzed approximately, disregarding all but the lowest space-charge mode, commonly called the dominant space-charge mode. This fact brings a thin electron beam into direct correspondence with the infinite-parallel-plane, one-dimensional beam model. Therefore, this approximation of the thin beam is often referred to as the "one-dimensional model." In the sequel we shall derive the equations for the dominant space-charge mode by physical reasoning which, it is hoped, will provide some insight into its nature.

The electron motion in the dominant space-charge mode of the thin beam is still governed by the force equation and the continuity equation. With proper precautions, Eqs. 3.21 and 3.22 of the infinite-parallel-plane analysis can be adapted to the analysis of the thin-beam case. The continuity equation, Eq. 3.22, holds unaltered, since it merely expresses the amount of current modulation produced by a given velocity modulation. The force equation, Eq. 3.21, preserves

its general meaning. On the left is (essentially) the a-c force of inertia.
On the right-hand side is the space-charge field. Yet, there is one
fundamental difference between the infinite-parallel-plane one-dimensional beam and the thin beam. Whereas in the infinite-parallel-plane
beam all electric-field lines are longitudinal, in a thin beam, the
bunched electrons produce also a radial fringing space-charge field, at
the expense of the longitudinal field. The longitudinal space-charge
repulsion forces produced by a given per cent current modulation
i/I_0 are less for the thin beam. The amount of fringing depends only
upon the distance between bunches relative to the beam radius
$\beta_e b$, and the drift-tube radius. Whereas the space-charge field corresponding to an infinite-parallel-plane beam is given by (compare
Eq. 3.21)

$$E_{sc} = j\beta_p{}^2 \frac{2V_0}{\beta_e |J_0|} J$$

we shall expect the longitudinal space-charge field for a thin beam to be

$$E_{sc} = j(p\beta_p)^2 \frac{2V_0}{\beta_e I_0} i$$

where p is a factor less than unity and a function† of $\beta_e b$ and $\beta_e a$.
(Note that I_0 is defined positive). The name plasma-frequency reduction factor has been applied to p. The combined expression $p\beta_p$ is
the so-called *reduced plasma propagation constant.*

$$\beta_q = p\beta_p \tag{3.48}$$

It can be expressed in terms of the beam voltage V_0, current I_0,
radius b, and plasma reduction factor p;

$$\beta_q = \frac{1}{(2\pi\epsilon_0)^{1/2} \left|2\frac{e}{m}\right|^{1/4}} \frac{pI_0{}^{1/2}}{V_0{}^{3/4}b} \tag{3.48a}$$

With these definitions, the fundamental Eqs. 3.21 and 3.22 of the
infinite-parallel-plane beam can be modified so as to apply to the case
of a thin beam as well.

$$\left(j\frac{\omega}{u} + \frac{d}{dz}\right) V = j\beta_q{}^2 \frac{2V_0}{\beta_e I_0} i \tag{3.49a}$$

$$\left(j\frac{\omega}{u} + \frac{d}{dz}\right) i = j\frac{\beta_e I_0}{2V_0} V \tag{3.49b}$$

† The factor p is plotted in Fig. 5.9, Chapter 5, for various values of $\beta_e a$ and $\beta_e b$.

Equations 3.49 apply strictly only for a freely drifting thin beam; i.e., for constant velocity u and beam radius b. However, it may be expected that Eqs. 3.49 are still applicable if both u and b vary slowly. Indeed, a variation of u and b does not affect the continuity equation (Eq. 3.49b) but modifies only the expression for the space-charge field on the right-hand side of Eq. 3.49a. If the beam characteristics change only by little within a distance between two successive current bunches (this distance is the electronic wavelength $\lambda_e = 2\pi/\beta_e$), Eq. 3.49a may still be expected to apply, provided β_q is considered to vary with distance. This variation will be due partly to the variation of β_p with distance; partly it will stem from the variation of p (see Eq. 3.48a).†

The Kinetic Power in the Unaccelerated Thin Beam

The solution of Eqs. 3.49 are particularly simple when the beam is freely drifting and u and β_q are distance-independent. We then have (compare with Eqs. 3.30 and 3.31)

$$V(z) = (U_+ e^{j\beta_q z} + U_- e^{-j\beta_q z})e^{-j\beta_e z} \qquad (3.50a)$$

$$i(z) = \frac{1}{W}(U_+ e^{j\beta_q z} - U_- e^{-j\beta_q z})e^{-j\beta_e z} \qquad (3.50b)$$

W is now given by

$$W = \frac{2V_0}{I_0}\frac{\beta_q}{\beta_e} \qquad (3.51)$$

Expressing β_q and β_e in terms of the beam voltage and current, beam radius b, and plasma reduction factor p, we have

$$W = 2\frac{\left|\dfrac{e}{2m}\right|^{1/4}}{(\pi\epsilon_0)^{1/2}}\frac{V_0^{3/4}p}{\omega I_0^{1/2}b} \qquad (3.51a)$$

U_+ and U_- in Eqs. 3.50 are the complex amplitudes of the fast and slow waves at $z = 0$. Again, it is found that the solution consists of two waves, with propagation constants $\beta_e \pm \beta_q$. The reduced plasma propagation constant, β_q, is always smaller than β_e, so that here at least the name "fast wave" is justified for the wave with the propagation constant $\beta_e - \beta_q$ (compare the statement in Section 1,

---

† Electron beams are surrounded in practice by conducting drift tubes which are below cut-off for electromagnetic propagation at the frequency of operation. Thus, in all practical cases no analog exists to the TEM propagation which was mentioned in connection with the infinite-parallel-plane, one-dimensional model.

page 90). The quantity β_q is in general a slowly varying function of frequency, whereas, for the infinite-parallel-plane beam of Section 1 the corresponding quantity, β_p, was independent of frequency.

Therefore, the group velocity of the two space-charge waves is

$$v_g = \frac{d\omega}{d\beta} = \frac{1}{d\beta/d\omega} = \frac{1}{1/u \pm d\beta_q/d\omega} \simeq u\left(1 \mp u\frac{d\beta_q}{d\omega}\right)$$

since $u(d\beta_q/d\omega)$ is, in general, a small quantity. From the above we see that the group velocity, i.e., the velocity of signal propagation, differs slightly from the electron velocity. This arises from the fact that the space-charge waves have a finite, although small, electromagnetic power flow.[29] A small part of the excitation is carried by the electromagnetic fields. The speed with which the signal is carried forward, therefore, does not have to correspond exactly to the beam velocity.

Turning now to the kinetic power, we note that in a thin beam the current density J and the kinetic voltage V in the two lowest-order space-charge waves are nearly uniform across the electron beam. We can carry out the integration of the kinetic power S_k across the beam cross section directly.

$$\int S_k \cdot d\mathbf{\sigma} = \tfrac{1}{2}\int V(\mathbf{r})\,J^*(\mathbf{r})\,d\sigma = \tfrac{1}{2}V(z)\,i^*(z)$$

If we introduce normalized wave amplitudes a_1 and a_2 according to Eq. 3.33,

$$a_1 = (2W)^{-\frac{1}{2}}U_+ \quad \text{and} \quad a_2 = -(2W)^{-\frac{1}{2}}U_- \tag{3.33}$$

we can write the real part of the kinetic power in a particularly simple form:

$$P_k = \tfrac{1}{2}\,\mathrm{Re}\,[V(z)\,i(z)^*] = |a_1|^2 - |a_2|^2 \tag{3.52}$$

The real part of the kinetic power, P_k, is independent of distance. This is to be expected on the basis of the kinetic-power theorem if no electromagnetic power is *extracted* from the beam.

Electromagnetic power can be extracted from a thin beam if it flows through a structure other than a drift tube. The helix of a traveling-wave tube is an example of a structure whose fields may impart to, or extract from the beam, electromagnetic power. In this instance it is convenient to adapt Eq. 3.47 for a thin beam; then the kinetic voltage $V(\mathbf{r})$ and the current-density modulation $J(\mathbf{r})$ are independent of the transverse co-ordinates. The integration in Eq. 3.47 can be carried over the cross section of the beam with the result (see Fig. 3.3):

$$\tfrac{1}{2}\,\mathrm{Re}\,[\oint E(\mathbf{r}) \times H(\mathbf{r})^* \cdot \overline{d\sigma}] = \tfrac{1}{2}\mathrm{Re}\,[V(z_1)\,i(z_1)^* - V(z_2)\,i(z_2)^*] \tag{3.47a}$$

where $i(z)$ is the RF current modulation in the beam at the cross section z. According to Eq. 3.47a, any time-average electromagnetic power extracted from the electron beam between two cross sections z_1 and z_2 is balanced by a decrease in the real kinetic power. On the other hand, if electromagnetic power is fed into the beam, and the integral on the left is negative, then the real kinetic power must increase correspondingly between the cross sections z_1 and z_2.

Now that the role of the kinetic-power flow is known, we can attempt to obtain a physical understanding of its meaning. Equation 3.52 shows that the real kinetic-power flow associated with the fast wave is positive. This follows from the fact that the kinetic voltage and the current in the fast wave are in phase, as is shown in Eqs. 3.50. Thus, if we view at a particular cross section z an electron beam propagating a fast wave only, we find that the kinetic voltage reaches its maximum at the same instant of time as the current modulation. A positive value of the kinetic voltage corresponds, according to the definition of Eq. 3.17, to a negative value of the velocity modulation v, since the electron charge e is negative. Thus, at the instant of time when the kinetic voltage is a maximum, the total velocity of the electrons passing the cross section z reaches the minimum value, $u - |v|$. The electrons passing the cross section at this instant of time travel more slowly than they would travel in the absence of an excitation. Simultaneously, the alternating current reaches its maximum instantaneous value $|i|$. A positive alternating current in a beam of negative charge occurs when there is a deficiency of negative particles. Thus, when the current swings into its maximum, the number of electrons passing the cross section is less than it would be in the absence of an excitation. Conversely, an excess velocity of the electrons occurring when the kinetic-voltage modulation swings negatively is accompanied by an excess of particle current. Thus, the number of electrons that passes the cross section with a velocity higher than u is larger than that passing the cross section with a velocity lower than u. We may therefore conclude that the electron beam carries, on the average, electrons with a higher kinetic energy in the presence of a fast wave than it carries in the absence of an excitation.

Conversely, we find that in the slow wave the kinetic-voltage and current modulations are 180 degrees out of phase. Thus, if only the slow wave is excited, the number of electrons passing a given cross section with a velocity higher than u is, on the average, smaller than the number of electrons with a velocity lower than u. On the average, the beam transports less kinetic energy when it propagates a slow wave than it would carry in the absence of such an excitation. This interpretation of the kinetic-power flow, although not quite rigorous in view of the limitations of small-signal theory, gives a useful physical picture. According to this picture, a negative kinetic-power flow does not signify a transport of energy in the negative z direction but rather a transport of a deficit of kinetic energy in the positive z direction.

The Accelerated Thin Beam

If the electron beam is accelerated, expanded, or is enclosed in a conducting tube of varying diameter, the parameters in Eqs. 3.49 are distance-dependent. Similar equations are encountered in the analysis of tapered transmission lines. To emphasize the similarity further, let us introduce the transformation of variables (compare Eqs. 3.23 and 3.24),

$$V = Ue^{-j\theta} \tag{3.53a}$$

$$i = Ie^{-j\theta} \tag{3.53b}$$

where

$$\theta = \omega \int_0^z \frac{dz}{u} \tag{3.54}$$

With this transformation we are able to put Eqs. 3.49a and 3.49b into the form

$$\frac{dU}{dz} = j\beta_q{}^2 \frac{2V_0}{\beta_e I_0} I \tag{3.55a}$$

$$\frac{dI}{dz} = j \frac{\beta_e I_0}{2V_0} U \tag{3.55b}$$

where β_q, V_0, and β_e may be all functions of z. We note that Eqs. 3.55a and 3.55b correspond to a *lossless* tapered transmission line. The power on this line

$$\tfrac{1}{2} \mathrm{Re}\, [UI^*] = \tfrac{1}{2} \mathrm{Re}\, [Vi^*] = P_k \tag{3.56}$$

is independent of distance. This is a confirmation of the kinetic-power theorem for a beam which is not delivering electromagnetic power. Equations 3.55a and 3.55b can be further simplified if we introduce the definition of the characteristic impedance W, Eq. 3.51. Clearly, W is now a function of distance. Expressing the distance in terms of the *"plasma transit angle"* ϕ,

$$\phi = \int_0^z \beta_q(z)\, dz \tag{3.57}$$

we obtain, for Eqs. 3.55a and 3.55b,

$$\frac{dU}{d\phi} = jW I \qquad (3.58a)$$

$$\frac{dI}{d\phi} = \frac{j}{W} U \qquad (3.58b)$$

Combining the two above equations, we get

$$\frac{d^2U}{d\phi^2} - \frac{1}{W}\frac{dW}{d\phi}\frac{dU}{d\phi} + U = 0 \qquad (3.59a)$$

$$\frac{d^2I}{d\phi^2} + \frac{1}{W}\frac{dW}{d\phi}\frac{dI}{d\phi} + I = 0 \qquad (3.59b)$$

Solutions of Eqs. 3.59a and 3.59b will be discussed in Chapter 5, Section 3, which deals with the design of low-noise guns.

Matrix Representation of Beam Transducers

The electron beam of a longitudinal beam amplifier is formed in an electron gun in which it is accelerated to anode potential. Following the anode there may be some accelerating or decelerating regions like those used in modern low-noise amplifiers (Fig. 3.4). These regions

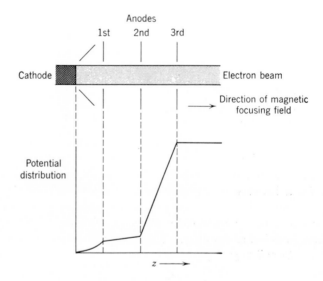

Fig. 3.4. Multi-electrode gun and its potential distribution on the beam axis.

are termed "beam transducers."[27] No exact analysis exists for an accelerated beam of finite diameter confined by a large magnetic field. Instead, the one-dimensional analysis is used, with a simple substitution of the reduced plasma frequency ω_q for the plasma frequency ω_p as discussed in the preceding subsection. Transmission-line equations (Eqs. 3.55a and 3.55b) result that are linear and have the property of being "lossless"; i.e., they conserve kinetic power defined in Eq. 3.44 (compare Eq. 3.56). Therefore, even without solving the transmission-line equations some general statements can be made about the interrelations between the beam voltage V_a and beam current i_a at an "input" cross section a, and the corresponding quantities V_b and i_b at an "output" cross section b. Linear relations must exist among these quantities. It is advantageous to write linear relations among sets of variables in matrix form. Particularly in the case of noise we shall find that the matrix formulation of the beam equations will provide simple proofs of general theorems. Defining the column matrices (column vectors)†

$$\mathbf{w}_a = \begin{bmatrix} V_a \\ i_a \end{bmatrix} \qquad \mathbf{w}_b = \begin{bmatrix} V_b \\ i_b \end{bmatrix} \tag{3.60}$$

we can write the linear relations in the form

$$\mathbf{w}_b = \mathbf{K}\mathbf{w}_a \tag{3.61}$$

The $\mathbf{K}$ matrix is sometimes called the "matrix of general circuit parameters"[30] or the $(ABCD)$ matrix. The $\mathbf{K}$ matrix is usually written as the following array of complex scalars (compare Eqs. 3.60 and 3.61 with Eqs. 3.37 to 3.41):

$$\mathbf{K} = \begin{bmatrix} A & B \\ C & D \end{bmatrix}$$

Conservation of Power. If no RF electromagnetic power is extracted from the beam in the region between the cross sections a and b, as is true for the beam transducer of Fig. 3.4, the real part of the kinetic power must be conserved.

$$4(P_{ka} - P_{kb}) = (V_a{}^*i_a + i_a{}^*V_a) - (V_b{}^*i_b + i_b{}^*V_b) = 0 \tag{3.62}$$

It is expedient to write Eq. 3.62 in matrix form. For this purpose we

---

† The operations and theorems of matrix algebra which we use here can be found in many texts on matrices or applied mathematics. See, for example, F. B. Hildebrand, *Methods of Applied Mathematics* (Prentice-Hall, New York, 1952).

introduce the permutation matrix

$$\mathbf{R} = \begin{bmatrix} 0 & 1 \\ 1 & 0 \end{bmatrix} \tag{3.63}$$

According to the rules of matrix multiplication we find that

$$\mathbf{RR} = \mathbf{I} \tag{3.64}$$

where $\mathbf{I}$ is the identity matrix. Equation 3.64 can also be written in the form

$$\mathbf{R} = \mathbf{R}^{-1}$$

indicating that $\mathbf{R}$ is equal to its own inverse. Further, we definx by $\mathbf{A}^{+}$ the Hermitian (complex) conjugate of the matrix $\mathbf{A}$. The Hermitian conjugate of a matrix $\mathbf{A}$ is obtained by taking the complex conjugate of all elements of $\mathbf{A}$ and then transposing them.

$$(\mathbf{A}^{+})_{ij} = A_{ji}{}^{*}$$

In particular, the Hermitian conjugate of a column matrix is a row matrix. Referring to definition 3.60, we have, for example,

$$\mathbf{w}_a{}^{+} = [V_a{}^{*}, i_a{}^{*}]$$

With the aid of these definitions we can write the real kinetic power P_{ka} of an excitation with the excitation matrix $\mathbf{w}_a$ at cross section a as

$$P_{ka} = \tfrac{1}{4}\mathbf{w}_a{}^{+}\mathbf{R}\mathbf{w}_a \tag{3.65}$$

Equation 3.62 can be rewritten using Eq. 3.65, and a corresponding equation for the real kinetic power at cross section b:

$$P_{ka} - P_{kb} = \tfrac{1}{4}\mathbf{w}_a{}^{+}\mathbf{R}\mathbf{w}_a - \tfrac{1}{4}\mathbf{w}_b{}^{+}\mathbf{R}\mathbf{w}_b = 0 \tag{3.66}$$

The excitation $\mathbf{w}_b$ at cross section b is related to that at cross section a by Eq. 3.61. We further note that the Hermitian conjugate of a product of two matrices $\mathbf{A}$ and $\mathbf{B}$ is equal to the product in reverse order of the Hermitian conjugates of the matrix factors.

$$(\mathbf{AB})^{+} = \mathbf{B}^{+}\mathbf{A}^{+} \tag{3.67}$$

Applying the rule, Eq. 3.67, to Eq. 3.61, we have

$$\mathbf{w}_b{}^{+} = \mathbf{w}_a{}^{+}\mathbf{K}^{+}$$

and thus we obtain, for the difference of the kinetic powers in Eq. 3.66,

$$\tfrac{1}{4}\mathbf{w}_a{}^{+}(\mathbf{R} - \mathbf{K}^{+}\mathbf{R}\mathbf{K})\mathbf{w}_a = 0$$

The above relation has to be fulfilled for an arbitrary choice of the excitation vector $\mathbf{w}_a$, that is, an arbitrary choice of the boundary conditions. This is possible if, and only if,

$$\mathbf{K}^+\mathbf{RK} = \mathbf{R} \tag{3.68}$$

This condition is the restriction imposed upon the $\mathbf{K}$ matrix by the requirement for the conservation of the real kinetic power, Eq. 3.62.

Alternate Form of the Power-Conservation Condition. For the analysis of noise in the electron beam it will be convenient to use another form of Eq. 3.68. In order to obtain that alternate form we shall make use of some additional definitions of matrix algebra.

A matrix $\mathbf{A}$ is termed nonsingular if its determinant det $(\mathbf{A})$ is not equal to zero. A nonsingular matrix $\mathbf{A}$ always has an inverse, $\mathbf{A}^{-1}$. Further, it follows from the properties of matrix multiplication and the multiplication of determinants that the determinant of a matrix product is equal to the product of the determinants of the matrix factors.

$$\det (\mathbf{AB}) = \det (\mathbf{A}) \det (\mathbf{B}) \tag{3.69}$$

With the aid of Eq. 3.69 we obtain, from Eq. 3.68,

$$\det (\mathbf{K}^+) \det (\mathbf{K}) = |\det (\mathbf{K})|^2 = 1 \tag{3.70}$$

By virtue of Eq. 3.70, the determinant of $\mathbf{K}$ is nonzero; correspondingly, $\mathbf{K}$ is nonsingular. The matrix $\mathbf{K}$ has an inverse. Multiplying Eq. 3.68 from the right by $\mathbf{K}^{-1}\mathbf{R}$, we have

$$\mathbf{K}^+\mathbf{RKK}^{-1}\mathbf{R} = \mathbf{RK}^{-1}\mathbf{R}$$

or, with the aid of Eq. 3.64,

$$\mathbf{K}^+ = \mathbf{RK}^{-1}\mathbf{R} \tag{3.71}$$

This is the equation that we shall use in the analysis of noise in electron beams.

Examples of Lossless Beam Transducers. A drift region is a simple example of a lossless beam transducer. With the aid of Eqs. 3.50 it is easy to show that the kinetic-voltage and current modulations, V_a and i_a, at the plane a, transform into corresponding modulations, V_b and i_b, at the plane b, through the following equations:

$$V_b = (V_a \cos \phi + i_a jW \sin \phi)e^{-j\theta} \tag{3.72}$$

$$i_b = \left(V_a \frac{j}{W} \sin \phi + i_a \cos \phi \right) e^{-j\theta}$$

where $\phi = \omega_q(z_b - z_a)/u$ is the transit angle measured in terms of the plasma period, $T = 2\pi/\omega_q$; and $\theta = \omega(z_b - z_a)/u$ is the conventional transit angle. We refer to ϕ as the *"plasma transit angle."* Equations 3.72 show that the $\mathbf{K}$ matrix of the transformation of voltage and current by a drift region is

$$
\mathbf{K} = \begin{bmatrix} \cos\phi & jW\sin\phi \\ \dfrac{j}{W}\sin\phi & \cos\phi \end{bmatrix} e^{-j\theta} \tag{3.73}
$$

Simple matrix manipulations show that $\mathbf{K}$ as given by Eq. 3.73 indeed satisfies the condition of conservation of the real part of kinetic power, Eq. 3.68.

A relation analogous to Eqs. 3.72 is given in the example of Section 1, page 94, for the voltage–current transformation by an infinite-parallel-plane accelerated electron beam. It is not difficult to confirm that the matrix $\mathbf{K}$, whose coefficients are given by Eqs. 3.38 to 3.41, satisfies the condition of power conservation, Eq. 3.68.

Wave Formalism. We earlier discussed another set of parameters that is also able to describe the excitation of a beam: the normalized amplitudes of the fast and the slow waves a_1 and a_2. Consider a lossless beam transducer that extends from cross section a to cross section b. Imagine that the transducer is preceded and followed by drift regions of characteristic impedance W_a and W_b, respectively. Then the normalized amplitudes a_1 and a_2 of Eq. 3.33 can be found uniquely in terms of the voltage V_a and current i_a by use of Eqs. 3.50, in which we set $z = 0$, thus choosing an appropriate origin of the co-ordinate system in the input drift region.

$$
a_1 = \frac{1}{2\sqrt{2W_a}}\,(W_a i_a + V_a)
$$
$$
\tag{3.74a}
$$
$$
a_2 = \frac{1}{2\sqrt{2W_a}}\,(W_a i_a - V_a)
$$

Similarly, we can choose the origin of the co-ordinate system in the output drift region to coincide with the cross section b. In order to avoid confusion, we denote the normalized amplitudes of the fast and slow waves in the output drift region by b_1 and b_2, respectively. Thus, we have

$$
b_1 = \frac{1}{2\sqrt{2W_b}}\,(W_b i_b + V_b)
$$
$$
\tag{3.74b}
$$
$$
b_2 = \frac{1}{2\sqrt{2W_b}}\,(W_b i_b - V_b)
$$

The linear relations among the kinetic voltages and currents at reference cross sections a and b, respectively, summarized in Eq. 3.61, imply corresponding linear relations among the normalized wave amplitudes. Introducing the column matrices

$$\mathbf{a} = \begin{bmatrix} a_1 \\ a_2 \end{bmatrix} \quad \text{and} \quad \mathbf{b} = \begin{bmatrix} b_1 \\ b_2 \end{bmatrix} \tag{3.75}$$

we may reformulate the transformation by a beam transducer, Eq. 3.61, in terms of the variables $\mathbf{a}$ and $\mathbf{b}$. This can be done in two ways. One way is to use the laws of transformation between the current–voltage variables and the wave variables as given in Eqs. 3.74. This method is reproduced in the appendix at the end of this chapter. Another way is to go through a separate derivation for the wave variables of the laws derived for the voltage–current variables. This method is mathematically simpler and, in addition, emphasizes the basic symmetry between the voltage–current formalism on one hand and the wave formalism on the other hand. This is why we follow this approach here in the text.

The relations among the waves b_1 and b_2 at the output of a beam transducer and the waves a_1 and a_2 at its input have to be linear within the assumptions of small-signal theory. These linear relations can be summarized in matrix form,

$$\mathbf{b} = \mathbf{Ma} \tag{3.76}$$

where $\mathbf{M}$ is a two-by-two matrix.

Equation 3.76 is the expression analogous to Eq. 3.61 for the voltage–current formalism.

The real kinetic power at the cross section a can be written according to Eq. 3.52 in the form

$$P_{ka} = |a_1|^2 - |a_2|^2 \tag{3.77}$$

Defining a matrix $\mathbf{P}$

$$\mathbf{P} = \begin{bmatrix} 1 & 0 \\ 0 & -1 \end{bmatrix} \tag{3.78}$$

we can cast the kinetic-power expression 3.77 into the matrix form

$$P_{ka} = \mathbf{a}^+ \mathbf{Pa} \tag{3.79}$$

The expression for the kinetic power in wave variables, Eq. 3.79, is

analogous to Eq. 3.65 derived for the voltage–current formalism.†
The matrix **P** plays the role of the matrix ¼**R**.

Again, a lossless beam transducer must fulfill the condition that the
net power flowing *into* the transducer must be zero. If we write this
condition in matrix form, we have

$$P_{ka} - P_{kb} = \mathbf{a}^+\mathbf{Pa} - \mathbf{b}^+\mathbf{Pb} = 0 \qquad (3.80)$$

We may express **b** in terms of **a** by the transducer equation, Eq. 3.76.
We obtain

$$\mathbf{a}^+(\mathbf{P} - \mathbf{M}^+\mathbf{PM})\mathbf{a} = 0 \qquad (3.81)$$

The requirement that Eq. 3.81 be fulfilled for an arbitrary excitation
a leads to the restriction on **M**,

$$\mathbf{M}^+\mathbf{PM} = \mathbf{P} \qquad (3.82)$$

Equation 3.82 is analogous to Eq. 3.68 of the voltage–current formalism.

In later work it will be profitable to relate the excitation matrices **a**
and $\mathbf{w}_a$, **b** and $\mathbf{w}_b$ of the two formalisms by means of a simple matrix
relation. This can be done by inspection of Eqs. 3.74. We define the
matrix $\mathbf{T}_a$ at the cross section a by

$$\mathbf{T}_a = \frac{1}{2\sqrt{2W_a}}\begin{bmatrix} 1 & W_a \\ -1 & W_a \end{bmatrix} \qquad (3.83a)$$

and the corresponding matrix at the cross section b

$$\mathbf{T}_b = \frac{1}{2\sqrt{2W_b}}\begin{bmatrix} 1 & W_b \\ -1 & W_b \end{bmatrix} \qquad (3.83b)$$

With the aid of these definitions we can summarize Eqs. 3.74 by

$$\mathbf{a} = \mathbf{T}_a\mathbf{w}_a \qquad (3.84a)$$

$$\mathbf{b} = \mathbf{T}_b\mathbf{w}_b \qquad (3.84b)$$

In conclusion, we may state that the wave formalism is applicable
even when the transducer under consideration is not preceded or
followed by a drift region. Then, Eqs. 3.84 are the definitions of
quantities **a** and **b**, which have but mathematical significance. The
impedance W in the definitions 3.83 is arbitrary but is conveniently
chosen to correspond to the characteristic impedance of a drift region

---

† The fact that Eq. 3.65 contains a factor of ¼ is an incidental consequence of
the use of the unnormalized voltage and current variables in **w**, whereas the wave
amplitudes have been properly normalized.

with a time-average voltage and a current density equal to those existing at the reference cross section.

Matrix Representation of Longitudinal-Beam Amplifiers

The kinetic-power concept formulated in Eq. 3.47a is common to all one-dimensional electron-beam systems. (To remind the reader: A one-dimensional beam is one whose excitation can be characterized in terms of only two parameters.) With the aid of Eq. 3.47a on page 101, an interesting formalism can be developed for all longitudinal-beam microwave amplifiers. Consider, for example, a traveling-wave tube as shown in Fig. 3.5. The excitation of the fast wave and the slow wave in the beam at the gun end we denote, as usual, by a_1 and a_2. The excitation of the same set of waves at the collector end we denote by b_1 and b_2. The normalized amplitude of the incident wave on the input transmission line of the amplifier we denote by a_3; the incident wave on the output transmission line by a_4. The reflected waves at the input and output are denoted by b_3 and b_4. The arrows in Fig. 3.5 indicate whether a particular wave carries power into or out of the amplifier. The wave amplitudes a_1 to a_4 can be adjusted by means external to the amplifier. Thus, we could, for example, excite the beam before it enters the amplifier. In this way a_1 and a_2 could be adjusted arbitrarily. Furthermore, the output transmission line of the amplifier could be terminated in a matched load, which corresponds to choosing $a_4 = 0$. Power could be fed through an attenuator matched to the input transmission line of the amplifier. The wave a_3 would thus be fixed in amplitude and phase. This example shows that the wave amplitudes a are under our control. The wave amplitudes

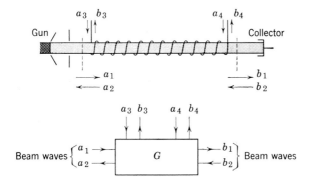

Fig. 3.5. Schematic of a traveling-wave tube.

b_1 to b_4 must then be related to the quantities a_1 to a_4 by linear relations (small-signal theory).

The amplifier can be characterized in terms of a four-by-four matrix **G**, so that

$$\mathbf{b} = \mathbf{Ga} \tag{3.85}$$

where

b is the column matrix $\begin{bmatrix} b_1 \\ b_2 \\ b_3 \\ b_4 \end{bmatrix}$ and **a** is the column matrix $\begin{bmatrix} a_1 \\ a_2 \\ a_3 \\ a_4 \end{bmatrix}$

The kinetic-power theorem, Eq. 3.47a imposes some conditions upon the elements of **G**. Indeed, according to this theorem, the sum of the electromagnetic power and of the kinetic power entering the amplifier must be equal to the sum of the powers leaving the amplifier. The electromagnetic power P_3 entering the circuit of the amplifier at the input is

$$P_3 = |a_3|^2 - |b_3|^2 \tag{3.86}$$

The electromagnetic power leaving the amplifier on the output transmission line is

$$P_4 = |b_4|^2 - |a_4|^2 \tag{3.87}$$

The kinetic power entering the amplifier at the input end is

$$P_{ka} = |a_1|^2 - |a_2|^2 \tag{3.88}$$

and that leaving the amplifier at the collector end is

$$P_{kb} = |b_1|^2 - |b_2|^2 \tag{3.89}$$

In general, electromagnetic power flows in the space around the excited beam and can leave the amplifier when the beam leaves the amplifier. This effect, however, is negligible in all practical cases. Indeed, it was remarked on page 101 that the electromagnetic power associated with an excited beam is much smaller than the kinetic power of the excitation.[29] Thus, the power balance corresponding to Eq. 3.47a is established satisfactorily by equating

$$P_3 + P_{ka} = P_4 + P_{kb} \tag{3.90}$$

Using Eqs. 3.86 to 3.89 in Eq. 3.90 and rearranging, we obtain

$$|b_1|^2 - |b_2|^2 + |b_3|^2 + |b_4|^2 = |a_1|^2 - |a_2|^2 + |a_3|^2 + |a_4|^2 \tag{3.91}$$

Introducing a "parity matrix" **P**

$$\mathbf{P} = \text{diag} \,(1, \, -1, \, 1, \, 1) \tag{3.92}$$

we can write Eq. 3.91 in a more elegant form:

$$\mathbf{b^+Pb = a^+Pa} \tag{3.93}$$

By the use of Eq. 3.85 we can express Eq. 3.93 as

$$\mathbf{a^+(G^+PG - P)a = 0} \tag{3.94}$$

The matrix equation 3.94 has to be satisfied for an arbitrary choice of the $\mathbf{a}$ matrix. This is possible if and only if

$$\mathbf{G^+PG = P} \tag{3.95}$$

The matrix equation 3.95 contains scalar equations of the form

$$|G_{13}|^2 - |G_{23}|^2 + |G_{33}|^2 + |G_{43}|^2 = 1 \tag{3.96}$$

and

$$G_{13}G_{14}{}^* - G_{23}G_{24}{}^* + G_{33}G_{34}{}^* + G_{43}G_{44}{}^* = 0 \tag{3.97}$$

Let us try to interpret the elements of $\mathbf{G}$ and the interrelations among them. The elements G_{33}, G_{34}, G_{43}, and G_{44} establish interrelations among the incident and reflected circuit waves a_3, b_3, a_4, and b_4, if the beam at the input of the amplifier is unexcited. This is in general true if noise is disregarded. Thus, the four elements of $\mathbf{G}$ listed above characterize the noise-free amplifier from the *circuit* point of view as a two-terminal-pair device. From our definitions of the waves a_3, b_3, a_4, and b_4 it follows that the matrix

$$\mathbf{S} = \begin{bmatrix} G_{33} & G_{34} \\ G_{43} & G_{44} \end{bmatrix} \tag{3.98}$$

is the scattering matrix of the amplifier.

In order to find a simple interpretation for Eq. 3.96, let us assume that, with the output transmission line matched to the load ($a_4 = 0$), the amplifier is matched to the input transmission line. Thus, $G_{33} = 0$ (but G_{44} is not necessarily zero, so that the output impedance of the amplifier is not necessarily matched to the load impedance.) Under this circumstance, the term $|G_{43}|^2$ has the meaning of (actual) power gain: i.e., the ratio of the actual output power to the actual input power. If the power gain of the amplifier is appreciably greater than unity, $|G_{43}|^2 \gg 1$, Eq. 3.96 implies that $|G_{23}|^2$, the only term preceded by a minus sign, must also be appreciably greater than unity. In other words, the input wave a_3 must couple strongly to the slow mode in the beam leaving the amplifier, if the amplifier has an appreciable gain. The electromagnetic power gain is obtained at the expense of kinetic power in the beam, which is correspondingly large, and negative, when the beam leaves the amplifier.

It is worth while noting that the term $|G_{43}|^2$ has a simple meaning in the most general case, even when $G_{33} \neq 0$. Let us assume only that the *source impedance* is matched to the *input transmission line*, and that the *load impedance* is matched to the *output transmission line*. Such an assumption does not restrict the generality of the results. Indeed, practical cases in which the source and load impedances have positive real parts (are passive) can be treated under this assumption by a simple stratagem.† Then, $|a_3|^2$ is the available power from the source, $|b_4|^2$ is the actual power flowing into the load since the wave reflected from the load a_4 is zero. Accordingly,

$$\frac{|b_4|^2}{|a_3|^2}\bigg|_{a_4=0} = |G_{43}|^2 = \frac{\text{actual output power}}{\text{available input power}}$$

Thus, $|G_{43}|^2$ is the so-called "transducer gain" of the amplifier.

If the amplifier contains elements with ohmic loss, Eq. 3.95 has to be modified. Because of added complications, it is believed that a discussion of such structures is not warranted here. The qualitative results of such an investigation agree in essence with those for lossless structures. Reference 21 gives the details.

3. Noise in Electron Beams

The preceding sections were devoted to the analysis of electron beams and the interaction of electron beams with RF structures under

† Any mismatch (or match) between the amplifier on one hand and the source and load impedances on the other hand can be simulated by a proper choice of the characteristic impedance of the input and output transmission lines. Assume, for example, that the output load impedance has the complex value Z_L, and that the characteristic impedance of the actual transmission line connected to the load is Z_0. Then, a voltage standing-wave ratio greater than unity is set up on the transmission line. At the voltage standing-wave maximum, a real impedance $R > Z_0$ is presented by the load. Now imagine that a fictitious transmission line with a characteristic impedance equal to R and half a wavelength long is inserted at the position of the voltage standing-wave maximum. Adding the fictitious line to the load does not affect the impedance presented by the load. The fictitious line can now be considered to be the output transmission line of the amplifier. The waves a_4 and b_4 as used here would then be the incident and reflected waves on the fictitious output transmission line. The scattering matrix of the amplifier is then renormalized with regard to the characteristic impedance of the fictitious output transmission line. Similar considerations allow a representation of an arbitrary passive source impedance as matched to the (perhaps fictitious) input transmission line. It should be noted that a change in normalization of the amplifier scattering matrix from one set of characteristic impedances of the input and output transmission lines to another set changes the elements of the scattering matrix (while not changing the actual mismatch between the amplifier and its terminations).

steady-state excitation at a frequency ω. The matrix equation 3.61 states that a steady-state, sinusoidal modulation of the electron beam is completely determined by the values of the kinetic-voltage and current modulations at one reference cross section of the beam. This result can be used as the starting point of the analysis of noise in electron beams.

Harmonic Analysis of Noise. Noise is a statistical process which must be analyzed by statistical methods. Here we shall make use of the extension of Fourier integral theory to the harmonic analysis of random functions.[31] We assume that the noise process in the electron beam is stationary and has no hidden periodic components. An observation of the noise process at the reference cross section a of the beam gives the kinetic noise voltage $V_a(t)$ and the current modulation $i_a(t)$ as functions of time. The subsequent analysis will be devoted to finding the statistical properties of the kinetic-voltage and current modulations at some other reference cross section b in terms of the statistical properties of $V_a(t)$ and $i_a(t)$.

The kinetic-voltage modulation $V_a(t)$ is not a periodic function of time and therefore cannot be analyzed by Fourier series methods. Neither are Fourier integral methods adequate for the analysis of $V_a(t)$, since it is intuitively obvious that the integral

$$\int_{-\infty}^{\infty} |V_a(t)|^2 \, dt$$

carried over an infinite time interval does not converge. We can choose, however, a function $V_{aT}(t)$ defined by

$$V_{aT}(t) = \begin{cases} V_a(t) & -T < t < T \\ 0 & T < |t| \end{cases} \tag{3.99}$$

The function $V_{aT}(t)$ satisfies the requirement of convergence of the integral

$$\int_{-\infty}^{\infty} |V_{aT}(t)|^2 \, dt$$

as long as T is chosen finite. It represents accurately the random function $V_a(t)$ over the interval $2T$. We can form the Fourier transform

$$V_{aT}(\omega) = \frac{1}{2\pi} \int_{-\infty}^{\infty} V_{aT}(t) e^{-j\omega t} \, dt$$

In a similar way we can define a function $i_{aT}(t)$ with the corresponding Fourier transform $i_{aT}(\omega)$. The function $i_{aT}(t)$ represents the random noise current $i_a(t)$ over the finite time interval $2T$.

Transformations of Noise. A linear-beam transducer changes the

Fourier transforms $V_{aT}(\omega)$ and $i_{aT}(\omega)$ of kinetic-voltage and current modulations applied to its input cross section to the corresponding Fourier transforms at its output b.

$$V_{bT}(\omega) = A V_{aT}(\omega) + B i_{aT}(\omega) \tag{3.100}$$

$$i_{bT}(\omega) = C V_{aT}(\omega) + D i_{aT}(\omega) \tag{3.101}$$

The coefficients A to D are, in general, functions of frequency; $V_{bT}(\omega)$ and $i_{bT}(\omega)$ are the Fourier transforms of the time functions $V_{bT}(t)$ and $i_{bT}(t)$. These, in turn, give the output of the beam transducer produced when the modulations $V_{aT}(t)$ and $i_{aT}(t)$ are applied to the input over a finite period of time $2T$. It may be expected that $V_{bT}(t)$ and $i_{bT}(t)$ will resemble the true noise output of the transducer $V_b(t)$ and $i_b(t)$ over a portion of the period $2T$. This portion encompasses the response of the transducer to the noise input $V_{aT}(t)$ and $i_{aT}(t)$ over the time during which it is possible to neglect the transients in the transducer set up at $t = -T$. The squares of the absolute values of Eqs. 3.100 and 3.101 are

$$|V_{bT}(\omega)|^2 = |A|^2 |V_{aT}(\omega)|^2 + |B|^2 |i_{aT}(\omega)|^2 + AB^* V_{aT}(\omega) i_{aT}(\omega)^* \\ + A^*B V_{aT}(\omega)^* i_{aT}(\omega) \tag{3.102}$$

$$|i_{bT}(\omega)|^2 = |C|^2 |V_{aT}(\omega)|^2 + |D|^2 |i_{aT}(\omega)|^2 + CD^* V_{aT}(\omega) i_{aT}(\omega)^* \\ + C^*D V_{aT}(\omega)^* i_{aT}(\omega) \tag{3.103}$$

The functions $V_{aT}(t)$ and $i_{aT}(t)$ do not represent the random functions $V_a(t)$ and $i_a(t)$ exactly as long as the interval $2T$ is held finite. If, however, the interval T is allowed to go to infinity, we see from the definition 3.99 that $V_{aT}(t)$ becomes indistinguishable from $V_a(t)$. The same statement can be made concerning $i_{aT}(t)$ and $i_a(t)$, $V_{bT}(t)$ and $V_b(t)$, as well as $i_{bT}(t)$ and $i_b(t)$. Generalized harmonic analysis proves that in the limit $T \to \infty$ the following quantities approach finite limits:

$$\lim_{T \to \infty} \frac{\pi}{T} \overline{|V_{aT}(\omega)|^2} \equiv \Phi_a(\omega) \tag{3.104}$$

$$\lim_{T \to \infty} \frac{\pi}{T} \overline{|i_{aT}(\omega)|^2} \equiv \Psi_a(\omega) \tag{3.105}$$

$$\lim_{T \to \infty} \frac{\pi}{T} \overline{V_{aT}(\omega) i_{aT}(\omega)^*} = \left[\lim_{T \to \infty} \frac{\pi}{T} \overline{V_{aT}(\omega)^* i_{aT}(\omega)} \right]^* \equiv \Theta_a(\omega) \tag{3.106}$$

The bar over the quantities on the left side of Eqs. 3.104, 3.105, and 3.106 indicates an ensemble average. To obtain such an average, consider a set (ensemble) of statistical processes of identical statistical character. In our particular case we can imagine that measurements are performed on a large number of identical electron beams. The average over such a set of measurements is then the ensemble average.

The quantity Φ_a is the self-power density spectrum (SPDS) of the kinetic noise-voltage modulation, Ψ_a is the SPDS of the noise-current modulation at the reference cross section a. The cross-power-density spectrum (CPDS) between the kinetic-voltage and current modulations is Θ_a. The frequency dependence of these quantities will be henceforth implied, and the parentheses (ω) will be omitted. If we take the limit $T \to \infty$ of Eqs. 3.102 and 3.103 multiplied by π/T, we find, after taking an ensemble average,

$$\Phi_b = |A|^2 \Phi_a + |B|^2 \Psi_a + AB^*\Theta_a + A^*B\Theta_a{}^* \qquad (3.107)$$

$$\Psi_b = |C|^2 \Phi_a + |D|^2 \Psi_a + CD^*\Theta_a + C^*D\Theta_a{}^* \qquad (3.108)$$

Multiplication of Eq. 3.100 by the complex conjugate of Eq. 3.101, transition to $T \to \infty$ of the resultant equation multiplied by π/T, and an ensemble average lead to the relation

$$\Theta_b = AC^*\Phi_a + BD^*\Psi_a + AD^*\Theta_a + BC^*\Theta_a{}^* \qquad (3.109)$$

Equations 3.107, 3.108, and 3.109 give the self- and cross-power-density spectra of the kinetic voltage and current of the noise at cross section b in terms of the corresponding quantities at cross section a. The three quantities, Φ_a, Ψ_a, Θ_a, the last of them complex, characterize the noise in an electron beam sufficiently for most practical purposes. Thus, a noise process at a particular frequency in a one-dimensional electron beam is specified by four real parameters.

The SPDS's are related by a factor $4\pi \, \Delta f$ to the more commonly used quantities, the "mean-square fluctuations within a frequency band Δf." Thus, for example, the SPDS of pure shot noise in a beam with a direct current I_0 is $\Psi = |e| I_0/2\pi$, whereas the mean-square fluctuations of the current within the frequency band Δf are known to be: $\overline{|i^2|} = 2|e| I_0 \, \Delta f$. One of the reasons for the deviation from conventional engineering use of the definition of the SPDS lies in the simplicity of the resulting relation between the mean-square value of the fluctuations and the frequency integral of the SPDS. We have, for example,

$$\lim_{T \to \infty} \frac{1}{2T} \int_{-T}^{T} i^2(t) \, dt = \int_{-\infty}^{\infty} \Psi(\omega) \, d\omega$$

We have a corresponding relation for the cross-power density spectrum:

$$\lim_{T \to \infty} \frac{1}{2T} \int_{-T}^{T} v(t)\, i(t)\, dt = \int_{-\infty}^{\infty} \Theta(\omega)\, d\omega = \int_{-\infty}^{\infty} \Theta^*(\omega)\, d\omega$$

The SPDS of the noise current, Ψ, at a cross section z of the electron beam is a measurable quantity. A high-Q cavity with a short gap at the position z has a power output proportional to the value $\Psi(\omega_0)$ at the resonant frequency ω_0 of the cavity. The classical experiment by Cutler and Quate[13] was a measurement of Ψ as a function of distance in a drifting beam performed in the described way.

Matrix Representation of Noise. Equations 3.107, 3.108, and 3.109 can be obtained in an alternate way by matrix methods. The advantage of such an approach lies partly in its elegance and partly in the fact that general theorems of matrix algebra can be applied to the noise problem. We define the column matrices

$$\mathbf{w}_{aT}(\omega) = \begin{bmatrix} V_{aT}(\omega) \\ i_{aT}(\omega) \end{bmatrix} \quad \text{and} \quad \mathbf{w}_{bT}(\omega) = \begin{bmatrix} V_{bT}(\omega) \\ i_{bT}(\omega) \end{bmatrix} \quad (3.110)$$

Equations 3.100 and 3.101 can be cast in matrix form.

$$\mathbf{w}_{bT}(\omega) = \mathbf{K}\mathbf{w}_{aT}(\omega) \quad (3.111)$$

where

$$\mathbf{K} = \begin{bmatrix} A & B \\ C & D \end{bmatrix}$$

We define the matrices

$$\mathbf{W}_a = \lim_{T \to \infty} \frac{\pi}{T}\, \overline{\mathbf{w}_{aT}(\omega)\, \mathbf{w}_{aT}(\omega)^+}$$

$$\mathbf{W}_b = \lim_{T \to \infty} \frac{\pi}{T}\, \overline{\mathbf{w}_{bT}(\omega)\, \mathbf{w}_{bT}(\omega)^+}$$

$$(3.112)$$

A study of the definitions 3.112 and the definitions of Eqs. 3.104, 3.105, 3.106, and 3.110 shows that the matrix $\mathbf{W}_a$ is composed of the SPDS's and the CPDS's as follows:

$$\mathbf{W}_a = \begin{bmatrix} \Phi_a & \Theta_a \\ \Theta_a{}^* & \Psi_a \end{bmatrix} \quad (3.113)$$

A similar expression holds for $\mathbf{W}_b$. The relation between the matrices $\mathbf{W}_a$ and $\mathbf{W}_b$ can be found by multiplying Eq. 3.111 from the right by its Hermitian conjugate and by π/T, and by transition to the limit

$T \to \infty$. Finally, taking an ensemble average, we obtain the result

$$\mathbf{W}_b = \mathbf{K}\mathbf{W}_a\mathbf{K}^+ \qquad (3.114)$$

Equations 3.107 to 3.109 are contained in the matrix equation 3.114, as can easily be demonstrated by matrix multiplication and the aid of Eq. 3.113. Although the matrix equation 3.114 is a short-hand expression for four equations, it contains only three distinct relations; the 12 element of the matrix equation 3.114 is equal to the complex conjugate of the 21 element. This fact is a direct consequence of the Hermitian character of the $\mathbf{W}$ matrices, namely:

$$\mathbf{W}_b = \mathbf{W}_b{}^+ \quad \text{and} \quad \mathbf{W}_a = \mathbf{W}_a{}^+$$

Transformation of Noise by Lossless Beam Transducers

An important class of beam transducers is "lossless," i.e., conserves the real part of the kinetic power. It is therefore of practical interest to devote special attention to noise transformations by means of such transducers.

The $\mathbf{K}$ matrix of a lossless transducer satisfies Eq. 3.71. The transformation of Eq. 3.114 can then be written in the alternate form:

$$\mathbf{W}_b = \mathbf{K}\mathbf{W}_a\mathbf{R}\mathbf{K}^{-1}\mathbf{R}$$

Multiplication of this equation from the right by $\mathbf{R}$ and the condition on the $\mathbf{R}$ matrix (Eq. 3.64) lead to the transformation

$$\mathbf{W}_b\mathbf{R} = \mathbf{K}\mathbf{W}_a\mathbf{R}\mathbf{K}^{-1} \qquad (3.115)$$

Equation 3.115 shows that *the noise matrix* $\mathbf{WR}$ *undergoes a similarity transformation when the beam is passed through a lossless transducer.* A similarity transformation leaves the trace and the determinant of a two-by-two matrix invariant. The $\mathbf{WR}$ matrix written out explicitly has the form

$$\mathbf{WR} = \begin{bmatrix} \Theta & \Phi \\ \Psi & \Theta^* \end{bmatrix} \qquad (3.116)$$

The trace of $\mathbf{WR}$ is

$$\text{trace } (\mathbf{WR}) = \Theta + \Theta^* = 2 \text{ Re } (\Theta) \qquad (3.117)$$

The determinant is

$$\det (\mathbf{WR}) = |\Theta|^2 - \Phi\Psi \qquad (3.118)$$

The physical meaning of the first invariant, the trace, is not hard to grasp. According to Eq. 3.106, Θ is proportional to the kinetic power carried in a narrow frequency band around the frequency ω. The real part of the kinetic power in a particular frequency band has to be conserved in a transition through a linear lossless transducer; hence the invariance of the trace of **WR**.

The meaning of the second invariant is less self-evident. Indeed, conventional network theory has no analog to this invariant. In order to demonstrate this, we consider, first, a random voltage $V(t)$ applied to a two-terminal network characterized by the admittance function $Y(\omega)$. Again, an equivalent Fourier transform of the voltage $V(t)$ based on a sample of length $2T$ can be constructed. Denote the Fourier transform by $V_T(\omega)$. The Fourier transform of the current flowing into the network under the influence of $V_T(\omega)$ is $i_T(\omega) = Y(\omega)\,V_T(\omega)$. Forming the CPDS Θ between the voltage and current, we obtain

$$\Theta(\omega) = \lim_{T \to \infty} \frac{\pi}{T} \overline{V_T(\omega)\,i_T(\omega)^*}$$

$$= \lim_{T \to \infty} \frac{\pi}{T}\, Y(\omega)^* \overline{\left|V_T(\omega)\right|^2}$$

$$= Y(\omega)^* \lim_{T \to \infty} \frac{\pi}{T} \overline{\left|V_T(\omega)\right|^2}$$

$$= Y(\omega)^*\, \Phi(\omega) \tag{3.119}$$

where, as before, we denote the self-power density spectrum of the voltage by $\Phi(\omega)$. In a similar way we obtain the self-power density spectrum of the current $\Psi(\omega)$;

$$\Psi(\omega) = \lim_{T \to \infty} \frac{\pi}{T} \overline{i_T(\omega)\,i_T(\omega)^*}$$

$$= \left|Y(\omega)\right|^2 \Phi(\omega) \tag{3.120}$$

Introducing Eqs. 3.119 and 3.120 into Eq. 3.118, we find that, for this process,

$$\det\,(\mathbf{WR}) = 0 \tag{3.121}$$

Consider, next, a cascade of linear two-terminal-pair transducers of conventional network theory. The admittance seen across any terminal pair within the cascade is determined uniquely by the admittance connected to the end of the cascade. Forming the matrix **W** of the SPDS's and CPDS's of the noise voltage across, and the noise current into, any terminal pair of the cascade, we find again that $\det\,(\mathbf{WR}) = 0$. Thus, noise propagating along a cascade of lossless two-terminal-pair networks has only one invariant parameter, namely, Re (Θ), which is proportional to the time-average power of the noise within a narrow-frequency band around the frequency ω fed into the termination of the cascade. Since a conventional transmission line can be considered as the limit of a cascade of an infinite number of infinitesimal two-terminal-pair networks, the above reasoning applies as well to noise propagation along conventional transmission lines.

The close analogy between lossless beam transducers and lossless two-terminal-pair networks makes one wonder whether or not the determinant of the matrix **WR** for a noise process in an electron beam can have a finite value. But there is a basic physical difference between a drift region and a transmission line. An electromagnetic wave of a given frequency incident upon a transmission-line termination is accompanied by a reflected wave, the phase and amplitude of which depend upon the termination. The voltage and current in the incident and reflected waves combine to satisfy the boundary condition imposed by the termination. The admittance at any cross section of the transmission line is determined by the terminating admittance. In an electron beam the roles of the incident and reflected waves are played by the fast and slow waves. Both these waves have a group velocity approximately equal in magnitude and direction to the time-average velocity of the electron beam. Thus, these waves can be excited only at the entry plane into a drift region, the plane passed first by the electrons. The phase relation between the fast and slow waves is determined by the method of excitation before the entry of the electron beam into the drift region. Noise mechanisms may excite a kinetic voltage and a current which are not in a definite ratio to each other, so that the kinetic voltage cannot be expressed in terms of the current by a simple impedance relation. We shall return to the question of noise excitation later. At this point it is sufficient to state that the kinetic noise voltage and noise current in the electron beam may or may not be in a definite ratio to each other, and therefore we must allow for the possibility that $|\Theta|^2 \neq \Phi\Psi$. It is known from statistical theory that the inequality holds. (See the definitions, Eqs. 3.104, 3.105, and 3.106).

$$|\Theta|^2 \leq \Phi\Psi \tag{3.122}$$

With the aid of this inequality and Eq. 3.118, we find, for a noise process in an electron beam,

$$\det (\mathbf{WR}) \leq 0 \tag{3.123}$$

According to Eq. 3.123, noise in an electron beam may well possess two invariants with regard to lossless beam transformations. Let us

choose distinct symbols for the two invariants of a noise process in an
electron beam, for they will prove of great importance. Since Re (Θ)
plays the role of power carried by noise propagation along a transmission line, we use the Greek letter Π to refer to it.

$$\text{Re } (\Theta) \equiv \Pi \tag{3.124}$$

The imaginary part of Θ we denote by

$$\text{Im } (\Theta) \equiv \Lambda \tag{3.125}$$

The determinant of $\mathbf{WR}$ can be written, according to Eqs. 3.118, 3.124,
and 3.125, as

$$\det (\mathbf{WR}) = \Pi^2 - (\Phi\Psi - \Lambda^2) \tag{3.126}$$

Since Π in itself is an invariant, we must conclude that the term
$\Phi\Psi - \Lambda^2$ must also be an invariant with regard to lossless transformations. We introduce here a symbol for it. We set

$$S = (\Phi\Psi - \Lambda^2)^{1/2} \tag{3.127}$$

The inequality 3.123 assures that S is always real. We shall choose S
positive by definition. We shall see later that the invariant S is more
easily explained in terms of physical quantities than the invariant
$\det (\mathbf{WR})$ itself.

The accelerating regions of a multi-electrode electron gun as shown
in Fig. 3.4 are lossless beam transducers. The noise in any of the
regions is determined by the noise at its input plane, the plane first
passed by the electrons. The parameters S and Π are invariant with
regard to lossless transformations and may be traced back to the input
of the first region. Where, then, should the input plane of the first
drift region be chosen, and what determines the values of the two noise
invariants? These two questions are intimately connected. The first
region of the multi-electrode gun is formed by a space-charge limited
diode. A space-charge-limited diode is a lossless transducer, provided
that the small-signal, single-velocity approximations are applicable.
These approximations hold as long as the range of velocities possessed
by the majority of the electrons is small compared to the average velocity of the electrons. The single-velocity approximation will hold at
potentials as low as a few volts above cathode potential. The input
plane for the first region can be picked in front of the cathode beyond
the potential minimum at a plane a few volts above the cathode potential. The values of S and Π at this plane are conserved throughout the
multi-electrode gun, under the assumption that no electromagnetic

power is extracted from the beam on the way. *The parameters S and II are thus entirely functions of the conditions in the potential minimum-cathode region in which the single-velocity assumption does not hold.*

A word of caution is in order. If the potentials applied to the successive electrodes of the gun differ widely, a strong electrostatic lens may result which may cause crossovers of the electron paths. A deviation from laminar electron flow is accompanied by an enhancement of noise over the values predicted by one-dimensional, single-velocity theory (Chapter 5, Section 4, pages 278ff.).

Little is now known about the effect of the region around the potential minimum upon the noise.† If we assume with Pierce[32] that the effect of this region upon the noise is negligible, then it is quite reasonable to suppose that the noise at the input reference plane, beyond which the single-velocity approximation is legitimate, consists of a current modulation of full shot-noise value, and an equivalent velocity modulation of the Rack value[33] (Chapter 1). The SPDS of the shot noise of a beam with a direct current I_0 is

$$\Psi = \frac{|e| I_0}{2\pi}$$

The SPDS of the kinetic voltage corresponding to the Rack equivalent velocity is

$$\Phi = \left(\frac{m}{e} u\right)^2 \frac{1}{4\pi \, \Delta f} \overline{v^2} = \left(1 - \frac{\pi}{4}\right) \frac{mkT_c}{\pi |e| I_0} u^2$$

where u is the average velocity of the electrons at the reference plane, T_c is the cathode temperature, and the value of the mean-square velocity fluctuations $\overline{v^2}$ is taken from Chapter 1, Eq. 1.74. If the reference plane is taken exactly at the potential minimum, a choice not very convincing in view of the warnings given above, we can take the value $2kT_c/m$ for u^2. Further, assuming that the velocity and current are uncorrelated at the first input plane, the II parameter is equal to zero. For the S parameter we obtain

$$S = (\Phi\Psi)^{\frac{1}{2}} = \left(1 - \frac{\pi}{4}\right)^{\frac{1}{2}} \frac{kT_c}{\pi} \tag{3.128}$$

An Interpretation of the S Parameter

A drift region is a lossless transducer with a **K** matrix given by

† See Chapter 1 for various analyses of this region. Also, compare the latest work by Siegman, Watkins, and Hsieh.[18]

Eq. 3.73. The matrix equation 3.114 carried out explicitly for a drift region gives, with the aid of a simple trigonometric identity,

$$\Phi_b = \tfrac{1}{2}(\Phi_a + W^2\Psi_a) + \tfrac{1}{2}(\Phi_a - W^2\Psi_a)\cos 2\phi + W\Lambda_a \sin 2\phi \quad (3.129)$$

$$\Psi_b = \frac{1}{2}\left[\left(\frac{1}{W}\right)^2 \Phi_a + \Psi_a\right] + \frac{1}{2}\left[\Psi_a - \left(\frac{1}{W}\right)^2 \Phi_a\right]\cos 2\phi$$
$$- \frac{1}{W}\Lambda_a \sin 2\phi \quad (3.130)$$

$$\Theta_b = \Pi_a + \frac{1}{2}j\left[W\Psi_a - \frac{1}{W}\Phi_a\right]\sin 2\phi + j\Lambda_a \cos 2\phi \quad (3.131)$$

Equation 3.131, split into its real and imaginary parts, and use of the definitions of Eqs. 3.124 and 3.125 lead to

$$\Pi_b = \Pi_a \quad (3.131a)$$

$$\Lambda_b = \frac{1}{2}\left[W\Psi_a - \frac{1}{W}\Phi_a\right]\sin 2\phi + \Lambda_a \cos 2\phi \quad (3.131b)$$

Equation 3.131a is simply an expression of the conservation of Re [Θ] through a transformation by a section of a drift region which is a lossless transducer. Equations 3.129 and 3.130 show that the SPDS of the kinetic noise-voltage and current modulations have the form of standing waves as functions of ϕ, the plasma transit angle. The maxima of the kinetic-voltage and current modulations lie at angles $\Delta\phi = 90°$ apart. (See Fig. 3.6.) The maximum of the current SPDS has

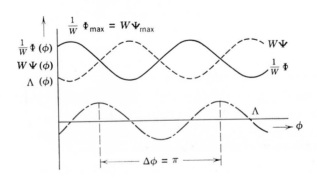

Fig. 3.6. Plot of the power density spectra as functions of plasma transit angle in a drift region.

the value

$$\Psi_{max} = \frac{1}{2}\left(\frac{1}{W^2}\Phi_a + \Psi_a\right) + \frac{1}{2}\left[\left(\frac{1}{W^2}\Phi_a - \Psi_a\right)^2 + 4\frac{1}{W^2}\Lambda_a{}^2\right]^{\frac{1}{2}}$$
(3.132)

The minimum is

$$\Psi_{min} = \frac{1}{2}\left(\frac{1}{W^2}\Phi_a + \Psi_a\right) - \frac{1}{2}\left[\left(\frac{1}{W^2}\Phi_a - \Psi_a\right)^2 + 4\frac{1}{W^2}\Lambda_a{}^2\right]^{\frac{1}{2}}$$
(3.133)

From Eqs. 3.132 and 3.133 we find that the product of the maximum and the minimum is

$$\Psi_{max}\Psi_{min} = \frac{1}{W^2}(\Phi_a\Psi_a - \Lambda_a{}^2) = \frac{S^2}{W^2}$$
(3.134)

Equation 3.134 shows that the product of the maximum and the minimum of the current SPDS is proportional to the square of the invariant S. The proportionality factor is the square of the characteristic admittance of the beam.†

The theoretically predicted conservation of the parameter S under lossless transformations can be checked experimentally. The noise standing-wave ratio in a drifting beam can be varied by adjustments of the voltages on the electrodes of a multi-electrode gun preceding the drift region. If the final potential of the drifting beam is left unchanged in the process, the characteristic impedance of the beam W is not changed. Theory then predicts that the product $\Psi_{max}\Psi_{min}$ has to stay invariant. Experiments were performed[34] which checked the theory satisfactorily. At standing-wave ratios Ψ_{max}/Ψ_{min} greater than 10 db, the experimentally observed product $\Psi_{max}\Psi_{min}$ exceeded that observed at lower standing-wave ratios. This effect was ascribed to the existence of higher-order space-charge modes which are not accounted for in the one-dimensional theory.[35]

The Equivalent Noise Admittance

Two noise processes with the same power-density spectra, that is, the same **WR** matrix, are transformed by a transducer with a given **K** matrix in the same way. This recognition enables one to construct a useful analog for propagation of noise along an electron beam. Consider a cascade of two-terminal-pair networks excited by noise and

---

† The invariance of the product $\overline{|i_{max}|^2}\,\overline{|i_{min}|^2}$, first proved by Pierce, is a special case of the invariance of S under lossless beam transformations.

terminated in some particular impedance. The noise matrix **WR** at
any point along the cascade has the property that det (**WR**) = 0, as
mentioned before. An admittance can be defined at any point, in
terms of the SPDS Ψ and the CPDS Θ by

$$Y = \frac{\Psi}{\Theta} \qquad (3.135)$$

Next, consider a noise process propagating along an electron beam.
Assume first that for this noise process the condition is fulfilled:
det (**WR**) = 0. It is then possible to choose the above network cascade,
its terminating impedance, and its noise excitation so that the matrix
WR, at any terminal pair of the network cascade is equal to the matrix
WR on the corresponding reference plane in the electron beam.[†] The
noise in the cascade of the networks (some of which may be uniform
transmission lines) can be used as an analog for the beam noise. The
analog noise permits the definition of an admittance, Eq. 3.135, which
can then be applied to the noise in the electron beam.

Once the admittance of the analog noise process is defined at any
reference terminal pair, the admittance at any other terminal pair is
uniquely determined according to the well-known laws of transformations of admittance by two-terminal-pair networks. Many properties
of the analog can be derived from the knowledge of its admittance.
Thus assume, for example, that the admittance $Y(\omega)$ of the analog
noise is given at a reference cross section of one of the transmission
lines. A particular value of Y is associated with a particular current
standing-wave ratio on the transmission line:

$$\left| \frac{i_{\max}}{i_{\min}} \right| = \frac{1 + \left| \dfrac{1 - WY}{1 + WY} \right|}{1 - \left| \dfrac{1 - WY}{1 + WY} \right|}$$

Since a noise process is defined by power-density spectra rather than
by Fourier amplitudes, this relation can be conveniently modified
so that it yields the ratio of the maximum to the minimum of the
current SPDS along the line. We have

$$\frac{\Psi_{\max}}{\Psi_{\min}} = \left[\frac{1 + \left| \dfrac{1 - WY}{1 + WY} \right|}{1 - \left| \dfrac{1 - WY}{1 + WY} \right|} \right]^2 \qquad (3.136)$$

[†] In general, nonreciprocal networks may be needed to accomplish this.

Equation 3.136 is an example of one of the many uses of the admittance concept for noise propagation along transmission lines. The admittance concept of the analog noise process can be applied directly to the corresponding noise process in the electron beam. The vast body of knowledge concerning admittance and impedance transformations can be brought to bear directly on the analysis of noise processes in electron beams for which an analog transmission-line noise process exists.

A noise process in an electron beam with det $(\mathbf{WR}) \neq 0$ does not possess a network-theory analog. The admittance concept is, therefore, not directly applicable. However, if the noise is transformed by lossless transducers only, it is still possible to define an analog admittance, as we shall now show.

The $\mathbf{WR}$ matrix of a general noise process can be written with the aid of Eq. 3.116 and the definitions of Eqs. 3.124 and 3.125 in the form

$$\mathbf{WR} = \begin{bmatrix} \Pi + j\Lambda & \Phi \\ \Psi & \Pi - j\Lambda \end{bmatrix} \tag{3.137}$$

Let us split the $\mathbf{WR}$ matrix of a general noise process with det $(\mathbf{WR}) \neq 0$ into two parts as shown below.

$$\mathbf{WR} = \mathbf{W'R} - \mathbf{I}\,\Delta\Pi \tag{3.138}$$

where $\mathbf{I}$ is the identity matrix and $\Delta\Pi$ is a real scalar given in terms of the invariants S and Π of the $\mathbf{WR}$ matrix.

$$\Delta\Pi = S - \Pi \tag{3.139}$$

According to Eqs. 3.137, 3.138, and 3.139, the explicit form of $\mathbf{W'R}$ is

$$\mathbf{W'R} = \begin{bmatrix} S + j\Lambda & \Phi \\ \Psi & S - j\Lambda \end{bmatrix} \tag{3.140}$$

The determinant of $\mathbf{W'R}$ is, from Eq. 3.127,

$$\det (\mathbf{W'R}) = 0 \tag{3.141}$$

Let us now study the transformation of the $\mathbf{WR}$ and $\mathbf{W'R}$ matrices by lossless transducers. Denoting the values of the $\mathbf{WR}$ and $\mathbf{W'R}$ matrices at the input to the transducer by the subscript a and at the output by the subscript b, we have, corresponding to Eq. 3.138,

$$\begin{aligned} \mathbf{W}_a\mathbf{R} &= \mathbf{W'}_a\mathbf{R} - \mathbf{I}\,\Delta\Pi \\ \mathbf{W}_b\mathbf{R} &= \mathbf{W'}_b\mathbf{R} - \mathbf{I}\,\Delta\Pi \end{aligned} \tag{3.142}$$

where the scalar $\Delta\Pi$ is the same in both equations because of the choice of Eq. 3.139 and the invariance of S and Π with regard to lossless transformations. The lossless transformation (Eq. 3.115) applied to Eq. 3.142 gives

$$\mathbf{W}_b\mathbf{R} = \mathbf{W'}_b\mathbf{R} - \mathbf{I}\,\Delta\Pi = \mathbf{K}\mathbf{W}_a\mathbf{R}\mathbf{K}^{-1} = \mathbf{K}\mathbf{W'}_a\mathbf{R}\mathbf{K}^{-1} - \mathbf{I}\,\Delta\Pi$$

or

$$\mathbf{W'}_b\mathbf{R} = \mathbf{K}\mathbf{W'}_a\mathbf{R}\mathbf{K}^{-1}$$
$$\mathbf{W}_b\mathbf{R} = \mathbf{K}\mathbf{W}_a\mathbf{R}\mathbf{K}^{-1}$$

$$(3.143)$$

Comparison of Eqs. 3.137 and 3.140 shows that the $\mathbf{W'R}$ matrix has the same off-diagonal elements and the same imaginary parts of its diagonal elements, as the $\mathbf{WR}$ matrix. According to Eqs. 3.143, the $\mathbf{W'R}$ matrix transforms in the same way as the $\mathbf{WR}$ matrix. Thus, the transformation of the off-diagonal elements, Φ and Ψ, and the transformation of the imaginary parts of the diagonal elements, $j\Lambda$, of $\mathbf{WR}$ can be studied with the aid of the transformation of $\mathbf{W'R}$. The latter satisfies the condition of Eq. 3.141 and thus has an analog noise process with the admittance

$$Y'(\omega) = \frac{\Psi'}{\Theta'} = \frac{\Psi}{S + j\Lambda} \qquad (3.135a)$$

Equation 3.130 shows that the SPDS of the current Ψ in a drift region depends only upon the parameters Φ_a, Ψ_a, and Λ_a of the noise process. These parameters are common to both $\mathbf{WR}$ and $\mathbf{W'R}$. Thus, all information concerning Φ, Ψ, and Λ can be obtained from a study of the analog noise process pertaining to the matrix $\mathbf{W'R}$. For example, the standing-wave ratio of the current SPDS for the analog noise process is given in terms of its admittance by Eq. 3.136. Thus, the standing-wave ratio of an arbitrary process may be found from Eq. 3.136, using the analog admittance Y' of Eq. 3.135a. It follows that *the admittance concept can be applied to any noise process, even those with det ($\mathbf{WR}$) $\neq 0$, if only transformations by lossless transducers are studied.*†

Transformations that extract power from the electron beam do not conserve S and Π. A split of the $\mathbf{WR}$ matrix according to Eq. 3.138 is not independent of the reference cross section. In general, no analog process with a $\mathbf{W'R}$ matrix which is such that det ($\mathbf{W'R}$) $= 0$ can be found whose Φ, Ψ, and Λ would transform identically with those of the $\mathbf{WR}$ matrix.

If we follow Pierce[32] in his assumptions for the noise near the poten-

---

† The admittance concept was first used in noise transformations by S. Bloom and R. W. Peter.[36]

tial minimum, we obtain a particularly simple expression for the equivalent admittance at the potential minimum. We have at the potential minimum

$$\Psi = \frac{|e|I_0}{2\pi}$$

$$\Lambda = 0$$

$$S = \sqrt{\left(1 - \frac{\pi}{4}\right)\frac{kT_c}{\pi}}$$

$$Y' = \frac{|e|I_0}{\sqrt{4 - \pi}\, kT_c} \qquad (3.144)$$

It should be noted that, aside from the uncertainty that exists as to the proper input conditions to be used at the potential minimum, an extension of the single-velocity analysis to the region of the potential minimum is in itself an approximation.

Alternate Representation of Noise

In many computations it is more convenient to use the wave formalism rather than the voltage–current representation of an excitation in an electron beam.[37] For this purpose it is necessary to describe a noise process in terms of normalized wave amplitudes. The Fourier transforms of the kinetic voltage and current in an electron beam of a noise process viewed during a time $2T$ were written in the form of a column matrix in Eq. 3.110. Normalized wave amplitudes can be assigned to these Fourier transforms according to Eqs. 3.84. We have

$$\mathbf{a}_T(\omega) = \mathbf{T}_a \mathbf{w}_{aT}(\omega)$$

$$\mathbf{b}_T(\omega) = \mathbf{T}_b \mathbf{w}_{bT}(\omega) \qquad (3.145)$$

The formation of the noise matrices $\mathbf{W}_a$ and $\mathbf{W}_b$ suggests similar noise matrices formed of the normalized wave amplitudes. We define

$$\mathbf{A} = \lim_{T \to \infty} \frac{2\pi}{T} \overline{\mathbf{a}_T(\omega)\,\mathbf{a}_T(\omega)^+} = 2 \lim_{T \to \infty} \frac{\pi}{T}\,(\mathbf{T}_a\,\overline{\mathbf{w}_{aT}(\omega)\,\mathbf{w}_{aT}(\omega)^+}\mathbf{T}_a^+) \qquad (3.146)$$

$$= 2(\mathbf{T}_a\mathbf{W}_a\mathbf{T}_a^+)$$

The factor of 2 was introduced for reasons of normalization. In order to explain the normalization involved in Eq. 3.146, we give an example. Imagine that a lossless transmission line is terminated at its two ends by matching resistors at a temperature T. The resistors exchange noise power over the transmission line. Denote by a_1 the normalized amplitude of the wave traveling from left to right, by a_2 the normalized amplitude of the wave from right to left. The values of the elements of the matrix **A** as defined by Eq. 3.146 would then be

$$A_{12} = A_{21} = 0 \qquad \text{since the two waves are uncorrelated}$$

and

$$A_{11} = A_{22} = \frac{kT}{4\pi} \tag{3.147}$$

according to the Nyquist formula,[38] which in its more conventional form gives the mean power carried in any one of the waves within the frequency band Δf as $kT \, \Delta f$. The definition of the **A** matrix thus differs by a normalization factor of $4\pi \, \Delta f$ from the conventional engineering use. This normalization is the same as that of the current SPDS discussed on page 117.

A similar "noise wave matrix" can be defined at reference cross section b by

$$\mathbf{B} = \lim_{T \to \infty} \frac{2\pi}{T} \overline{\mathbf{b}_T(\omega) \, \mathbf{b}_T(\omega)^+} = 2\mathbf{T}_b \mathbf{W}_b \mathbf{T}_b^+ \tag{3.148}$$

A beam transducer between the reference cross sections a and b is conveniently characterized in terms of the transformation of Eq. 3.76. This transformation establishes a relation between the noise wave matrices **A** and **B**. We have

$$\mathbf{B} = \lim_{T \to \infty} \frac{2\pi}{T} \overline{\mathbf{b}_T(\omega) \, \mathbf{b}_T(\omega)^+} = \lim_{T \to \infty} \frac{2\pi}{T} \mathbf{M} \, \overline{\mathbf{a}_T(\omega) \, \mathbf{a}_T(\omega)^+} \mathbf{M}^+ = \mathbf{MAM}^+$$

Thus

$$\mathbf{B} = \mathbf{MAM}^+ \tag{3.149}$$

If, in particular, the transducer is lossless, we must have, according to Eq. 3.82,

$$\mathbf{M}^+\mathbf{PM} = \mathbf{P} \tag{3.82}$$

or

$$\mathbf{M}^+ = \mathbf{PM}^{-1}\mathbf{P} \tag{3.150}$$

Introducing Eq. 3.150 into Eq. 3.149, and multiplying from the right by P (which is equal to its own inverse), we find

$$\mathbf{BP} = \mathbf{MAPM}^{-1} \tag{3.151}$$

Equation 3.151 is analogous to the transformation of the **WR** matrix by lossless transducers, Eq. 3.115.

Invariants. Equation 3.151 leads to the conclusion that the transformation by a lossless transducer must leave the trace and the determinant of the **AP** matrix equal to that of the **BP** matrix. Naturally, the invariance of the trace and determinant of the **AP** matrix must be related to the invariance of the same quantities pertaining to the **WR** matrix. In order to find how these are related, let us study more carefully the transformation of Eq. 3.146. Multiplying Eq. 3.146 from the right by **P**, and using Eq. 3A.5a of the Appendix for **P**, we obtain

$$\mathbf{AP} = \tfrac{1}{2}\mathbf{T}_a\mathbf{W}_a\mathbf{T}_a{}^+(\mathbf{T}_a{}^+)^{-1}\mathbf{RT}_a{}^{-1} = \tfrac{1}{2}\mathbf{T}_a\mathbf{W}_a\mathbf{RT}_a{}^{-1} \qquad (3.152)$$

The **AP** matrix and the **WR** matrix are related by a similarity transformation. Accordingly, there is a direct relation between the traces and the determinants of the two matrices.

$$\text{trace } (\mathbf{AP}) = A_{11} - A_{22} = \tfrac{1}{2}\text{ trace } (\mathbf{WR}) = \Pi \qquad (3.153)$$

$$\det (\mathbf{AP}) = |A_{12}|^2 - A_{11}A_{22} = \tfrac{1}{4}\det (\mathbf{WR}) = \tfrac{1}{4}(\Pi^2 - S^2) \qquad (3.154)$$

The trace of the **AP** matrix is equal to the real part of the CPDS between the kinetic-voltage and current fluctuations of the noise process.

The inverse of Eq. 3.152, which gives $\mathbf{W}_a\mathbf{R}$ in terms of **AP**, is also of interest. It is obtained from Eq. 3.152 by premultiplication by $\mathbf{T}_a{}^{-1}$ and postmultiplication by $\mathbf{T}_a$.

$$\mathbf{W}_a\mathbf{R} = 2\mathbf{T}_a{}^{-1}\mathbf{APT}_a \qquad (3.155)$$

The relations between the elements of the **WR** matrix, as given in Eq. 3.137, and the elements of the **AP** matrix are written out in detail below. The set of Eqs. 3.156 to 3.158 is obtained by carrying out the matrix operations in Eq. 3.152; the set of Eqs. 3.159 to 3.162 follows from Eq. 3.155.

$$A_{11} = \frac{1}{4}\left(\frac{1}{W}\Phi + W\Psi\right) + \frac{1}{2}\Pi \qquad (3.156)$$

$$-A_{12} = \frac{1}{4}\left(\frac{1}{W}\Phi - W\Psi\right) - \frac{1}{2}j\Lambda \qquad (3.157)$$

$$-A_{22} = -\frac{1}{4}\left(\frac{1}{W}\Phi + W\Psi\right) + \frac{1}{2}\Pi \qquad (3.158)$$

$$\Pi = A_{11} - A_{22} \qquad (3.159)$$

$$\Lambda = j(A_{12}{}^* - A_{12}) = -j(A_{12} - A_{21}) \qquad (3.160)$$

$$\Phi = W(A_{11} + A_{22} - A_{12} - A_{12}{}^*) \qquad (3.161)$$

$$\Psi = \frac{1}{W}(A_{11} + A_{22} + A_{12} + A_{12}{}^*) \qquad (3.162)$$

The subscript a of the elements of the **WR** matrix has been omitted in these equations, thus emphasizing their applicability at any reference plane.

Transformations of the Wave Matrix **A**.   The transformation of the **A** matrix by a section of a drift region is much simpler than that of the **WR** matrix.   A section of a drift region of length ϕ, measured in terms of the plasma transit angle ϕ and the transit angle θ, has the **M** matrix:

$$\mathbf{M} = \begin{bmatrix} e^{-j(\theta-\phi)} & 0 \\ 0 & e^{-j(\theta+\phi)} \end{bmatrix} \tag{3.163}$$

The expression for **M**, Eq. 3.163, applied to the transformation of Eq. 3.149, gives (written in detail)

$$B_{11} = A_{11}$$
$$B_{12} = A_{12}e^{2j\phi} \tag{3.164}$$
$$B_{22} = A_{22}$$

This transformation is easy to comprehend.   Along a drift region the fast and the slow waves change in phase only.   This accounts for the invariance of the diagonal elements A_{11} and A_{22}.   Within a drift region of plasma transit angle ϕ, the fast and the slow wave get out of phase by an angle 2ϕ, which accounts for the relationship between B_{12} and A_{12}.

A slightly different interpretation of Eq. 3.164 is possible.   Let us assume that the reference plane a in a drift region is continuously varied.   A movement of the reference plane forward by a plasma transit angle ϕ corresponds to a change of the argument of A_{12} by an amount 2ϕ.   The reference plane can be so chosen that A_{12} is real and positive.   Equation 3.162 shows that this reference plane coincides with the maximum of the current SPDS in the drift region. On the other hand, if the argument of A_{12} is not equal to zero, Eqs. 3.162 and 3.164 indicate that the plasma transit angle from the maximum of the SPDS to the reference plane is $\Delta\phi = \arg (A_{12})/2$.

If the off-diagonal element of the **A** matrix, A_{12}, is equal to zero, the current SPDS is independent of distance, according to Eqs. 3.162 and 3.164.   The fast wave and the slow wave in the beam are uncorrelated. The **A** matrix is diagonal.   Equation 3.157 shows that this is the case when

$$\Phi = W^2\Psi \quad \text{and} \quad \Lambda = 0 \tag{3.165}$$

The following question will prove of interest: given a general noise process at a reference plane a with some noise wave matrix **A**, is it possible to pass this noise process through a lossless beam transducer so that the noise matrix **B** at its output b is diagonal? A general proof of the possibility of such a transformation and how it is achieved is given in reference 21. Here we shall answer the question by simple physical reasoning which will help toward an understanding of noise transformations.

We have shown in Section 3, page 128, that it is possible to find an equivalent admittance for an arbitrary noise process as long as transformations of noise by lossless beam transducers are considered. The transformation of the current SPDS of the analog noise process with the admittance $Y'(\omega)$ of Eq. 3.135a on page 128 is the same as that of the general noise process. The question above can be recast into the terminology of admittance transformations. The requirement that the noise wave matrix be diagonal at the cross section b is tantamount to (see Eq. 3.165)

$$W^2\Psi_b = \Phi_b \quad \text{and} \quad \Lambda_b = 0 \tag{3.166}$$

The admittance of the analog noise process at cross section b must be, according to Eq. 3.135a and the definition of Eq. 3.127,

$$Y'_b = \frac{1}{W}$$

The equivalent admittance at reference cross section a is

$$Y'_a = \frac{\Psi_a}{S + j\Lambda_a}$$

Thus, the original question may be formulated as follows: Is there a lossless transducer that transforms the admittance $1/W$ at its output into Y'_a at the input? The admittance at the output of the transducer Y'_b is positive real; the admittance at the input has a positive real part. But it is always possible to find a lossless transducer that transforms any admittance with a positive real part into any other admittance with positive real part. Thus, we can conclude that *it is always possible to find a lossless beam transducer, which, inserted between the reference cross sections a and b, transforms an arbitrary noise wave matrix into a diagonal noise wave matrix.* Whether such a beam transducer is physically realizable is an entirely different problem. We know that a great variety of noise transformations can be achieved with a multi-electrode gun[25] as shown in Fig. 3.4. In the subsequent discussion we shall postulate that we can always find the beam transducer that is theoretically required.

When the noise matrix $\mathbf{A}$ is brought into a diagonal form $\mathbf{B}$ by a lossless beam transducer, the trace and the determinant have to be conserved. According to Eqs. 3.153 and 3.154, we must have

$$B_{11} - B_{22} = \Pi$$

$$B_{11}B_{22} = \tfrac{1}{4}(S^2 - \Pi^2)$$

with the result that

$$B_{11} = \tfrac{1}{2}(S + \Pi), \qquad B_{22} = \tfrac{1}{2}(S - \Pi) \qquad (3.167)$$

If $\mathbf{M}$ is the particular transducer that brings the noise wave matrix $\mathbf{A}$ into the diagonal form $\mathbf{B}$, we have

$$\mathbf{MAM}^+ = \mathbf{B} \qquad (3.168)$$

with $\mathbf{B}$ diagonal. Premultiplying this equation by $\mathbf{M}^{-1}$, and postmultiplying it by $(\mathbf{M}^+)^{-1}$, we obtain

$$\mathbf{A} = (\mathbf{M}^{-1})\mathbf{B}(\mathbf{M}^{-1})^+ \qquad (3.169)$$

Since $\mathbf{M}$ satisfies the condition of power conservation

$$\mathbf{M}^+\mathbf{PM} = \mathbf{P} \qquad (3.82)$$

we find, after premultiplying Eq. 3.82 by $(\mathbf{M}^+)^{-1}$, and postmultiplying by $\mathbf{M}^{-1}$,

$$(\mathbf{M}^{-1})^+\mathbf{P}(\mathbf{M}^{-1}) = \mathbf{P} \qquad (3.170)$$

Equation 3.170 shows that the matrix $\mathbf{M}^{-1}$ is the matrix of a lossless transducer. Using this knowledge we can interpret Eq. 3.169 as follows: *Any noise process with a matrix $\mathbf{A}$ can be represented by a diagonal noise process B followed by a lossless transducer with the matrix* $\mathbf{M}^{-1}$. The matrix $\mathbf{M}^{-1}$ is the inverse of the matrix $\mathbf{M}$ required to diagonalize $\mathbf{A}$ by the operation of Eq. 3.168.

4. The Minimum Obtainable Noise Figure

Sections 1 and 2 developed a matrix formalism for the signal excitation of microwave structures containing a one-dimensional electron beam (a beam whose excitation is fully described by two parameters at any one cross section). Section 3 applied the formalism to the study of noise propagation along electron beams. These results will now be exploited to find the minimum obtainable noise figure of a microwave amplifier. The generality of the formalism used will ensure a corresponding generality of the results. The minimum of the noise figure will be found to be common to all longitudinal-beam amplifiers with a given gain, using a beam with given noise parameters S and Π.

The noise-figure minimization will be carried out in two ways. Each approach will help understand the reasons behind the existence of a lower limit on the noise figure. We shall also find that the form of the minimum-noise-figure expression will suggest a definition of "noisiness" of a longitudinal electron beam.

The general formalism will be applied to practical cases such as a lossless traveling-wave tube, a traveling-wave tube with severed helix, a klystron, and a backward-wave amplifier.

The Noise-Figure Expression

The generally adopted measure of the sensitivity of an amplifier is the noise figure.[39,40] Here we shall deal exclusively with the spot-noise figure. The spot-noise figure can be evaluated as the ratio of the total noise output of the amplifier N, within the narrow frequency band Δf, over the hypothetical noise output N_o which the amplifier would have if no additional noise were introduced by the amplifier. Both N and N_0 are evaluated for a given impedance connected to the amplifier input. The impedance has an internal noise corresponding to the equilibrium temperature T. The noise figure F is

$$F = \frac{N}{N_o}$$

The frequency band Δf is selected small enough so that the amplifier characteristics can be assumed to be constant within the band Δf. The noise output N of the amplifier can be ascribed to two sources: the contribution N_i caused by the noise internal to the amplifier, and the noise N_o from the input circuit which would be present even if the amplifier were noise-free. The two contributions N_o and N_i are, in general, uncorrelated. Thus, the total noise output N can be written as the sum of N_o and N_i. The noise figure can be written in the form

$$F = 1 + \frac{N_i}{N_o} \tag{3.171}$$

Figure 3.7 shows a schematic of a microwave longitudinal-beam amplifier. The excitation on the transmission line feeding the amplifier can be expressed in terms of the normalized amplitudes of the incident and reflected waves a_3 and b_3. The excitation on the output transmission line is given in terms of the incident- and reflected-wave amplitudes a_4 and b_4. The electron beam enters the amplifier at the gun end with an excitation of the fast and slow waves a_1 and a_2, respectively. The excitation in the beam leaving the amplifier at the collector end is characterized by the normalized amplitudes of the fast and slow waves b_1 and b_2. The amplifier can be described by the

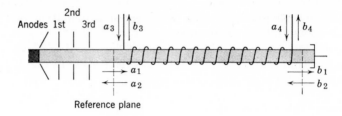

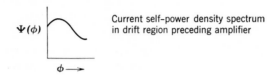

Fig. 3.7. Schematic of microwave longitudinal-beam amplifier with multi-electrode gun.

four-by-four matrix **G** which gives the linear relation between the column matrices **a** and **b**, each formed of the four normalized wave amplitudes:

$$\mathbf{b} = \mathbf{Ga} \qquad (3.85)$$

Without loss of generality we can assume that the input source is matched to the input transmission line and that the output load is matched to the output transmission line. For the noise leaving the amplifier with the output transmission line matched to the load, and thus a_4 equal to zero, we have†

$$N = \lim_{T \to \infty} \frac{2\pi}{T} \overline{b_{4T} b_{4T}^{*}} \, 4\pi \, \Delta f$$

The fourth element of the matrix relation, Eq. 3.85, gives the amplitude b_4.

$$b_4 = G_{41} a_1 + G_{42} a_2 + G_{43} a_3 \qquad (3.172)$$

The noise power N_o is found by disregarding the noise contribution of the electron beam: i.e., by setting $a_1 = a_2 = 0$.

$$N_o = \lim_{T \to \infty} \frac{2\pi}{T} \left| G_{43} \right|^2 \overline{a_{3T} a_{3T}^{*}} 4\pi \, \Delta f = \left| G_{43} \right|^2 A_{33} 4\pi \, \Delta f \qquad (3.173)$$

† Note the remarks on the normalization of Eq. 3.146.

The noise power carried by the incident wave is given by the Nyquist formula (compare Eq. 3.147):

$$A_{33} 4\pi\, \Delta f = kT\, \Delta f \qquad (3.174)$$

where k is Boltzmann's constant, and T is the temperature of the input circuit. The noise power N_i caused by the noise internal to the amplifier comes from the noise in the electron beam. We have

$$N_i = \lim_{T\to\infty} \frac{2\pi}{T} \overline{(G_{41}a_{1T} + G_{42}a_{2T})(G_{41}a_{1T} + G_{42}a_{2T})^*}\, 4\pi\, \Delta f$$

$$= (|G_{41}|^2 A_{11} + |G_{42}|^2 A_{22} + G_{41}G_{42}{}^* A_{12}$$
$$+ G_{41}{}^* G_{42} A_{12}{}^*) 4\pi\, \Delta f \qquad (3.175)$$

Combining Eqs. 3.171, 3.173, 3.174, and 3.175, we find the noise figure of the amplifier.

$$F = 1 + \frac{4\pi}{kT} \frac{1}{|G_{43}|^2}\, (|G_{41}|^2 A_{11} + |G_{42}|^2 A_{22} + G_{41}G_{42}{}^* A_{12}$$
$$+ G_{41}{}^* G_{42} A_{12}{}^*) \qquad (3.176)$$

For a given amplifier, the noise figure is a function of the beam noise at the input reference plane a. The noise parameters A_{11}, A_{22}, and A_{12} can be varied by lossless beam transformations in transducers preceding the reference plane. The noise parameters cannot be varied independently. We shall now study the minimization of the noise figure subject to the constraints imposed on the A's. (Note that no restrictions of losslessness have been imposed on the **G** matrix. Therefore, Eq. 3.176 and the results immediately following are applicable to amplifiers employing lossy as well as lossless microwave structures.)

Minimization of the Noise Figure

Let us imagine that the beam passes through a multi-electrode gun as shown in Fig. 3.7 before it enters the amplifier. The multi-electrode gun is a lossless transducer by which the noise can be adjusted at the reference cross section at the gun end of the amplifier. In many amplifiers the beam passes through a drift region before it enters the amplifier. We shall assume this to be the case, merely because the adjustment of the standing wave of the current SPDS in this drift region gives a good visual representation of the noise transformations by the multi-electrode gun. We shall assume that the multi-electrode gun is versatile enough to be able to give an arbitrary standing-wave ratio of the current SPDS within the drift region. Stated in mathematical language this assumption is equivalent to the requirement that A_{11}, A_{22}, and A_{12} at the input reference cross section can be adjusted arbitrarily, subject to the condition of conservation of S and Π by lossless transformations. We have, from Eqs. 3.153 and 3.154,

$$A_{11} - A_{22} = \Pi = \text{const}$$
$$(A_{11} + A_{22})^2 - 4|A_{12}|^2 = S^2 = \text{const} \tag{3.177}$$

The requirement of the invariance of S and Π leaves two of the four noise parameters adjustable. One of these parameters is the argument of A_{12}, which does not enter into the conditions of Eq. 3.177. The choice of arg (A_{12}) which minimizes the expression of Eq. 3.176 is, obviously,

$$\text{arg } (A_{12}) = \text{arg } (G_{42}) - \text{arg } (G_{41}) + \pi \tag{3.178}$$

This condition is achieved when the minimum of the current SPDS in the drift region preceding the amplifier is located at a plasma transit angle,

$$\phi_0 = \frac{1}{2} \left[\text{arg } (G_{42}) - \text{arg } (G_{41})\right] = \frac{1}{2} \text{arg } \left(\frac{G_{42}}{G_{41}}\right) \tag{3.179}$$

in front of the reference plane in the beam (see Fig. 3.7). The first minimization leads to the noise-figure expression

$$F = 1 + \frac{4\pi}{kT} \frac{1}{|G_{43}|^2} \left(|G_{41}|^2 A_{11} + |G_{42}|^2 A_{22} - 2|G_{41}G_{42}| \, |A_{12}|\right) \tag{3.180}$$

This expression allows one further minimization. We can express A_{11} and A_{22} in terms of $|A_{12}|$ with the aid of Eq. 3.177

$$A_{11} = \tfrac{1}{2}[(S^2 + 4|A_{12}|^2)^{1/2} + \Pi]$$
$$A_{22} = \tfrac{1}{2}[(S^2 + 4|A_{12}|^2)^{1/2} - \Pi] \tag{3.181}$$

where the square roots are to be taken positive.

These expressions can be introduced into the noise-figure expression (Eq. 3.180), which becomes a function of the single variable $|A_{12}|$. Minimization with regard to this variable is achieved when

$$|A_{12}| = \left| \frac{G_{41}G_{42}}{|G_{41}|^2 - |G_{42}|^2} \right| S \tag{3.182}$$

The minimum noise figure is

$$F_{\min} = 1 + \frac{2\pi}{kT} |D| \left[S - \frac{D}{|D|} \Pi \right] \tag{3.183}$$

where

$$D = \frac{|G_{42}|^2 - |G_{41}|^2}{|G_{43}|^2}$$

The standing-wave ratio of the current SPDS required for the minimization of the noise figure is (see Eq. 3.162), from Eqs. 3.181 and 3.182,

$$\frac{\Psi_{\max}}{\Psi_{\min}} = \frac{A_{11} + A_{22} + 2|A_{12}|}{A_{11} + A_{22} - 2|A_{12}|} = \left(\frac{|G_{41}| + |G_{42}|}{|G_{41}| - |G_{42}|}\right)^2 \qquad (3.184)$$

The minimum noise figure of the amplifier obtainable by means of a lossless transducer in the beam depends, according to Eq. 3.183, partly on the two noise invariants S and Π, partly on the amplifier structure through the constant D. In practice, it is often necessary to find an appropriate multi-electrode gun that will give the best possible noise figure when used in a given amplifier. In view of our limited theoretical knowledge of the noise input conditions beyond the potential minimum in the electron gun, a semi-empirical approach to this problem may be used. We shall illustrate this approach with the aid of an example.

Application to Traveling-Wave Tubes

For a traveling-wave tube the parameters G_{41}/G_{43} and G_{42}/G_{43} are, in Pierce's[1] notation (see reference 37, appendix II, where a similar notation has been employed):

$$\frac{G_{41}}{G_{43}} = j(\delta_2 + \delta_3)(QC)^{\frac{1}{4}} - \frac{1}{2}(\delta_2\delta_3 - 4QC)(QC)^{-\frac{1}{4}}$$

$$\frac{G_{42}}{G_{43}} = -j(\delta_2 + \delta_3)(QC)^{\frac{1}{4}} - \frac{1}{2}(\delta_2\delta_3 - 4QC)(QC)^{-\frac{1}{4}} \qquad (3.185)$$

The parameter QC is Pierce's space-charge parameter. The delta's are the incremental propagation constants which are solutions of the determinantal equation

$$(\delta^2 + 4QC)(j\delta + jd - b) = 1 \qquad (3.186)$$

where b is Pierce's velocity parameter and d is the loss parameter.

The root of the above determinantal equation corresponding to the growing wave (i.e., the root with a positive real part) is denoted by the subscript 1. The root corresponding to the decaying wave which, in the absence of helix loss, would be equal to $-\delta_1{}^*$ is denoted by the subscript 2. It should be noted that the roots of Eq. 3.186 depend upon the space-charge parameter QC, the loss parameter d, and the velocity parameter b. If the beam voltage is adjusted for maximum gain, the corresponding velocity parameter has a value dependent upon QC and d. Thus, in the case of maximum gain, δ_2 and δ_3 are functions of QC and d only. Correspondingly, the functions G_{41}/G_{43} and G_{42}/G_{43} depend only upon QC and d, provided it is assumed that the beam velocity is adjusted for maximum gain. The position of the current standing-wave minimum and the noise standing-wave ratio for optimum noise figure are thus functions of QC and d only. Plots of these will be presented in Chapter 5. There the design and adjustment of a low-noise gun will be treated in greater detail. Here we shall only discuss briefly the measures that have to be taken for noise minimization.

Once the noise standing wave required for optimum noise figure is found from Eqs. 3.179 and 3.184, a multielectrode gun (see Fig. 3.4) can be tested for the standing wave of the noise-current SPDS. This can be done by a sliding-cavity beam tester used by several laboratories.[13,34,41] The standing-wave ratio can be adjusted to the desired value (Eq. 3.184) by changes in the electrode potentials. If the level of the product $\Psi_{max}\Psi_{min}$ as given by Eq. 3.134 does not change in the process, we can assume that the gun performs as ideally predicted by the theory. Should the level rise, the gun design is suspect and must be changed. Experience shows that gun designs with the smoothest possible potential distribution—as shown, for example, in Fig. 3.4—perform closest to the idealized theory. When the proper current standing wave is achieved, the gun can be built into the amplifier. Small final adjustments in the potentials of the electrodes may be necessary in order to achieve the best performance.

The adjustments described above do not require the knowledge of Π. The value of Π serves only to determine the final value of the optimized noise figure.

Deviations of the noise behavior from that predicted by the idealized theory are ascribed, today, to two effects. First, there is the lens action of the various electrodes in the multi-electrode gun.† The lens action may result in crossovers of the electron trajectories. Such crossovers are believed to be harmful to the noise figure, since they may cause a transformation of the random motion in the transverse direction (which is always present under physically realizable magnetic focusing fields) into motion in the longitudinal direction. Second, an electron beam propagates, aside from the dominant space-charge waves included in the one-dimensional theory (see remark on p. 125), higher-order space-charge waves. These higher-order space-charge waves are excited by the input noise and are picked up by a sliding-cavity beam tester. Thus, a sliding-cavity beam tester observes not only the standing wave of noise current set up by the dominant space-charge waves but also the noise-current pattern of the higher-order space-charge waves. Their effect is particularly pronounced at the standing-wave minima of the noise current in the dominant space-charge waves and may influence a reading of the standing-wave ratio of the noise current in the dominant space-charge waves. Thus, the measured standing-wave ratio of the dominant wave may differ from the true value. It has been found experimentally, on a number of low-noise,

---

† See the discussion on lens effect in Chapter 5, Section 4, pages 278ff.

multi-electrode guns[34,42] that the product $\Psi_{max}\Psi_{min}$ is conserved through noise transformations that give standing-wave ratios of noise current of less than 10 db. These findings have been taken as the experimental proof that the effect of the higher-order space-charge waves upon a reading of the noise-current standing-wave ratio is negligible for current standing-wave ratios of less than 10 db. The value of the standing-wave ratio of noise current to be used in the computations of the noise figure of an amplifier is that of the dominant space-charge waves. This follows directly from the fact that the theory is based on a two-wave picture and can, therefore, account only for noise carried in two waves. Since the coupling of slow-wave structures to the higher- order space-charge waves is weak in a thin beam, the noise figure of a traveling-wave tube with a thin beam is not believed to be strongly affected by the existence of the higher-order space-charge waves.

Magnitude of the Parameter *D* for a Lossless Amplifier

The subsequent analysis will be limited to lossless amplifier structures. (The effect of loss in the amplifier is treated in general terms in reference 21.) It may be stated here that the minimum noise figure of an amplifier with loss can only be higher than or, at best, equal to that of a lossless amplifier.

Equation 3.95 imposes upon the elements of the **G** matrix a condition that has a profound influence upon the characteristic constant *D*. From Eq. 3.95 it follows that

$$\left|\det \mathbf{G}\right|^2 = 1$$

Thus, the **G** matrix has a reciprocal, $\mathbf{G}^{-1}$. Premultiplying Eq. 3.95 by **GP**, and postmultiplying it by $(\mathbf{PG})^{-1}$, we get

$$\mathbf{GPG}^+\mathbf{PG}(\mathbf{PG})^{-1} = \mathbf{GPP}(\mathbf{PG})^{-1}$$

and, since **P** is its own reciprocal, we have

$$\mathbf{GPG}^+ = \mathbf{P} \tag{3.187}$$

Superficially, Eq. 3.187 looks like Eq. 3.95. But it implies relations among the rows of the **G** matrix, as, for example,

$$\left|G_{41}\right|^2 - \left|G_{42}\right|^2 + \left|G_{43}\right|^2 + \left|G_{44}\right|^2 = 1 \tag{3.188}$$

From Eq. 3.188 we find for the amplifier constant *D* (see Eq. 3.183),

$$D = \frac{\left|G_{42}\right|^2 - \left|G_{41}\right|^2}{\left|G_{43}\right|^2} = 1 - \frac{1 - \left|G_{44}\right|^2}{\left|G_{43}\right|^2} \tag{3.189}$$

Since $\left|G_{43}\right|^2$ is the transducer gain of the amplifier, it must always be greater than unity. Thus, *D* is always positive. Two classes of amplifiers have to be distinguished. One class has $\left|G_{44}\right| < 1$; the other has $|G_{44}| > 1$. Let us recall the significance of the matrix element G_{44}. The fourth element of the matrix relation, Eq. 3.85, reads

$$b_4 = G_{41}a_1 + G_{42}a_2 + G_{43}a_3 + G_{44}a_4$$

With no excitation in the electron beam, $a_1 = a_2 = 0$, and, with the input transmission line matched to a passive load, $a_3 = 0$, we find that the reflection coefficient measured in the output transmission line is $G_{44} = b_4/a_4$. A reflection coefficient of magnitude less than unity is caused by a passive load; a reflection coefficient of magnitude greater than unity is produced by an impedance with a negative real part. For $|G_{44}| > 1$, the output impedance of the amplifier with the input transmission line matched has a negative real part. The amplifier is only conditionally stable with regard to end loading.

Conversely, if $|G_{44}| < 1$, the output impedance of the amplifier has a positive real part. In this case it is always possible to find a lossless two-terminal-pair network which matches the output impedance of the amplifier to the line. When the output is matched, the output power for a given input power is maximized. The quantity $|G_{43}|^2$ becomes the "available power gain" G of the amplifier: i.e., the ratio of the available output power over the available input power. The element G_{44} is reduced to zero. The parameter D becomes

$$D = 1 - \frac{1}{G} \tag{3.190a}$$

The class of amplifiers with $|G_{44}| > 1$ cannot be matched to the output transmission line. The quantity $|G_{43}|^2$ still has the meaning of transducer gain G_t, the ratio of the actual output power over the available input power. For this class of amplifiers,

$$D \geq 1 - \frac{1}{G_t} \tag{3.190b}$$

Comparison of Eqs. 3.190a and 3.190b shows that amplifiers with an output impedance with a negative real part are, in general, less desirable from the point of view of noise performance.

A third possibility should not be excluded: namely, amplifiers with $G_{44} = 1$. For these amplifiers Eq. 3.189 gives directly

$$D = 1 \tag{3.190c}$$

independent of the gain.

Since D is always positive, the minimum obtainable noise figure as given by Eq. 3.183 depends only upon the *difference* of the noise invariants, $S - \text{II}$. (This result does not depend upon the assumption of zero loss, as shown in general in reference 21.)

The lowest possible noise figure with a given gain is achieved with amplifiers that have $|G_{44}| < 1$ and have been subsequently matched to the output transmission line. The minimum noise figure of this class of amplifiers has the form

$$F_{\min} = 1 + \left(1 - \frac{1}{G}\right)\frac{2\pi}{kT}(S - \text{II}) \qquad (3.191)$$

It should be noted that, aside from the losslessness of the amplifier structure, nothing has been assumed concerning the details of the structure in deriving Eq. 3.191. Thus, no lossless structure, no matter how elaborate, can have a noise figure lower than that of Eq. 3.191.

Cascaded Amplifiers. Equation 3.191 shows that the noise figure of an amplifier can, in general, be reduced to unity at a corresponding sacrifice of gain. Such a result is not surprising. Indeed, one can almost always decouple the microwave structure from the electron beam, thus preventing any beam noise from entering the structure, with a resulting noise figure of unity. Naturally, the gain is then reduced to unity also. The dependence of the minimum noise figure upon the gain may suggest another scheme for achieving amplification with a small noise figure. Assume that a set of n electron guns is available, all with the same lowest possible value of $S - \text{II}$. Then, construct n amplifiers, using these guns, each of the amplifiers with a low gain and low-noise figure corresponding to Eq. 3.191. Is it then possible to achieve a noise figure lower than that given by Eq. 3.191 at some large available gain G by cascading the n amplifiers? To find an answer to this question let us assume that all n amplifiers have the same available gain as the first amplifier g. The gain of the cascade is

$$G = g^n$$

The noise figure of the cascade can be found from the well-known formula for the noise figure of a cascade of amplifiers:

$$F = F_1 + \frac{F_2 - 1}{g} + \cdots + \frac{F_n - 1}{g^{n-1}} \qquad (3.192)$$

where the subscripts refer to the order of the arrangement of the amplifier. In our case we have assumed that all of the amplifiers have the same gain, and, according to Eq. 3.191, they have the same noise figure. For the case of n such amplifiers we find, from Eq. 3.192,

$$F = 1 + (F_1 - 1) \frac{1 - 1/g^n}{1 - 1/g}$$

where F_1 and g refer to the noise figure and gain of the first amplifier. Introducing Eq. 3.191 for the noise figure of the first amplifier, and using the expression for the available gain G of the amplifier cascade, we find

$$F = 1 + \left(1 - \frac{1}{G}\right) \frac{2\pi}{kT} (S - \Pi) \qquad (3.193)$$

The over-all noise figure is identical with the noise figure obtainable with a single amplifier at a corresponding gain G. Thus, the cascading scheme cannot lead to any improvement over the noise figure of Eq. 3.191.

The influence of the electron beam upon the amplifier noise figure appears always in terms of the beam-noise parameter $S - \Pi$. We have also found that cascading of amplifiers using different beams with the same noise parameter $S - \Pi$ leads to the same noise figure at high gain as would have been achieved with one single high-gain amplifier using a beam with a corresponding noise parameter $S - \Pi$. It appears, therefore, that the parameter $S - \Pi$ is a measure of the basic *"noisiness"* of an electron beam.

Applications

The general theory that led to Eq. 3.183 can best be illustrated with the aid of examples. We shall study the minimum-noise-figure expressions of the lossless traveling-wave tube, the klystron amplifier, the traveling-wave tube with a severed helix, and the backward-wave amplifier.

Lossless Traveling-Wave Tube. In principle, a lossless traveling-wave tube, as shown schematically in Fig. 3.7, can be matched to the input and output transmission lines. Then, no reflection occurs when power is fed into the output with the input transmission line terminated by a matched resistor, $G_{44} = 0$. Equation 3.191 for the minimum noise figure applies directly. Thus, the lossless traveling-wave tube is a microwave-beam amplifier that theoretically achieves the lowest possible noise figure. If we assume, following Pierce, that S is given by Eq. 3.128, and $\Pi = 0$, we find that the limiting noise figure is

$$F_{\min} = 1 + \left(1 - \frac{1}{G}\right) (4 - \pi)^{1/2} \frac{T_c}{T} \qquad (3.194)$$

This expression was obtained by various authors in the limit of large gain.[36,43,44] With the choice of $T = 300°$ K, $T_c = 1200°$ K, and $G \to \infty$, we have

$$F_{\min} \approx 6 \text{ db}$$

Klystron Amplifier. Next let us consider a klystron amplifier. In order to be consistent with the assumptions of the preceding theory, we must postulate that *the cavities of the klystron are lossless.* If, in addition, the cavity gaps are very short, the beam loading admittance is theoretically zero. The output cavity viewed from the output transmission line looks like a reactive termination; thus $G_{44} = 1$. According to Eq. 3.190c, $D = 1$, and the minimum noise figure is[45]

$$F_{\min} = 1 + \frac{2\pi}{kT} (S - \Pi) \tag{3.195}$$

In a more realistic model of a klystron with lossy cavities (but still with negligible gap transit angle), the noise figure is greater for two reasons. First, to obtain sufficient gain, the generator is usually matched to the buncher cavity. If the gap impedance as seen by the electron beam is then retained at its optimum value, as computed from the lossless klystron, a reduction of $\sqrt{2}$ in gap voltage results. This increases $F - 1$ by 3 db. Secondly, the internal thermal noise in the cavity contributes an additive term to $F - 1$.

Severed-Helix Traveling-Wave Tube. The formalism that led to Eq. 3.191 for the minimum noise figure was developed under the assumption of a lossless microwave structure. It is not hard to modify it so that it is applicable to several practical cases with loss. One case of practical interest is the traveling-wave tube with a severed helix. Somewhere between the input and output, the helix of the traveling-wave tube is interrupted by a lossy section so that regenerative feedback from reflections from the output is prevented. One can represent this effect by assuming that the traveling-wave tube has two lossless helices, one end of each terminated in a matched load (Fig. 3.8). The traveling-wave tube is now characterized by a lossless six-by-six **G** matrix, relating the column vectors **a** and **b**, each of sixth order. The numbering of the waves is shown in Fig. 3.8. The fast and slow waves in the beam at the input end are still a_1 and a_2. The input and output of the traveling-wave tube are denoted by the subscripts 3 and 4. The two new terminal pairs that are terminated in matched resistances are terminals 5 and 6. The 4,4 element of the matrix relation, Eq. 3.187, with a **G** matrix of sixth order and

$P = \text{diag}(1, -1, 1, 1, 1, 1)$, reads

$$|G_{41}|^2 - |G_{42}|^2 + |G_{43}|^2 + |G_{44}|^2 + |G_{45}|^2 + |G_{46}|^2 = 1 \quad (3.196)$$

If the second helix is matched to the output, we have $G_{44} = 0$. Further, a wave traveling in a direction opposite to the flow of the beam cannot excite the beam. Hence, the wave a_6 cannot couple to the

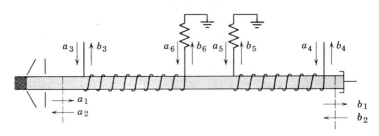

Fig. 3.8.   Schematic of traveling-wave tube with severed helix.

output, $G_{46} = 0$. The quantity $|G_{43}|^2$ is the available gain of the over-all amplifier G; the term $|G_{45}|^2$ would be the available gain G' if only the output helix section were used as an amplifier. If the contribution to the noise by the matching resistor 5 is neglected, Eq. 3.189 for the parameter D still holds unchanged. These conditions introduced into Eq. 3.186 give

$$D = 1 - \frac{1}{G} + \frac{G'}{G} \quad (3.197)$$

Comparison of Eq. 3.197 with Eq. 3.190a shows that D is not necessarily much larger than the D of a lossless traveling-wave tube. If the gain in the first section of the severed helix is appreciable, G'/G can be kept small. The noise figure then gets established essentially in the first section of the amplifier. The over-all traveling-wave tube acts somewhat like a cascade of two amplifiers in which the noise figure of the second amplifier does not affect the over-all noise figure if the gain in the first amplifier is sufficiently large.

   Backward-Wave Amplifier. If we consider a backward-wave amplifier as an example, we find that its formalism is identical with that of a traveling-wave tube. A backward-wave amplifier with a lossless microwave structure can be described by a lossless four-by-four **G** matrix. The input terminals 3 are now situated at the collector end; the output terminals 4 are at the gun end. Aside from this difference in physical appearance; the mathematical formalism is identical with that of a traveling-wave tube, and the minimum-noise-figure expression (Eq. 3.191) applies to the backward-wave amplifier as well.

An Alternate Derivation of the Minimum Noise Figure

In the derivation of the minimum-noise-figure expression 3.183, we considered a microwave amplifier characterized by a given **G** matrix. We adjusted the noise at the beam-input reference plane of the amplifier for the lowest noise figure. Another way of optimizing the noise figure is to keep the noise at the beam-input reference plane fixed and to adjust the amplifier (vary **G**) for minimum noise figure. This second process of optimization brings out some interesting physical facts. We shall, therefore, study it in greater detail.

Let us assume that the noise waves at the beam-input reference plane of Fig. 3.7 are decorrelated through a proper adjustment of the beam transducer preceding the amplifier. We then ask: What is the best possible amplifier that gives the lowest noise figure? In answering this question we shall gain a better understanding of the noise parameter $S - \Pi$.

The noise figure expression (Eq. 3.176) can be applied directly to the problem by merely identifying the elements of the **A** matrix with those of the diagonal **B** matrix of Section 3, page 134. We have

$$F = 1 + \frac{4\pi}{kT} \frac{1}{|G_{43}|^2} \left(|G_{41}|^2 B_{11} + |G_{42}|^2 B_{22} \right) \qquad (3.198)$$

Now we shall optimize F by a proper choice of the amplifier. Since both B_{11} and B_{22} are positive, Eq. 3.198 is clearly optimized if $|G_{41}|^2$ and $|G_{42}|^2$ are selected as small as possible, and $|G_{43}|^2$ is selected as large as possible. Condition 3.188 shows that this is the case when $G_{41} = G_{44} = 0$, and

$$\frac{|G_{42}|^2}{|G_{43}|^2} = 1 - \frac{1}{|G_{43}|^2} = 1 - \frac{1}{G} \qquad (3.199)$$

But, according to Eq. 3.167, $B_{22} = (S - \Pi)/2$. Combining Eqs. 3.167, 3.198, and 3.199, we find expression 3.191 for the minimum noise figure.

This simple minimization brings out the connection between the *gain mechanism* in a microwave tube and the *optimum noise performance* that can be achieved with a beam of given "noisiness" $S - \Pi$.

An amplifier has *electromagnetic power gain* at the expense of kinetic power in the electron beam. Electromagnetic power can be extracted from the electron beam only if the negative kinetic-power content in the electron beam is increased. In other words, an amplifier has to couple to the slow wave in the electron beam which carries negative kinetic power. This phenomenon was recognized when the magnitude of the term $|G_{23}|^2$ in Eq. 3.96 was discussed.

Noise is carried in the beam both by the fast and by the slow wave. As long as there is partial correlation between the two waves, the noise in one can be used to cancel part of the noise in the other. Such a scheme fails in the absence of correlation. The *optimum noise performance* is achieved when the amplifying structure is coupled to the slow wave only. Such a coupling is necessary for the operation of the amplifier. The fast wave should not couple to the output ($G_{41} = 0$), since it carries positive kinetic power and thus cannot be used as a source of electromagnetic power. Coupling to the fast wave would only introduce additional noise.

The minimization as described above applies to traveling-wave tubes with large QC. A large value of QC implies widely different phase velocities for the fast and the slow space-charge wave in the electron beam. (Indeed, $\beta_q = \beta_e \sqrt{4QC^3}$.) A circuit wave in synchronism with the slow beam wave has negligible coupling with the fast wave, provided the coupling parameter C is not too large. Thus, the traveling-wave tube with large QC is a realization of a device within which only coupling to the slow wave takes place. The noise performance of such a device is optimized when the noise of the fast wave is decorrelated from that of the slow wave in the beam entering the tube. As a consequence, according to Eq. 3.162, the noise input conditions to the tube are optimized when the noise-current SPDS is independent of distance; in other words, the standing-wave ratio $\Psi_{\max}/\Psi_{\min}$ is unity.

Conclusions

Based on the assumptions of the one-dimensional, single-velocity, small-signal theory, the preceding investigation has established a lower limit to the noise figure for high-gain microwave-beam amplifiers. This lower limit is a function solely of the noise process in the region of the potential minimum. The value of the basic noise parameter $S - \Pi$, established in the region of the potential minimum, gives the minimum noise figure at high gain $F_{\min}$.

$$F_{\min} = 1 + \frac{2\pi}{kT}(S - \Pi)$$

It was also shown that a traveling-wave tube with a microwave structure of zero (i.e., small) loss achieves the theoretical minimum noise figure solely with the aid of a conventional multi-electrode gun.

Future work on low-noise microwave tubes will, therefore, have to be concerned with two major questions: (a) What is the lowest possible value of the noise parameter $S - \Pi$? (b) How can the minimum noise figure given by Eq. 3.193 be lowered by schemes that fall outside the realm of validity of the assumption of one-dimensional, single-velocity theory?

The first question has been approached theoretically by P. K. Tien[17] and Siegman et al.[18] Saito[46] has reported experimental measurements of the noise parameter $S - \Pi$ essentially in agreement with the theory of Siegman et al. Undoubtedly, more theoretical and experimental work will be concentrated around this intriguing question in the future.

The second possibility mentioned above is concerned with a means of circumventing the lower limit on the microwave-tube noise figure as established in Eq. 3.193. This would have to be accomplished by schemes for which the single-velocity, one-dimensional assumptions do not hold. One might try to influence the emission from the cathode, or potential minimum, so that the noise parameter $S - \Pi$ would be reduced. Or one might look for microwave tubes other than those of the longitudinal-beam type, which would possess a limit on noise figure lower than that of the longitudinal-beam tube. Among these, the transverse-field tube has been considered,[47] because it seemed to allow a reduction of noise by beam collimation, a scheme that promised to be simpler than are noise-reduction schemes that influence the cathode emission in longitudinal tubes. No useful results have been reported.

Finally, we note that the one-dimensional assumptions used in the present chapter imply that the beam can propagate only two waves, the dominant space-charge waves. If, on the other hand, other waves are, or are made to be, important, the present theory is not applicable. The more elaborate theory of reference 21 would have to be used.

APPENDIX

The Transformation
from the Voltage-Current Formalism
into the Wave Formalism

The transformations between the wave variables and voltage-current variables are given in Eqs. 3.74. They are summarized in matrix form in Eqs. 3.83 and 3.84.

$$\mathbf{a} = \mathbf{T}_a \mathbf{w}_a \tag{3.84a}$$

$$\mathbf{b} = \mathbf{T}_b \mathbf{w}_b \tag{3.84b}$$

It is easy to show that the matrices $\mathbf{T}_a$ and $\mathbf{T}_b$ have an inverse. Physically, this means that the excitation is fully described in terms of either the wave amplitudes or the voltage and current amplitudes. The inverse relations to Eqs. 3.84 are

$$\mathbf{w}_a = \mathbf{T}_a^{-1} \mathbf{a} \tag{3A.1a}$$

$$\mathbf{w}_b = \mathbf{T}_b^{-1} \mathbf{b} \tag{3A.1b}$$

Equation 3.61 of the text can be put into the form

$$\mathbf{w}_b = \mathbf{K} \mathbf{T}_a^{-1} \mathbf{a} \tag{3A.2}$$

Multiplying Eq. 3A.2 from the left by $\mathbf{T}_b$, we obtain

$$\mathbf{b} = \mathbf{M} \mathbf{a} \tag{3A.3}$$

where

$$\mathbf{M} = \mathbf{T}_b \mathbf{K} \mathbf{T}_a^{-1} \tag{3A.4}$$

Equation 3A.4 gives the relation between the transducer matrix in the wave formalism $\mathbf{M}$, as defined in Eq. 3.76, and the $\mathbf{K}$ matrix of the voltage-current formalism.

In a similar way we may relate the matrices $\frac{1}{4}\mathbf{R}$ and $\mathbf{P}$ which are characteristic of the kinetic-power expressions in the two formalisms. We have, from Eq. 3.65,

$$P_{ka} = \tfrac{1}{4} \mathbf{w}_a{}^+ \mathbf{R} \mathbf{w}_a \tag{3.65}$$

Using Eq. 3A.1a, we have

$$P_{ka} = \tfrac{1}{4} \mathbf{a}^+ (\mathbf{T}_a{}^+)^{-1} \mathbf{R} (\mathbf{T}_a)^{-1} \mathbf{a}$$

According to Eq. 3.79, the above expression has to be equal to

$$\mathbf{P}_{ka} = \mathbf{a}^+ \mathbf{P} \mathbf{a}$$

for all choices of the excitation $\mathbf{a}$. This is possible if, and only if,

$$\mathbf{P} = \tfrac{1}{4} (\mathbf{T}_a{}^+)^{-1} \mathbf{R} (\mathbf{T}_a)^{-1} \tag{3A.5a}$$

In a similar way one finds

$$\mathbf{P} = \tfrac{1}{4} (\mathbf{T}_b{}^+)^{-1} \mathbf{R} (\mathbf{T}_b)^{-1} \tag{3A.5b}$$

It is easy to verify by direct computation, using Eqs. 3.83, that the expressions 3A.5a and 3A.5b lead to the expression for **P**

$$\mathbf{P} = \begin{bmatrix} 0 & 0 \\ 0 & -1 \end{bmatrix} \tag{3.78}$$

as it was originally defined.

REFERENCES

1. J. R. Pierce, *Traveling-Wave Tubes*, D. Van Nostrand Co., New York (1950).
2. D. R. Hamilton, J. K. Knipp, and J. B. H. Kuper, "Klystrons and Microwave Triodes," *Rad. Lab. Ser.*, vol. 7, McGraw-Hill Book Co., New York (1948).
3. J. C. Slater, *Microwave Electronics*, D. Van Nostrand Co., New York (1950).
4. W. Kleen, *Einführung in die Mikrowellen Elektronik*, S. Hirzel, Zürich (1952).
5. C. K. Birdsall and J. R. Whinnery, "Waves in an Electron Stream with General Admittance Walls," *J. Appl. Phys.*, **24**, 314 (1953).
6. L. M. Field, P. K. Tien, and D. A. Watkins, "Amplification by Acceleration and Deceleration of a Single-Velocity Stream," *Proc. IRE*, **39**, 194 (1951).
7. A. V. Haeff, "The Electron-Wave Tube—a Novel Method of Generation and Amplification of Microwave Energy," *Proc. IRE*, **37**, 4 (1949).
8. C. K. Birdsall, "Rippled Wall and Rippled Stream Amplifiers," *Proc. IRE*, **42**, 1628 (1954).
9. R. Kompfner and N. T. Williams, "Backward-Wave Tubes," *Proc. IRE*, **41**, 1602 (1953).
10. D. A. Watkins, "Noise Reduction in Beam-Type Amplifiers," *Stanford Univ. Electronics Research Lab. Tech. Rept. 31* (1951).
11. G. Ecker, "Gesamtheiten mit kollektiver Wechselwirkung," parts I and II, *Z. Physik*, **140**, 274 (1955).
12. H. M. Mott-Smith, "Change of Electron Temperature in an Electron Beam," *J. Appl. Phys.*, **24**, 249 (1953).
13. C. C. Cutler and C. F. Quate, "Experimental Verification of Space-Charge and Transit-Time Reduction of Noise in Electron Beams," *Phys. Rev.*, **80**, 875 (1950).
14. D. A. Watkins, "The Effect of Velocity Distribution in a Modulated Electron Stream," *J. Appl. Phys.*, **23**, 568 (1952).
15. A. M. Clogston and L. R. Walker, unpublished memos, Bell Telephone Co.; also, "The Dispersion Formula for Plasma Waves," *J. Appl. Phys.*, **25**, 131 (1954).
16. H. A. Haus, "Propagation of Noise and Signals along Electron Beams at Microwave Frequencies," Sc.D. thesis, MIT (1954).
17. P. K. Tien, "A Dip in the Minimum Noise Figure of Beam-Type Microwave Amplifiers," *Proc. IRE*, **44**, 938 (1956).
18. A. E. Siegman, D. A. Watkins, and Hsung-Cheng Hsieh, "Density-Function Calculations of Noise Propagation on an Accelerated Multivelocity Electron Beam," *J. Appl. Phys.*, **28**, 1138 (1957).
19. W. C. Hahn, "Small-Signal Theory of Velocity-Modulated Electron Beams," *GE Rev;* **42**, 258 (1939).
20. S. Ramo, "Space Charge and Field Waves in an Electron Beam," *Phys. Rev.*, **56**, 276 (1939).

21. F. N. H. Robinson and H. A. Haus, "Analysis of Noise in Electron Beams," *J. Electronics*, ser. 1, **IV**, 373 (1956).

22. L. J. Chu, "A Kinetic Power Theorem," paper delivered at the IRE conference, PGED, Durham, N. H. (June 1951).

23. F. B. Llewellyn, *Electron Inertia Effects*, Cambridge University Press, Cambridge (1941).

24. F. B. Llewellyn and L. C. Peterson, "Vacuum-Tube Networks," *Proc. IRE*, **32**, 144 (1944).

25. R. W. Peter, "Low-Noise Traveling-Wave Amplifier," *RCA Rev.*, **13**, 344 (1952).

26. B. Agdur and H. A. Haus, "A One-Dimensional Electron Beam as a Four-Terminal Network, *MIT Research Lab. Electronics Quart. Prog. Rept. 30* (April 15, 1954).

27. S. Bloom and R. W. Peter, "Transmission-Line Analog of a Modulated Electron Beam," *RCA Rev.*, **15**, 95 (1954).

28. J. Müller, "Untersuchungen über Elektronenströmungen," *Zeit. angew. Math. u. Physik*, **5**, part I, 203; part II, 409 (1954).

29. W. H. Louisell and J. R. Pierce, "Power Flow in Electron Beam Devices," *Proc. IRE*, **43**, 425 (1955).

30. E. A. Guillemin, *Communication Networks*, vol. II, John Wiley & Sons, New York (1935).

31. Y. W. Lee, "Application of Statistical Methods to Communication Problems," *MIT Research Lab. Electronics Tech. Rept. 181* (1950).

32. J. R. Pierce, "A Theorem Concerning Noise in Electron Streams," *J. Appl. Phys.*, 931 (954).

33. A. J. Rack, "Effect of Space Charge and Transit Time on the Shot Noise in Diodes," *Bell System Tech. J.*, **17**, 592 (1938).

34. L. D. Smullin and C. Fried, "Microwave Noise Measurements on Electron Beams," *Trans. IRE* (PGED), **ED-1**, no. 4, 168 (1954).

35. H. E. Rowe, "Shot Noise in Electron Beams at Microwave Frequencies," Sc. D. thesis, MIT (1952); also *MIT Research Lab. Electronics Tech. Rept. 239* (1952).

36. S. Bloom and R. W. Peter, "A Minimum Noise Figure for the Traveling-Wave Tube," *RCA Rev.*, **15**, 252 (1954).

37. H. A. Haus and F. N. H. Robinson, "The Minimum Noise Figure of Microwave Beam Amplifiers," *Proc. IRE*, **43**, 981 (1955).

38. H. Nyquist, "Thermal Agitation of Electric Charge in Conductors," *Phys. Rev.*, **32**, 110 (1928).

39. H. T. Friis, "Noise Figure of Radio Receivers," *Proc. IRE*, **32**, 419 (1944).

40. K. Fränz, "Messung der Empfängerempfindlichkeit bei kurzen elektrischen Wellen," *Z. Elektr. Elektroak.*, **59**, 105 (1942).

41. "Research and Development on Microwave Generators, Mixing Devices, and Amplifiers," *U.S. Signal Corps Contract DA-36-039-SC-64443, Quart. Rept. 3* (Aug. 1955).

42. C. E. Muehe, Jr., "Noise Figure of Traveling-Wave Tubes," S. M. thesis, MIT (1952); also *MIT Research Lab. Electronics Tech. Rept. 240* (1952).

43. J. R. Pierce and W. E. Danielson, "Minimum Noise Figure of Traveling-Wave Tubes with Uniform Helices," *J. Appl. Phys.*, **25**, 1163 (1954).

44. F. N. H. Robinson, "Microwave Shot Noise in Electron Beams and the Minimum Noise Factor of Travelling Wave Tubes and Klystrons," *J. Brit. IRE*, **14**, 79 (1954).

45. H. A. Haus, "Noise in One-Dimensional Electron Beams," *J. Appl. Phys.*, **26**, 560 (1955).
46. S. Saito, "New Method of Measuring the Noise Parameters of the Electron Beam, Especially the Correlation between Its Velocity and Current Fluctuations," *MIT Research Lab. Electronics Tech. Rept. 333* (1957); a shorter version has been submitted to the *Trans. IRE.* (PGED).
47. G. Wade, K. Amo, and D. A. Watkins, "Noise in Transverse-Field Traveling-Wave Tubes," *J. Appl. Phys.*, **25**, 1514 (1954).

CHAPTER 4

Noise in Grid-Control Tubes

T. E. Talpey

1. Sources of Noise in Grid-Control Tubes

Soon after the first application of electron tubes as amplifiers, it was discovered that the output of high-gain amplifiers contained a randomly fluctuating component as well as the desired signal component. The fluctuating component is commonly called noise, referring to its physiological effect when converted into audible energy. It was found that short-circuiting the input terminals of the amplifier reduced the noise output somewhat, but did not eliminate it entirely. On the other hand, the noise practically disappeared when the input tube was removed from its socket. It was thus concluded that a considerable part of the output noise originated inside the first tube of the amplifier.

Schottky[15] in 1918 gave the first theoretical treatment of fluctuation phenomena in electron tubes, and his work, which has been verified experimentally countless times, shows that the corpuscular nature of the electron is the basic cause of tube noise.

The type of fluctuation studied by Schottky, which is called shot noise, is not the only source of noise in modern amplifying tubes. For the purpose of discussion it is convenient to divide the various types of tube noise into six categories. They are:

1. Shot noise (including space-charge-reduced shot noise).
2. Induced grid noise.

3. Partition noise.

4. Cathode noise.

5. Secondary-emission noise.

6. Anomalous sources of noise, due to faulty construction or improper functioning of the tube.

Each of these sources is discussed in detail on pages 156–188, and references to literature concerning them can be found at the end of this chapter.

A convenient means of characterizing the noise in a grid-control tube is shown in Fig. 4.1, where the noise sources are represented by random noise generators (both voltage and current). The type of noise represented by each generator is shown in the legend to the figure. A portion of $\overline{i_g^2}$ is correlated with $\overline{i_p^2}$ and $\overline{i_{g2}^2}$; this is the so-called correlated component of induced grid noise.

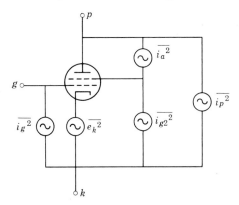

Fig. 4.1.   The noise equivalent circuit of a tetrode.

$\overline{i_g^2}$ = induced grid noise

$\overline{i_p^2}$ = shot noise in plate current

$\overline{i_{g2}^2}$ = shot noise in screen–grid current

$\overline{i_a^2}$ = partition noise

$\overline{e_k^2}$ = cathode-coating noise

The diagram in Fig. 4.1 is that of a tetrode or of a pentode with its suppressor connected to the cathode. The corresponding diagram for a triode is formed by omitting the two noise generators connected to the second grid. If secondary emission is present in the tube (perhaps intentionally as in a secondary-emission multiplier), an additional noise-current generator is necessary, shunting the two electrodes between which the exchange of secondaries takes place.

There will often be present in actual tubes one or more sources of noise which are not shown in Fig. 4.1. A list of these noises would include noise due to the presence of gas in the tube, noise accompanying leakage currents across the mica supports, and noise due to the motion of charges on the walls of the glass envelope. Although these sources of noise can be very troublesome, they can all be made negligible by proper design and manufacturing techniques. A brief discussion of these noise sources is included at the end of this section.

Shot Noise in Temperature-Limited Currents

The study of shot noise can be divided into two main categories, corresponding to temperature-limited and space-charge-limited anode currents. Although grid-control tubes employ space-charge-limited currents, much of the basic theoretical work on shot noise has dealt with temperature-limited currents. For this reason we shall start with a summary of the study of shot noise in temperature-limited diodes, extending this later to space-charge-limited currents and grid-control tubes.

The anode current of a grid-control tube or of a diode is not a smooth function of time, but consists of a large number of superimposed current pulses, each one caused by the flight of an electron from cathode to anode. These pulses are randomly distributed in time, since the emission of electrons from the cathode is a random process. The result is an anode current with minute fluctuations about its average (d-c) value. The magnitude of these fluctuations is less in space-charge-limited currents than in temperature-limited currents because of a smoothing effect exerted by the space charge near the potential minimum.

The theoretical work of Schottky[15] applies to temperature-limited currents in diodes operating at frequencies low enough so that the transit angle (cathode to anode) is much less than one radian. By assuming that the number of electrons emitted from the cathode fluctuates randomly in time about an average value, and that the electrons require negligible time to reach the anode, Schottky predicted that the mean-square value of the fluctuating component of anode current contained in a band of frequencies between f and $f + \Delta f$ would be given by

$$\overline{i^2} = 2eI\,\Delta f \text{ (amperes)}^2 \tag{4.1}$$

where e = charge on an electron = 1.6×10^{-19} coulomb
I = anode current in amperes
Δf = bandwidth in cycles per second

When measurements of the shot effect are performed at *higher* frequencies it is found that the above formula gives a value larger than that obtained experimentally. At a transit angle of one radian, for example, measured values of shot noise are about 20% lower than Eq. 4.1 predicts (see Fig. 4.2).

In 1928, Ballentine[8] presented an analysis taking into account the finite transit time of an electron traveling from cathode to anode. His theory predicted a decrease in shot noise with frequency, but to a smaller extent than actually observed experimentally. Ballentine's theory shows that, for a planar diode, the shot-noise formula, Eq. 4.1 should be multiplied by a factor

$$S(\theta) = \frac{4}{\theta^4} \left| 1 - e^{-j\theta} - j\theta e^{-j\theta} \right|^2$$

$$= \frac{4}{\theta^4} [\theta^2 + 2(1 - \cos\theta - \theta\sin\theta)] \tag{4.2}$$

where θ is the transit angle $\omega\tau$. In extending high-frequency noise analysis to the space-charge-limited case Spenke[17] in 1937 and Rack[14] in 1938 obtained independent verifications of Eq. 4.2. This equation is plotted for comparison in Fig. 4.2. It is seen that the theory is still unable to account for the low values near one radian.

Duvall[10] in 1948 predicted the effect of secondary electrons on the shape of the curve in Fig. 4.2. His calculations show that the low values of shot noise near one radian can be explained in terms of the secondary electrons emitted from the anode. A primary electron

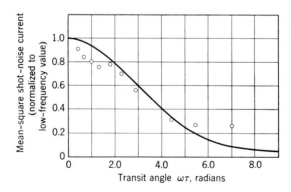

Fig. 4.2. Comparison of theory and experiment for temperature-limited shot noise at high frequencies. The solid curve is Ballentine's theory (Eq. 4.2), and the points are typical experimentally determined values, taken from Duvall.[10]

which produces a secondary will cause a current pulse in the external circuit similar to that shown in Fig. 4.3. The length and height of the tail on this current pulse depend on the energy of the secondary electron. Duvall showed that, with certain assumptions regarding the energy distribution of the secondary electrons, a theory could be formed which agrees with experimental results fairly well. Precise calculations are not possible for an actual tube because the energy distributions and the angular distributions of the secondary electrons can only be estimated.

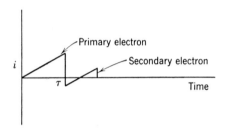

Fig. 4.3. Current pulse produced in the external leads of a temperature-limited diode by an electron which causes one secondary electron to be emitted from the anode.

Shot Noise in Space-Charge-Limited Currents at Low Frequencies

The low-frequency theory of shot noise in space-charge-limited currents was developed independently and more or less simultaneously by Schottky and Spenke,[16] North,[13] and Rack[14] (cf. Chapter 1). When the anode current of a diode is determined by space-charge rather than cathode temperature, the observed fluctuations in the plate current are considerably lower than those predicted by Eq. 4.1. When currents are space-charge-limited, there exists a potential minimum just off the surface of the cathode. If the cathode emission should momentarily increase slightly, the potential minimum will become more negative, a greater percentage of electrons will be returned to the cathode, and the plate current will not increase by as large a factor as the emission current. Thus a given fluctuation in emission produces a smaller fluctuation in plate current. In temperature-limited currents the space charge is practically negligible, and this compensating effect does not take place. Hence the fluctuations in space-charge-united currents are expected to be less than the fluctuations in temperature-united currents. The mean-square shot-noise current in space-charge-limited currents at low frequencies is commonly expressed as follows:

$$\overline{i^2} = \Gamma^2 \cdot 2eI \, \Delta f \tag{4.3}$$

This is simply the temperature-limited shot noise of Eq. 4.1 multiplied by a factor Γ^2, which has been termed the space-charge reduction factor. In most cases it has a value lying between 0.05 and 0.15, varying considerably with the structure and operating conditions of the tube. In terms of the diode conductance g and the cathode temperature T_c,[13]

$$\Gamma^2 = \frac{2kT_c g\theta}{eI} \tag{4.4}$$

where k = Boltzmann's constant = 1.37×10^{-23} joules per °K and θ is a factor† which in most practical cases is very nearly equal to its asymptotic value of $3(1 - \pi/4) = 0.644$. If Eq. 4.4 is substituted in Eq. 4.3, we obtain the interesting result

$$\overline{i_p{}^2} = 4kT_c g\theta \,\Delta f \tag{4.5}$$

This equation states that the magnitude of the space-charge-reduced shot-noise current at the plate can be computed by assuming that it arises as thermal noise‡ in a conductance g at a temperature equal to θT_c. The concept is misleading, however, for shot noise and thermal noise are not the same from a thermodynamical point of view. North[13] and Thompson[18] take pains to point this out, but for years before their papers were published there was general confusion on this point. Williams[19] has pointed out that it is not possible to apply thermodynamic reasoning to a situation in which the drift velocities of the electrons are greatly in excess of their thermal velocities. The thermal-agitations viewpoint is useful in some applications, but it should not serve as a substitute for an analysis based on a microscopic examination of electron behavior in the presence of space charge.

When attempts were made to verify the above theory of space-charge-reduced shot noise in diodes, it was found that experimental values were always considerably higher than predicted by the theory. In a convincing discussion, supported by experimental confirmation, North attributes this excess to a small percentage of electrons which are elastically reflected at the anode and hence return to the vicinity

† Note that θ does not stand for a transit angle in this case.

‡ A resistance R at a temperature T will deliver a short-circuit noise current whose mean-square value is

$$\overline{i^2} = \frac{4kT\,\Delta f}{R} = 4kTG\,\Delta f \tag{4.6}$$

This result was derived by Nyquist[50] in 1928 and was first verified experimentally by Johnson[46] in the same year. Thermal noise is sometimes called "Johnson noise."

of the potential minimum, modifying the space-charge reduction factor.

By a slight modification, however, the theory can be made to predict the plate noise in triodes, and in this form it has been verified experimentally.[13]

In a triode the negative grid wires deflect the reflected electrons as they return from the anode, and thus prevent many of them from reaching the potential minimum. In addition, at the higher anode voltage used in a triode, the anode reflection factor is considerably lower than in a diode. Consequently, excess plate noise due to reflected electrons in a triode is small compared with the shot noise, and verification of the theory of space-charge-reduced shot noise is possible. The modification in theory which is necessary leads to the substitution of g_m/σ for g in Eq. 4.4, yielding

$$\Gamma^2 = \frac{2kT_c g_m \theta}{\sigma e I} \tag{4.7}$$

The equivalent form of Eq. 4.5 for triodes becomes

$$\overline{i_p^2} = \frac{\theta}{\sigma} \cdot 4kT_c g_m \, \Delta f \tag{4.8}$$

The quantity σ is related to the amplification factor and electrode spacings and depends, to a minor extent, on electrode potentials. For conventional tubes it has a value between 0.5 and 1.0. Formulas for its calculation can be found in textbooks† and in North's paper.[13] It appears in the formula for the effective grid-plane potential V_e of a triode:

$$V_e = \sigma \left(V_c + \frac{V_b}{\mu} \right) \tag{4.9}$$

where V_c is the grid bias voltage, V_b is the plate voltage, and μ is the amplification factor of the tube.

A quantity often encountered in connection with amplifier tube noise is the equivalent noise resistance, R_{eq}. It is defined as being that resistance (at normal room temperature, T) which, when placed in series with the control grid, produces fluctuations in the plate current equal to the shot noise of the tube. Mathematically, we require that

$$\overline{i_p^2} = 4kTR_{eq} \, \Delta f \cdot g_m^2 \tag{4.10}$$

† See, for example, W. G. Dow, *Fundamentals of Engineering Electronics*, 2nd ed., p. 139, John Wiley & Sons, New York (1952).

Equating Eq. 4.8 and Eq. 4.10, we find

$$R_{\text{eq}} = \frac{\theta T_c}{\sigma T g_m} \tag{4.11}$$

For conventional triodes with oxide-coated cathodes Eq. 4.11 yields approximately

$$R_{\text{eq}} = \frac{2.5}{g_m} \tag{4.12}$$

In some cases however, R_{eq} may be as much as 50% greater than the value of Eq. 4.12, owing to a smaller value of σ. This occurs, for example, when the spacing between grid wires is large compared to the cathode–grid spacing.

Consideration of Eqs. 4.7 and 4.3 indicates that, the higher the ratio of g_m to I_b in a triode, the more noise current will be present in its output. This is not the complete picture, however, for, the higher the g_m of a tube, the more output signal current it can produce for a given input voltage. The noise performance is measured in terms of the signal-to-noise (power) ratio at the output of the tube and not merely the amount of output noise. In this respect the quantity R_{eq} is more suitable for judging the noise performance, since it transforms the plate noise into an equivalent noise in the input circuit where it can be compared with the applied signal conveniently. For cases in which the fluctuation current in the plate circuit is due mainly to shot noise, any design change that reduces R_{eq} will also improve the signal-to-noise ratio at the output. Equation 4.11 indicates that R_{eq} can be reduced by increasing g_m, which in turn can be accomplished in the design by increasing the cathode area, decreasing cathode–grid spacing or decreasing the pitch of the grid (increasing grid TPI) although there are practical limits to the spacings and dimensions that can be employed with present mechanical techniques. Changing the electrode spacings and grid pitch will also have an effect on the parameter σ, tending to decrease σ when g_m is increased. The product σg_m in general increases with g_m so that the net effect is still a decrease in R_{eq}.

Shot Noise in Space-Charge-Limited Currents at High Frequencies

As investigations of space-charge-reduced shot noise are carried to higher and higher frequencies, it is to be expected that Eq. 4.4 will not remain valid for diodes, nor Eq. 4.7 for triodes. This follows from a consideration of the fact that the development of Eqs. 4.4 and 4.7 involves the assumption that the duration of a fluctuation in emission current is long compared with the transit time of an electron.

A theoretical study by Rack[14] in 1938 indicated that the equivalent noise-voltage generator of a space-charge-limited diode should be given approximately by

$$\overline{e^2} = 3\left(1 - \frac{\pi}{4}\right) \cdot 4kT_c r_p \,\Delta f \cdot S(\theta) \tag{4.13}$$

where r_p is the low-frequency plate resistance of the diode, and $S(\theta)$ is as defined by Eq. 4.2. This is the same factor that appears in connection with temperature-limited shot noise at high frequencies.

Attempts to obtain experimental confirmation of the high-frequency diode theory under space-charge-limited conditions have been neither extensive nor outstandingly successful. Measurements by Duvall[10] and Kompfner et al.[12] show considerable disagreement with theory. Aside from the difficulties involved in measurement techniques at large transit angles, the effects of elastically reflected electrons prevent good experimental checks, just as they do in the low-frequency example discussed by North.[13]

A further point to be considered is the effect that fluctuations in the depth of the potential minimum may have on the length and shape of the anode-current pulses. Diemer and Knol[9] have pointed out that fluctuations in the depth of the potential minimum (caused by fluctuations in cathode emission) will produce long tails on the anode-current pulses, and have shown qualitatively that the result will be an increase in the diode noise at very high frequencies.

There have been some attempts to extend the diode noise theory to triode structures, with the thought that experimental confirmation of the theory could then be obtained in much the same manner as that employed by North in his studies at low frequencies. Although there is lack of sufficient investigation to warrant any definite conclusions in this respect, van der Ziel[6] has given a brief discussion of triodes from a theoretical point of view, and Duvall[10] has published some experimental data obtained on large-sized tubes.

Calculation of Noise at High Frequencies

A few remarks concerning the methods employed for calculating high-frequency noise seem to be in order here. In general two methods of attack have been used: (1) a method that makes use of the Fourier spectrum of the current pulse produced by a single electron and (2) a method that makes use of the Llewellyn–Peterson[49] electronics equations.

The first method can be applied rigorously only to temperature-limited currents or to cases that involve negligible space charge. An illustration of the method as applied to the calculation of shot noise in temperature-limited diodes can be found in an article by Fraser,[11] and an excellent description of the mathematical derivation of the method has been given by Beck.[1] The essential features of this derivation can be summarized as follows. The number of electrons flowing between two electrodes is assumed to fluctuate randomly about an average value which is measured in terms of the direct current I associated with the stream of electrons. It is assumed that each electron produces an identical current pulse $f(t)$ in the external circuit connected between the two electrodes. The current pulses are randomly distributed in time, however. The application of Campbell's theorem[43] allows us to write

$$\overline{i^2} = \frac{I}{e} \int_{-\infty}^{+\infty} f^2(t)\, dt \tag{4.14}$$

where $\overline{i^2}$ is the mean-square fluctuation current in the external circuit. It is common practice when a complicated time function is encountered to express it in terms of its Fourier spectrum, giving the amplitude of each frequency component. If the Fourier transform of $f(t)$ is represented by

$$F(\omega) = \frac{1}{2\pi} \int_{-\infty}^{+\infty} f(t) e^{-j\omega t}\, dt \tag{4.15}$$

then it can be shown (cf. Rice[54]) that

$$\int_{-\infty}^{+\infty} f^2(t)\, dt = 2\pi \int_{-\infty}^{+\infty} |F(\omega)|^2\, d\omega \tag{4.16}$$

This equation is an alternate form of Parseval's theorem.[58]

Combining Eqs. 4.14 and 4.16, we obtain an expression which can be used for the calculation of fluctuation currents

$$\overline{i^2} = \frac{4\pi I}{e} \int_0^\infty |F(\omega)|^2\, d\omega \tag{4.17}$$

In many applications we are interested in the noise current present in a narrow band of frequencies, $\Delta f = \Delta\omega/2\pi$, over which the integrand in Eq. 4.17 can be considered constant. The mean-square noise current in this narrow band is then given by

$$\overline{i^2} = \frac{4\pi I\, \Delta\omega}{e} |F(\omega)|^2 \tag{4.18}$$

In order to make use of Eq. 4.18 in any given case it is necessary to calculate the Fourier power spectrum $|F(\omega)|^2$ of the current pulse $f(t)$ produced in the external circuit by the passage of a single electron.

Because of induced charges, current starts to flow in the external circuit as soon as the electron enters the space between the two electrodes. The induced current is given by

$$i(t) = -e\left(\frac{E_0}{V_0}\right)v(t) \tag{4.19}$$

where $v(t)$ is the velocity of the electron as a function of time and E_0 is the component of the electric field which would exist along the path of the electron if a potential V_0 were applied between the electrodes and if no space charge were present.†   Equation 4.19 holds, however, even when space charge is present.   The effect of the space charge is taken care of entirely by the velocity term $v(t)$.   For plane-parallel electrodes separated by a distance d, Eq. 4.19 reduces to

$$i(t) = \frac{ev}{d} \tag{4.20}$$

Van der Ziel[7] has pointed out that the method embodied in Eq. 4.17 has been employed incorrectly for the calculation of noise in space-charge-limited currents.   The method requires that the current pulses be independent random events; this is not true when space charge is present and there is interaction between the electrons.   For this reason the method cannot be employed rigorously for the calculation of fluctuations in space-charge-limited currents, although it may be useful in some instances for predicting general behavior and orders of magnitude.

The method for calculating high-frequency noise by means of the Llewellyn–Peterson electronic equations[49] possesses the advantage that it can be applied when space charge is present as well as in the absence of space charge.   It is necessary with space-charge-limited currents, however, to neglect the region between cathode and potential minimum where electrons are moving in both directions, and to assume that all the electrons leaving the cathode at any instant have the same velocity rather than a Maxwellian distribution of velocities.

The Llewellyn–Peterson equations refer to two parallel planes a and b, and are commonly written‡

$$V_b - V_a = A^*I + B^*J_a + C^*v_a$$
$$J_b = D^*I + E^*J_a + F^*v_a \tag{4.21}$$
$$v_b = G^*I + H^*J_a + I^*v_a$$

---

† The validity of this relation was demonstrated by Ramo[53] and extended by Jen[45] to the case where space charge is present.

‡ Note that in reference 49 a different sign convention is used for the convection current density, there denoted by q ($q = -J$).

where $(V_b - V_a)$ is the fluctuation in potential between planes a and b, I is the fluctuation in total current density, J_b is the fluctuation in conduction current density at plane b, v_b is the fluctuation in velocity at b, and J_a and v_a are the impressed fluctuations in conduction current density and velocity at plane a. The coefficients $A*$ through $I*$ are tabulated in books by Beck[1] and van der Ziel[7] as well as in Llewellyn and Peterson's article.

The use of these equations in noise calculations will be illustrated by outlining their application to a space-charge-limited diode. In this case we need only the first of the three equations, identifying $(V_b - V_a)$ with the noise voltage appearing between the potential minimum and the anode. It can be shown that the term $B*J_a$ is negligible in this instance so that we obtain

$$V_b - V_a = A*I + C*v_a \qquad (4.22)$$

This equation can be interpreted as stating that the fluctuating voltage appearing across the diode (neglecting the region between cathode and potential minimum) is equal to the alternating voltage drop $A*I$ across the internal impedance of the diode, plus a noise voltage whose rms value is $C*v_a$. The coefficient $A*$ is simply the high-frequency impedance of a unit area of the diode. The open-circuit mean-square noise voltage is thus

$$\overline{e^2} = \overline{(V_b - V_a)^2} = |C*|^2 \overline{v_a^2} \qquad (4.23)$$

The quantity $C*$ can be evaluated from Llewellyn and Peterson's tabulation,

$$C* = \frac{J_0}{\omega^2 \epsilon_0} (1 - e^{-j\theta} - j\theta e^{-j\theta}) \qquad (4.24)$$

where J_0 is the d-c diode-current density, ϵ_0 is the dielectric constant of free space, and $\theta = \omega\tau$, where τ is the transit time from potential minimum to anode. The quantity $\overline{v_a^2}$ represents the mean-square fluctuation in velocity of the electrons passing the potential minimum (the procedure for its evaluation is outlined in Chapter 1).

$$\overline{v_a^2} = \frac{4kT_c e}{m J_0 A_c} \left(1 - \frac{\pi}{4}\right) \Delta f \qquad (4.25)$$

where T_c is the temperature of the cathode, and A_c is the cathode area.

At low frequencies it can be shown that

$$\lim_{\omega \to 0} A* = A_c r_p = \frac{e J_0 \tau^4}{12 m \epsilon_0^2} \qquad (4.26)$$

where r_p is the low-frequency plate resistance of the diode. Combining Eqs. 4.23 to 4.26, we obtain

$$\overline{e^2} = 4kT_c r_p \, \Delta f \left(1 - \frac{\pi}{4} \right) \frac{12}{\theta^4} \left| 1 - e^{-j\theta} - j\theta e^{-j\theta} \right|^2 \qquad (4.27)$$

Equation 4.27 can also be written

$$\overline{e^2} = 3 \left(1 - \frac{\pi}{4} \right) \cdot 4kT_c r_p \, \Delta f \cdot S(\theta) \qquad (4.13)$$

as given on page 162, using Eq. 4.2 as the definition of $S(\theta)$.

The short-circuit noise current i ($i = A_c I$) can be found by setting $V_b - V_a = 0$ in Eq. 4.22† and solving for I, with the result

$$\overline{i^2} = A_c{}^2 \overline{I^2} = A_c{}^2 \frac{|C^*|^2}{|A^*|^2} \overline{v_a{}^2} = \frac{A_c{}^2 \overline{e^2}}{|A^*|^2} \qquad (4.28)$$

The quantity $1/A^*$ is the high-frequency admittance per unit area of the diode; A^* is given by

$$\begin{aligned}
A^* &= \frac{eJ_0\tau^4}{m\epsilon_0{}^2} \cdot \frac{1}{\theta^4} \left[2(1 - e^{-j\theta}) - j\theta \left(1 + \frac{\theta^2}{6} + e^{-j\theta} \right) \right] \\
&= A_c r_p \cdot \frac{12}{\theta^4} \left[2(1 - e^{-j\theta}) - j\theta \left(1 + \frac{\theta^2}{6} + e^{-j\theta} \right) \right]
\end{aligned} \qquad (4.29)$$

The above calculation of the shot noise in a space-charge-limited diode is intended to serve as an illustration of the application of the Llewellyn–Peterson equations to noise problems in electron tubes. As shown by van der Ziel[7] and by Peterson,[51] the method can readily be extended to triodes and multigrid tubes. The basic formulas needed are the Llewellyn–Peterson equations and the relation for the velocity fluctuations, Eq. 4.25.

Induced Grid Noise

Induced grid noise is an important consideration in modern miniature tubes operating at frequencies above about 15 Mc per sec. It becomes more and more dominant as the operating frequency is increased, and at frequencies over 100 Mc per sec it becomes the principal limiting factor in low-noise amplifier design.

The mechanism by which induced grid noise is produced can be

† This is equivalent to short-circuiting the anode to the potential minimum. This can be done provided the fluctuations in the potential minimum voltage are negligible with respect to $C^* v_a$.

explained by considering the flight of an electron from cathode to plate in a triode. As the electron approaches the control grid, it will induce a charge on the grid wires, causing a motion of charge through the impedance connected between grid and cathode. The induced grid current, which is the rate at which this charge is induced, depends on the velocity of the electron and the electrode geometry. As the electron passes through the grid plane, the induced current reverses sign, because the velocity of the electron with respect to the grid is reversed. When the electron reaches the plate, the induced grid current stops. Thus for every electron that passes the grid plane, a doublet pulse of grid current is produced (Fig. 4.4). This pulse has equal areas above and below the zero axis, since a negative charge e flows off the grid while the electron is approaching, and an equal positive charge flows onto the grid when the electron moves away. Since electrons pass by the grid at randomly spaced time intervals, the actual induced grid current consists of a large number of randomly distributed current pulses superimposed on each other. Although the average value of each pulse, and hence of the total induced current, is zero, there is a fluctuating current due to the fact that more pulses may occur during one interval of time than during a succeeding interval. This fluctuating component flowing through the grid–cathode impedance, produces a fluctuating grid voltage which is amplified by the tube and appears as a noise current in the plate circuit in addition to the shot noise.

The first mention of induced grid noise in the literature is found in an article by Ballentine,[8] in 1928, in which the prediction is made that noise from this source will become important at high frequencies. Ballentine makes no attempt to calculate the induced grid noise, but indicates how it might be done if the shape of the induced grid-current pulses were known and the space charge negligible.

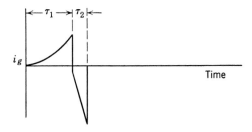

Fig. 4.4. Current pulse induced in the grid circuit of an ideal space-charge-limited triode by the passage of a single electron.

The first reports of experimental work on induced grid noise appeared in 1941 in articles by Bakker[20] and by North and Ferris.[23] Theoretical studies by Bakker indicated that the mean-square value of the induced grid noise current in an ideal planar triode should be given by

$$\overline{i_g{}^2} = \left(\frac{\omega\tau_1}{3}\right)^2 \cdot \overline{i_p{}^2} \tag{4.30}$$

where τ_1 = transit time from potential minimum to grid plane

$\overline{i_p{}^2}$ = mean-square value of the fluctuations in the plate current

This result was also obtained by Schremp.[5] It is valid only for small transit angles $\omega\tau_1$, and assumes that the grid noise has its origin in the same fluctuations of cathode emission that give rise to the plate noise. In the development of this equation, it is necessary to neglect the effect of electrons in the region between cathode and potential minimum, to neglect the grid–plate transit time, and to assume that all electrons cross the potential minimum with a uniform velocity.

Bakker presents measured values of induced grid noise and finds reasonably good agreement with calculated values. His calculations cannot be considered a direct verification of Eq. 4.30, however, since they are based on a modification of this equation in which measured values of the input conductance of the tube are used.

The article of North and Ferris[23] presents the same theoretical relationship between grid noise and input conductance as that derived by Bakker: namely,

$$\overline{i_g{}^2} = 1.43 \cdot 4kT_cG_g\,\Delta f \tag{4.31}$$

$$= 4k\beta TG_g\,\Delta f$$

where T_c = temperature of the cathode in degrees Kelvin

T = room temperature $\cong 293°$ K

$\beta = 1.43T_c/T \approx 5.0$ for oxide-coated cathodes

G_g = input conductance due to transit-time effects (excluding lead inductance loading)

This amounts to treating the induced grid noise *as if* it originated in a grid–cathode conductance equal to the actual input conductance of the tube at a temperature of approximately five times room temperature. At high frequencies it is usually simpler to measure the induced grid noise directly than to measure the input conductance of the tube. For this reason it is usually preferable to associate the factor β with the input conductance, defining the quantity βG_g as the equivalent grid-noise conductance of the tube.

Using the Llewellyn–Peterson equations,[49] van der Ziel[26] has extended Eq. 4.30 to take into account the grid–plate transit time τ_2. The result is

$$\overline{i_g^2} = \overline{i_k^2} \left(\frac{\omega\tau_1}{3}\right)^2 \left(1 + 2\frac{\tau_2}{\tau_1}\right)^2 \tag{4.32}$$

where τ_1 = transit time from potential minimum to grid plane
τ_2 = transit time from grid plane to plate
$\overline{i_k^2}$ = mean-square fluctuations in the cathode current
$= 2eI_b\Gamma^2 \Delta f$

In practice $\overline{i_k^2}$ can be replaced by $\overline{i_p^2}$, the mean-square fluctuation current at the plate, since $\overline{i_p^2} = \overline{(i_k - i_g)^2}$ and $\overline{i_g^2}$ is less than 1% of $\overline{i_k^2}$ for the range of $\omega\tau_1$ in which Eq. 4.32 is valid.

Bell[22] has shown that the induced grid noise can also be calculated from the relationship

$$\overline{i_g^2} = \left(\frac{\omega C_e}{g_m}\right)^2 \cdot \overline{i_k^2} \tag{4.33}$$

where C_e is the component of input capacitance due to space-charge effects, that is, the change in input capacitance as the current is increased from zero to its operating value.

Using the Llewellyn–Peterson equations, van der Ziel[26] has confirmed Bell's formula for a planar triode, but finds that Eq. 4.33 should be multiplied by a factor somewhat greater than unity, giving

$$\overline{i_g^2} = \left(\frac{\omega C_e}{g_m}\right)\overline{i_k^2} \left(\frac{1 + 2\dfrac{\tau_2}{\tau_1}}{\dfrac{1}{2} + \dfrac{3}{2}\dfrac{d_m}{d_1} + 2\dfrac{\tau_2}{\tau_1}}\right) \tag{4.34}$$

where d_m is the distance from the cathode to the potential minimum, and d_1 is the cathode–grid spacing.

All the formulas presented above were derived under the assumption that the induced grid noise has its origin in the same fluctuations in cathode emission that produce the shot-noise current in the plate circuit. This means that the induced grid-noise current from this source is completely correlated with the shot-noise current of the tube. It has been shown by a number of investigators[24,27] that there is *not* complete correlation between the induced grid noise and shot noise in an actual tube. In some cases it has been found[27] that the uncorrelated component of mean-square grid-noise current represents as much as 70% of the total grid noise, leaving only a 30% correlation between grid noise and plate noise.

Three explanations have been offered to account for this uncorrelated component of induced grid noise. We shall refer to them as

1. Transit-angle fluctuations.
2. Reflected electrons.
3. Total emission noise.

It appears likely that each of the three sources is partly responsible for the excess (uncorrelated) noise component, the relative contribution of each depending on the construction of the tube and the operating conditions. A thorough comparative investigation of each of these theories is needed before it can be stated that a complete and accurate understanding of induced grid noise exists.

1. Transit-Angle Fluctuations. The transit angle $\omega\tau_1$ in Eqs. 4.30 and 4.32 represents the portion of a cycle of the signal voltage that is required for an electron to travel from the potential minimum to the grid plane. These equations imply a single-valued transit angle— a concept that obviously applies only to an idealized structure. In an actual tube, an electron passing near a negative grid wire will exhibit a longer transit time than one passing midway between two grid wires. As the number of electrons passing near grid wires relative to those passing midway between grid wires can be expected to fluctuate randomly in time,† the *average* transit angle of all electrons can be said to fluctuate. Thus, not only are there fluctuations in the number of electrons passing the grid (these fluctuations in numbers cause the shot noise and the correlated component of grid noise), but there are also fluctuations in the average transit angle of the electrons. These transit-angle fluctuations are not taken into account in Eqs. 4.30 to 4.34. Since fluctuations in the average transit angle are independent of fluctuations in the numbers of electrons, the noise generated by transit-angle fluctuations is not correlated with the shot noise in the plate circuit.

The first suggestion that transit-angle fluctuations could be the cause of the uncorrelated component of induced grid noise was made by Bell[21] in 1948. After presenting the theory, Bell's article discusses the results of some computations which were made on a particular tube type in an attempt to estimate the magnitude of noise from

† The trajectory of an electron depends on the point along the cathode from which it leaves (i.e., behind a grid wire or between grid wires) and its emission velocity. Both of these are purely random phenomena.

transit-angle fluctuations.† A quotation relative to his findings follows: "Results indicate that even under the most favorable conditions— the approach of temperature limitation—the effect does not account for more than 30 per cent of the total grid noise, and that under normal conditions, involving considerable smoothing out of field irregularities by the space charge, the effect becomes very small."

More recently, however, there has been some re-examination[6,7] of Bell's calculations, particularly with regard to how the results might differ when applied to tubes with different geometries. Van der Ziel[7] mentions briefly some measurements on tubes with a coarse grid structure (relatively large ratio of grid pitch to grid–cathode spacing), which appear to lend support to Bell's theory by exhibiting a large uncorrelated component of induced grid noise. The use of a high negative grid bias constrains the electrons to flow through smaller areas in the grid plane and hence reduces the noise due to trajectory variations and the accompanying transit-angle fluctuations.

2. Reflected Electrons. It has been shown[25] that the elastic reflection of a small number of electrons from the plate of a tube can cause a large increase in the induced grid noise. A reflection coefficient of 2 or 3% is sufficient to cause a several-fold increase in induced grid noise. Unfortunately it is not possible to determine the reflection coefficient of the plate of a given tube, so that it is difficult to obtain a direct verification of the theory.

The mechanism by which reflected electrons produce grid noise can be summarized as follows. Electrons that are elastically reflected from the plate travel back toward the grid, inducing pulses of current in the grid circuit. Many of these reflected electrons succeed in passing back between grid wires and into the grid–cathode region before they lose their cathode-directed energy and are finally drawn back through the grid to the plate. Since electron reflections at the plate are independent random events, the pulses of current induced in the grid circuit by reflected electrons occur at randomly spaced time intervals and give rise to a component of noise in the grid circuit. This noise component is not correlated with shot noise since fluctuations in the numbers of reflected electrons are independent of fluctuations in the cathode current.

In comparing the noise due to reflected electrons with the correlated component of grid noise, we note that reflected electron noise is not smoothed by space charge. Consequently, a relatively small reflected electron current can produce a component of induced grid noise which

† Bell has called this component of noise "negative grid partition noise" or "induced partition noise."

is comparable to the space-charge-smoothed correlated component of grid noise. In addition, the current pulses produced by reflected electrons are of longer duration than those produced by electrons that traverse the tube only once, so that the noise-power spectrum of grid-current pulses due to reflected electrons has a larger amplitude than the noise power spectrum of pulses produced by electrons that are not reflected. These two considerations lead to the conclusion[25] that a reflection coefficient at the plate of 2 or 3%† is sufficient to give rise to an uncorrelated component of grid noise of about the observed magnitude. At the same time the additional component of noise induced in the plate circuit by these reflected electrons is negligible compared with the shot noise in the plate current.

In tubes with a large ratio of grid pitch to grid–cathode spacing (i.e., a relatively open grid structure), an electron reflected from the plate is more likely to get back between grid wires than in a tube with a finer grid structure, so that reflected electrons will produce a larger uncorrelated component of grid noise in a tube with a coarse grid. The use of a high negative grid bias and a high plate voltage will make it more difficult for reflected electrons to get back past the grid, and hence will reduce noise from this source.

3. Total Emission Noise. The designation "total emission noise" has been given[28,56] to the noise produced in the grid–cathode circuit by those electrons that do not have sufficient energy to get past the potential minimum, and hence are returned to the cathode. An electron of this class will induce a doublet-type pulse of current in the grid–cathode circuit. If the frequency is high enough and if the grid-cathode spacing is small enough to make the transit angle associated with a returned electron comparable to the transit angle of an electron which gets by the potential minimum, then noise from this source will be appreciable. When the grid–cathode spacing is made very small (of the order of 0.001 in. as in some recent tube types), the distance from the cathode to the potential minimum represents a considerable portion of the cathode–grid distance, and total emission noise may constitute an appreciable part of the uncorrelated component of induced grid noise. Further experimental work is needed to determine just how important the effect may be in close-spaced tubes.

It has been stated[55] that total emission noise can be calculated by assuming that the conductance representing the total emission loading

---

† Experimental work by Farnsworth[44] on electron reflection indicates that these values are reasonable for the reflection coefficient of a triode under normal operating conditions.

generates noise as if it were at cathode temperature. The reasoning is that the electrons which start out toward the potential minimum and return to cathode can be considered to be in thermal equilibrium with the cathode, and hence should produce noise similar to resistance noise. A calculation of total emission noise based on a study of electron dynamics in the region between cathode and potential minimum is needed to verify this assumption.

The presence of a small amount of grid current in a tube can also cause an appreciable amount of uncorrelated grid noise. This possibility is often neglected in grid-noise studies, perhaps because it arises from a conductive current rather than from an induced current. Several microamperes of grid current are enough to cause a doubling of the grid noise in a typical grid-control tube. This current could be caused either by gas in the tube or by primary emission from barium deposited on the grid wires. Such a current will produce noise according to the relation

$$\overline{i_g^2} = 2eI_c \, \Delta f \tag{4.35}$$

where I_c is the direct grid current. The operation of a grid-control tube with too small a value of negative grid bias can also result in a grid current produced, in this case, by the faster-moving electrons striking grid wires.

There are two quantities that are useful in discussions of the magnitudes and measurement of induced grid noise and in network calculations involving grid noise. These are the equivalent noise diode current and the equivalent noise conductance.

The equivalent noise diode current is defined as the direct anode current of a temperature-limited diode which would produce a short-circuit noise current equal to that produced by the noise source in question. Below 100 Mc per sec the equivalent noise diode currents for the induced grid noise of typical receiving tubes lie in the range from 2 to 50 μa. In terms of the equivalent noise diode current I_{eq}, the induced grid noise is given by

$$\overline{i_g^2} = 2eI_{eq} \, \Delta f \tag{4.36}$$

The equivalent noise conductance is defined as that conductance at room temperature (293° K) which would produce a short-circuit noise current equal to that produced by the noise source in question. In terms of the equivalent noise conductance G_{eq}, the induced grid noise of a tube is given by

$$\overline{i_g^2} = 4kTG_{eq} \, \Delta f \tag{4.37}$$

It is to be emphasized that the equivalent noise conductance, like the equivalent shot-noise resistance, is not a dissipative element but rather represents a noise source. When it is shown in a network diagram, it is to be interpreted as a noise generator and not as a load. Table 4.1 gives some measured values of G_{eq} at 30 Mc per sec. The equivalent noise conductance varies as the square of frequency so that at 60 Mc per sec, for example, the values in the table must be multiplied by four. Comparison of Eqs. 4.36 and 4.37 shows that, when numerical values are substituted for the constants,

$$G_{eq} = 20 I_{eq} \qquad (4.38)$$

It will be noticed also that G_{eq} is the same as the quantity βG_g in Eq. 4.31.

A few general comments can be made concerning the effect of changes in tube design on the magnitude of the various components of induced grid noise.

First, the mean-square induced grid-noise current is proportional to the number of electrons flowing past the grid, other factors remaining unchanged. This means, for example, that, in a planar tube operating at a fixed current density, the induced grid noise $\overline{i_g{}^2}$ is proportional to the cathode area. Thus increasing the cathode area of such a tube increases the g_m, decreases the shot-noise equivalent resistance, R_{eq} (as pointed out in this section, page 161), and increases G_{eq}. Consequently, changes in cathode area have very little effect on the minimum narrow-band noise figure which the tube will exhibit as an RF amplifier input tube. (Optimum narrow-band noise figure is approximately equal to $1 + 2\sqrt{R_{eq} \cdot G_{eq}}$; see Eq. 4.131.)

Second, changes in grid *pitch* will have several effects on the induced grid noise of a grid-control tube. Equation 4.30 states that the correlated component of grid noise is given by

$$\overline{i_g{}^2} = \left(\frac{\omega \tau_1}{3}\right)^2 \cdot \overline{i_p{}^2} \qquad (4.30)$$

Consider two triodes, identical in all respects except that the control grid of tube 2 is wound with twice as many turns per inch as the control grid of tube 1. At a given frequency the transit angle $\omega \tau_1$ of the input sections of these tubes depends primarily on the cathode–grid spacing and the equivalent grid-plane potential. The cathode–grid spacings are the same, and, since the current densities are assumed to be equal, the equivalent grid-plane potentials are equal. The transit angles are thus essentially the same for both tubes, and, according to Eq. 4.30, the correlated component of grid noise in this case is proportional only

TABLE 4.1

Measured Values of Plate Noise and Induced Grid Noise at 30 Mc per sec

Tube Type[†]	Number Examined	Plate Current I_b, ma	Transconductance g_m, micromhos	Plate Noise: Equivalent Noise Resistance R_{eq}, ohms	Induced Grid Noise: Equivalent Noise Conductance G_{eq}, Micromhos	Reference[‡]
6AC7	2	10	9,000	590	160–280	27
6AC7	1	10	9,000	520	140	24
6AG5	21	7	6,000	480	140	25
6AK5	30	10	5,500	460	45	25
6AS6§	4	10	5,800	450	45	25
6AU6	17	12	6,600	420	210	25
6BC5	6	8	5,800	590	130	25
6BC6§	10	12	7,300	410	175	25
6J4	2	15	12,000	320	220	25
6J4	4	10	—	315	50–105	27
6J6‖	10	8	4,400	720	60	25
6V6	1	30	4,000	860	560	24
7AF7	2	16	4,000	725	120–230	24
396A/2C51‖	8	8	5,400	550	40	25
404A	9	15	16,000	240	70	25

† All tubes are triode-connected.
‡ See page 217.
§ Suppressor connected to plate and screen.
‖ Values are for a single section.

to the plate noise.　As shown by Eq. 4.8, the plate noise is proportional to g_m/σ so that the tube with the higher g_m/σ ratio will have a larger correlated component of induced grid noise.　It is well known that the ratio of transconductance to plate current of a tube can be increased by decreasing the pitch of the grid wires (winding more turns per inch on the grid).　Tube 2 consequently will have the higher transconductance and hence† a larger correlated component of induced grid noise than tube 1.　At the same time the uncorrelated component of induced grid noise in tube 2 with its closer-wound grid wires will be less than the uncorrelated component in tube 1.　As pointed out above, this can be explained either in terms of transit-angle fluctuations or in terms of reflected electrons.　Thus, although tube 2 has a larger correlated component of grid noise than tube 1, it has a smaller uncorrelated component.　Stated in another way, tube 2 has a higher grid-noise correlation coefficient than tube 1.　It is shown in Section 2 that in certain applications it is possible to suppress the effects of the correlated component of grid noise so that in general a high correlation coefficient is desirable.

Changes in the cathode–grid spacing will also affect the correlation coefficient and the magnitude of the induced grid noise.　The effects of a given change in spacing are more difficult to evaluate, however, for a change in spacing will also alter the relative geometry of the input section.　If the grid is moved closer to the cathode, for example, the spaces between grid wires will become wider relative to the input spacing, and an increase in the uncorrelated component of grid noise will result.　A change in the correlated component of grid noise is also to be expected, since a change in input spacing will also affect transit time and transconductance (see Eq. 4.30).　The net effect of reducing the input spacing will be a decrease in the correlation coefficient. If the distance between cathode and grid is made very small, total emission noise may become appreciable, further reducing the correlation coefficient.

In any given tube it is possible to reduce the uncorrelated component of induced grid noise by operating with a high negative bias on the control grid.　In order to maintain the plate current at a value high enough to obtain sufficient gain, it is necessary to use a high plate voltage with this high negative bias.　Under these conditions the total induced grid noise is lowest and the correlation coefficient highest, the limitations of the procedure depending mostly on the allowable plate dissipation of the tube.

---

† The quantity σ is affected only to a minor extent by changes in grid pitch.

Partition Noise

When a grid-control tube contains one or more positive grids to which a direct current flows (as the screen grid in a tetrode or pentode), a type of noise called partition noise will also be present. This noise arises because of random fluctuations in the division of current between the positive grid and the plate. Assume a smooth stream of electrons flowing through a positive grid toward the plate of a vacuum tube. Some of the electrons will strike the grid and give rise to a direct current to the grid. During one interval of time, more than the average number of electrons may strike the grid, while fewer than average may strike the grid in the next (equal) interval of time. The current to the grid will thus show minute fluctuations about its average value, and the current to the plate will show equal fluctuations of the opposite sign. In a grid-control tube, these fluctuations in current division reduce the effectiveness of the space-charge smoothing of the shot noise in the plate current.

The first calculations of the magnitude of the fluctuations due to the partition effect were carried out more or less simultaneously by Schottky,[31] Bakker,[29] and North,[30] using various methods of analysis. The results as applied to a screen–grid tube can be summarized by the following three equations, representing the noise currents in the cathode lead, screen–grid lead, and plate lead, respectively:

$$\overline{i^2}_{\text{cath}} = \Gamma^2 \cdot 2eI_k \, \Delta f \qquad (4.39)$$

$$\overline{i^2}_{\text{screen}} = \frac{\Gamma^2 I_{c2} + I_b}{I_k} \cdot 2eI_{c2} \, \Delta f \qquad (4.40)$$

$$\overline{i^2}_{\text{plate}} = \frac{\Gamma^2 I_b + I_{c2}}{I_k} \cdot 2eI_b \, \Delta f \qquad (4.41)$$

where Γ^2 is the space-charge-reduction factor, I_k is the direct cathode current, I_{c2} the direct screen–grid current, and I_b the direct plate current.

The validity of Eqs. 4.39 to 4.41 can be justified in the following manner. In a grid-control tube, the mechanism of space-charge smoothing of the shot noise can be represented in terms of compensating currents which have their origin in the compensating action of the space charge near the potential minimum. A shot-noise fluctuation, $i = \sqrt{2eI_b \, \Delta f}$, in the plate current of a triode is linearly reduced by a compensating current† $i = (1 - \Gamma) \sqrt{2eI_b \Delta f}$. This gives a net fluctuation

$$i_p = \sqrt{2eI_b \Delta f} \left[1 - (1 - \Gamma) \right] \tag{4.42}$$

which yields the familiar space-charge-reduced shot-noise formula

$$\overline{i_p{}^2} = \Gamma^2 \cdot 2eI_b \Delta f \tag{4.43}$$

In a screen–grid tube the compensating current divides between the plate and screen in proportion to the average plate and screen currents. Thus a shot-noise fluctuation in the plate current is counteracted by a fraction I_b/I_k of the compensating current. This produces a net fluctuation in plate current whose Fourier component‡ is given by

$$i_{p1} = \sqrt{2eI_b \Delta f} \left[1 - \frac{I_b}{I_k} (1 - \Gamma) \right] \tag{4.44}$$

In addition, the plate receives a portion I_b/I_k of the compensating current which accompanied the shot-noise fluctuation in the screen current. This portion amounts to

$$i_{p2} = \sqrt{2eI_{c2} \Delta f} \left[-\frac{I_b}{I_k} (1 - \Gamma) \right] \tag{4.45}$$

The total noise in the plate current is thus given by

$$\overline{i^2}_{\text{plate}} = \overline{\lfloor i_{p1}{}^2} + \overline{i_{p2}{}^2} \tag{4.46}$$

$$= 2eI_b \Delta f \left[1 - \frac{I_b}{I_k} (1 - \Gamma) \right]^2 + 2eI_{c2} \Delta f \left[\frac{I_b}{I_k} (1 - \Gamma) \right]^2$$

Algebraic reduction of Eq. 4.46 yields

$$\overline{i^2}_{\text{plate}} = 2eI_b \Delta f \left(\Gamma^2 \frac{I_b}{I_k} + 1 - \frac{I_b}{I_k} \right) \tag{4.47}$$

Making use of the fact that $I_k = I_b + I_{c2}$, we obtain Eq. 4.41:

$$\overline{i^2}_{\text{plate}} = 2eI_b \Delta f \left(\frac{\Gamma^2 I_b + I_{c2}}{I_k} \right) \tag{4.48}$$

Equation 4.40 can be verified in a similar manner.

The noise currents given by Eqs. 4.39 to 4.41 are partially correlated

---

† Mathematical rigor is sacrificed here in the interest of brevity. Actually the calculation should be carried out by considering separately the effects of compensating currents for each velocity class of electrons and performing a summation over all the velocity classes.

‡ See footnote, page 190.

with each other. In applications involving screen–grid tubes, it is frequently more convenient to deal with noise-current generators that are uncorrelated. We proceed to derive these current generators.

Referring to the notation of Fig. 4.1, page 165, we postulate three independent noise-current generators: $\overline{i_p^2}$ connected between plate and cathode, $\overline{i_{g2}^2}$ connected between screen–grid and cathode, and $\overline{i_a^2}$ connected between screen–grid and plate. In terms of the Fourier components of these currents†

$$i_{\text{plate}} = i_p + i_a \tag{4.49}$$

$$i_{\text{screen}} = i_{g2} - i_a \tag{4.50}$$

$$i_{\text{cath}} = i_{g2} + i_p \tag{4.51}$$

The mean-square value of each of these equations contains the mean of a cross product of two of the currents. Since we postulated that the current generators $\overline{i_p^2}$, $\overline{i_{g2}^2}$, and $\overline{i_a^2}$ represent independent or uncorrelated noise sources, the mean of each cross product must be zero. Thus we obtain

$$\overline{i^2}_{\text{plate}} = \overline{i_p^2} + \overline{i_a^2} \tag{4.52}$$

$$\overline{i^2}_{\text{screen}} = \overline{i_{g2}^2} + \overline{i_a^2} \tag{4.53}$$

$$\overline{i^2}_{\text{cath}} = \overline{i_{g2}^2} + \overline{i_p^2} \tag{4.54}$$

The simultaneous solution of these three equations yields

$$\overline{i_p^2} = \tfrac{1}{2}(\overline{i^2}_{\text{cath}} - \overline{i^2}_{\text{screen}} + \overline{i^2}_{\text{plate}}) \tag{4.55}$$

$$\overline{i_{g2}^2} = \tfrac{1}{2}(\overline{i^2}_{\text{cath}} + \overline{i^2}_{\text{screen}} - \overline{i^2}_{\text{plate}}) \tag{4.56}$$

$$\overline{i_a^2} = \tfrac{1}{2}(-\overline{i^2}_{\text{cath}} + \overline{i^2}_{\text{screen}} + \overline{i^2}_{\text{plate}}) \tag{4.57}$$

Substituting Eqs. 4.39 through 4.41 into Eq. 4.55 and reducing terms, we obtain

$$\overline{i_p^2} = \Gamma^2 \cdot 2eI_b\,\Delta f \tag{4.58}$$

Similar substitutions in Eqs. 4.56 and 4.57 yield

$$\overline{i_{g2}^2} = \Gamma^2 \cdot 2eI_{c2}\,\Delta f \tag{4.59}$$

$$\overline{i_a^2} = (1 - \Gamma^2) \cdot 2e\,\frac{I_{c2}I_b}{I_k}\,\Delta f \tag{4.60}$$

---

† The signs in Eqs. 4.49 to 4.51 are chosen under the assumption that currents produced by the generators in Fig. 4.1 flow upward. Since they will be assumed uncorrelated, this actually makes no difference in the final result.

Equations 4.58 through 4.60 define a set of three uncorrelated noise-current generators which can be used to represent the noise sources in a screen–grid tube.

It is clear that the partition noise $\overline{i_a{}^2}$ vanishes when $\Gamma = 1$. This is a consequence of the fact that partition noise exists because the division of current between the plate and the screen–grid reduces the effectiveness of the compensating action of the space charge. When no space-charge smoothing of the shot noise takes place, that is, when $\Gamma^2 = 1$, partition noise is not present.

In the usual applications of screen–grid tubes the screen is by-passed to the cathode, short-circuiting the noise-current generator $\overline{i_{g2}{}^2}$. The noise current in the plate lead is given by Eq. 4.41 or the sum of Eqs. 4.58 and 4.60. It is often convenient to express this noise in terms of an equivalent noise resistance in series with the control grid, similar to the triode equivalent noise resistance defined in this section, page 161. The equivalent noise resistance is defined in such a way as to produce the same noise current in the plate lead as the tube itself produces. Thus

$$\frac{\Gamma^2 I_b + I_{c2}}{I_k} \cdot 2eI_b\,\Delta f = 4kTR_{eq} \cdot \Delta f \cdot g_m{}^2 \tag{4.61}$$

where g_m is the transconductance from control grid to plate. Solving Eq. 4.61 for R_{eq} and substituting Eq. 4.7 for Γ^2, we obtain

$$R_{eq} = \frac{\theta T_c}{\sigma T g_m} \cdot \frac{I_b}{I_k} + \frac{2e}{4kT} \cdot \frac{I_b I_{c2}}{I_k g_m{}^2} \tag{4.62}$$

When we use the approximation $T_c = 1000°\,K$, $\theta = 0.644$, $T = 293°\,K$, and $\sigma = 0.88$, this becomes

$$R_{eq} \cong \frac{2.5}{g_m} \cdot \frac{I_b}{I_k} + \frac{20}{g_m{}^2} \cdot \frac{I_b I_{c2}}{I_k} \tag{4.63}$$

If the tube is connected as a triode by connecting or by-passing the screen to the plate, the transconductance will be increased in approximately the direct ratio of cathode current to plate current. Thus

$$g_{m,tri} = g_{m,pent} \cdot \frac{I_k}{I_b} \tag{4.64}$$

and Eq. 4.63 can be written

$$R_{eq,pent} = R_{eq,tri}\left(1 + \frac{8I_{c2}}{g_m}\right) \tag{4.65}$$

The effects of the tube design changes on the noise of a tetrode or pentode can be illustrated by reference to Eq. 4.62. The equivalent noise resistance can be decreased most effectively by increasing the transconductance, since it appears in the denominator of both terms in Eq. 4.62. Any design change which increases transconductance without increasing the plate current, such as decreasing cathode to control-grid spacing, will thus decrease R_{eq}. Some reduction in noise can also be effected by decreasing the screen-grid current. This can be done by electron-optical means, although the added complexity in construction may not always be justifiable.

Secondary-Emission Noise

When an electron strikes a positive electrode such as the plate of a vacuum tube, it may liberate one or more low-energy secondary electrons. Each primary electron does not liberate exactly the same number of secondaries, and some primary electrons may not liberate any secondary electrons. Consequently, for a given primary current the secondary-electron current shows fluctuations in time about an average value. If this noise current flows to another electrode, then the noise equivalent circuit of the tube should include a noise-current generator connected between the two electrodes between which the exchange of secondaries takes place. In certain applications this will cause an additional component of noise to be present in the output circuit.

In a tetrode, for example, there may be an exchange of secondary electrons between plate and screen under certain operating conditions. With the screen grid by-passed to the cathode, there will be a secondary-emission noise current in the plate lead in addition to the shot-noise and partition-noise currents. All these noise currents add quadratically, giving

$$\overline{i^2}_{tot} = \overline{i_p^2} + \overline{i_a^2} + \overline{i^2}_{sec} \tag{4.66}$$

where $\overline{i^2}_{sec}$ is the mean-square noise current due to secondary emission, and $\overline{i_p^2}$ and $\overline{i_a^2}$ are the shot noise and partition noise, respectively.

Secondary-emission noise in most grid-control tubes can be eliminated or reduced to negligible proportions by causing the secondaries liberated at the plate to return to the plate. In a triode this takes place on its own accord since the plate is usually at a much higher potential than any other electrode. In a tetrode the secondaries can be returned to the plate by adding a suppressor grid (converting the tube to a pentode) or by making the spacing between screen grid and plate large enough so that a space-charge cloud forms between the two electrodes, depressing the potential sufficiently to return the secondary electrons to the plate. Operating the plate at a potential 20 to 30 volts above the potential of the screen will also force most of the secondaries back to the plate.

At very high frequencies, low-energy secondaries have considerable influence on output noise, even when they return to the electrode from which they are liberated. This effect has been discussed briefly in this section, page 157, where the work of Duvall[10] on high-frequency diode noise was mentioned.

The magnitude of low-frequency secondary-emission noise is given to a very rough approximation by[35]

$$\overline{i^2}_{\text{sec}} \approx 2eI_s \, \Delta f \qquad (4.67)$$

where I_s is the average value of the secondary-electron current. This relation is predicated on the assumption that the probability that a given primary electron will liberate exactly n secondaries follows a Poisson distribution. Ziegler[34] has shown that the secondary-emission noise current is often considerably higher than Eq. 4.67 would indicate, depending on the energy of the primary electrons. He points out that this must mean that the standard deviation in the number of secondaries emitted per primary is greater than would be predicted by a Poisson distribution. Kurrelmeyer and Hayner[32] have shown that a theory which provides good agreement with a considerable amount of experimental data can be formulated by dividing the primary electrons into three classes: primary electrons that are reflected, primary electrons that produce no secondaries, and primary electrons that liberate true secondary electrons. The effect of each class is considered separately in calculating the total noise due to secondary emission.

For the purpose of estimating the amount of secondary-emission noise to be expected in a grid-control tube, we refer to some calculations made by Shockley and Pierce.[33] Their calculations indicate that, for the operating voltages commonly employed, the secondary-emission noise can be expected to lie between 85% and 150% of the value given by Eq. 4.67.

The use of electron multiplication by secondary emission has been proposed[47,48] as a means of enhancing the transconductance of a grid-control tube. The electron-multiplier stage of such a tube amplifies both the noise and the signal currents which are present in the primary-electron stream. In addition the output contains secondary-emission noise generated in the electron multiplier. Thus the output noise current is given by

$$\overline{i^2} = N^2 \overline{i^2}_{\text{pri}} + \overline{i^2}_{\text{sec}} \qquad (4.68)$$

where N is the amplification ratio of the secondary-emission multiplier, $\overline{i^2}_{\text{pri}}$ is the noise in the input current to the multiplier stage, and $\overline{i^2}_{\text{sec}}$ is the noise generated by secondary emission. The noise figure of a single-stage amplifier employing an electron multiplier tube is always higher than that which can be obtained by using a conventional tube operating at the same stage gain.

Cathode Noise

The subject of cathode noise is included in this chapter merely to complete the picture of noise in grid-control tubes. Theories, references, and measurements of cathode noise are not discussed since they are treated in detail in Chapter 2. A few general remarks here will suffice.

A type of low-frequency noise which is important in some applications of grid-control tubes is that which has been termed flicker effect. Flicker noise has a power spectrum whose amplitude decreases with increasing frequency. At frequencies above 100 kc per sec, flicker noise is usually negligible with respect to the shot noise in the cathode current. Flicker noise is suppressed considerably by the space charge near the potential minimum in a grid-control tube. There appears to be some connection between flicker effect and the interface† resistance of a grid-control tube, although flicker noise cannot be accounted for solely in terms of this interface resistance. A short discussion of proposed flicker-noise theories and of circuit considerations pertaining to flicker noise has been published by van der Ziel.[7]

A simplified equivalent circuit which is often used to represent the electrical characteristics of a cathode is shown in Fig. 4.5. R_1 represents the interface resistance, and C_1 represents the capacitance between the cathode base metal and the inner surface of the cathode coating. R_2 represents the bulk resistance of the cathode coating between inner and outer surfaces. It is thought that most of this resistance occurs near the outer surface of the cathode coating. At high frequencies, the capacitance C_1 effectively by-passes R_1, and the interface resistance has no effect. If the resistance R_2 is appreciable, however, it will cause a reduction in the effective transconductance of the tube. Furthermore, it is to be expected that R_2 will act as a

† Interface is a term that is used to describe a high-resistance compound which forms between the cathode base metal and the oxide coating of conventional cathodes.

source of noise.   The mean-square voltage can be expressed as

$$\overline{e_k^2} = \alpha \cdot 4kT_cR_2\,\Delta f \tag{4.69}$$

where T_c is the temperature of the cathode, and α is a factor (equal to or greater than unity) which takes into account the possibility that R_2 may not produce pure thermal noise.   To our knowledge no experimental investigations of bulk resistance as a source of noise have yet been reported.   It may be that ordinary values of R_2 are so small

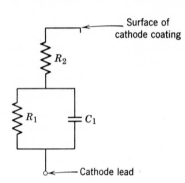

Fig. 4.5.   Simplified equivalent circuit for an electron-tube cathode.

that the effect is of importance only in very high transconductance tubes.   As an illustration, assume $\alpha = 1$, $T_c = 1000°$ K, and $R_2 = 10$ ohms.   The noise generated by R_2 would then effectively cause an increase of about 30 ohms in the apparent value of the equivalent noise resistance of the tube.   This increase would probably not be noticed in a triode whose transconductance is less than 10,000 micromhos.

Anomalous Sources of Noise

There are several types of noise which may exist in a grid-control tube which do not fit into the same category as those noise sources which have been discussed up to this point.   With proper precautions the noise from these anomalous sources can be reduced to negligible proportions.   Included in this classification are noise due to

1. Gas ions.
2. Primary emission from the control grid.
3. Leakage conductances.
4. Charges on the walls of the glass envelope.

Each of these sources of noise is discussed briefly in this section.

A number of factors may cause free gas molecules to be present in a vacuum tube.   During the manufacture of a tube, adsorbed gases are driven off the glass envelope by baking, and occluded gases are removed from the metal parts by RF induction heating. It is not economically feasible or even possible to remove all the gas in this manner, so that, if the tube elements are overheated during the subsequent operation of the tube, there is a possibility that gas may be liberated at a rate that exceeds the capabilities of the getter. Insufficient baking or heating, insufficient evacuation, improper flashing of the getter, and other irregularities in the manufacturing process can also lead to the presence of free gas molecules inside the tube. With proper manufacturing techniques, a new tube will not contain enough free gas molecules to produce any appreciable amount of noise. If a tube has been operated above its ratings, however, the internal gas pressure may become high enough so that it becomes important to be able to estimate the magnitude of the gas ion noise.

As an electron is accelerated toward the plate of an electron tube, it may hit a free gas molecule with sufficient velocity to knock off another electron and ionize the molecule. Gas ions produced in this manner cause plate-current fluctuations in several ways. Some of the ions travel to the control grid where their charge is neutralized by electrons flowing onto the grid through the external circuit. The d-c value of this grid current provides a means for estimating the noise produced in this manner. The formation of ions and the flow of ions to the grid are both random processes. The grid current I_{c1} due to gas ions exhibits shot-noise fluctuations given by

$$\overline{i_{g'}^{2}} = 2eI_{c1}\,\Delta f \tag{4.70}$$

This fluctuating current flowing through the grid–cathode impedance produces a fluctuating grid voltage which is amplified by the tube and manifests itself as a noise current in the plate circuit. This component of noise is not of great importance except for very high grid-circuit impedances or for abnormally high grid currents.

Plate-current fluctuations are also produced directly by the electrons which are liberated by the random formation of ions in the grid–plate region. These electrons induce current pulses in the plate circuit as they move toward the plate and cause an increase in the plate noise. The ions that are formed by these collisions have an even greater effect on plate noise. When an ion is formed in the grid–plate region, it is accelerated toward the grid. Since the grid is made of fine wires, a large fraction of the ions so formed will pass through the grid and travel on toward the potential minimum, neutralizing part of the space charge in that region, and giving rise to an increase in plate current. Pulses of plate current produced in this manner are of relatively long duration, owing to the fact that the gas ions remain in the vicinity of the potential minimum for a relatively long time. A detailed analysis of this source of noise has been carried out by Thompson and North.[57] The results of their study provide a rough estimate of the plate-noise current due to gas ions by means of the following formula:

$$\overline{i_{p'}^{2}} \approx \frac{6.4 \times 10^{-16} I_b I_{c1}}{g_m} \Delta f \tag{4.71}$$

where $\overline{i_{p'}^{2}}$ is expressed in (amperes)^2 if I_b and I_{c1} are in amperes, g_m in mhos, and Δf in cycles per second.

It will be noticed that both Eqs. 4.70 and 4.71 express gas-noise components in terms of the direct ion current to the control grid I_{c1}. This current will thus provide a convenient means for estimating the amount of noise produced by free gas ions in the tube. Usually, if I_{c1} is less than 0.01 μa, the gas noise will be negligible with respect to other tube noises. This condition is not at all difficult to attain with modern manufacturing techniques.

During the operation of an electron tube, especially at high heater voltages, barium atoms may be driven off the cathode and deposited on the control-grid wires, lowering the work function of the grid. In some cases the work function is reduced enough so that the grid actually begins to emit electrons itself. This phenomenon is called primary emission by the control grid. The electrons emitted by the grid travel to the plate and constitute a small electron current flowing from control grid to plate inside the tube. Since the emission of these electrons is a random process, and since they travel through a region where there is no space-charge depression of the potential, these electrons should produce pure shot noise in the plate and grid circuits,

$$\overline{i_{pe}^{2}} = 2eI_{c1}\,\Delta f \tag{4.72}$$

where I_{c1} now refers to the d-c value of the primary-emission current. (This current is in the same direction as the grid current owing to gas ions and produces noise in the same proportion; cf. Eq. 4.70. It is frequently difficult to distinguish between the two.) This source of noise is not of great importance since it becomes appreciable only for grid currents of the order of 1 μa, at which point the primary emission begins to affect noticeably other characteristics of the tube.

Between any two electrodes inside an electron tube, there may often be considerable leakage conductance. This leakage may take place across the surface of the mica supports, or across the glass surface between stem pins. Barium atoms from the cathode deposited on the mica insulation between cathode and grid is one of the chief causes of grid–cathode leakage. Dirt inside the tube envelope can also be listed as one of the causes of high leakage conductance. An asbestos fiber or a piece of lint or dust lodged between grid and cathode, for example, can cause as much as one micromho or more of leakage conductance between these two elements.

The noise generated in a leakage conductance G should be given approximately by a formula of the form

$$\overline{i^2} = \alpha \cdot 4kTG\,\Delta f \qquad (4.73)$$

where T is an "average" temperature of the conductance, and α is a factor equal to or greater than unity which takes into account the possibility that the leakage conductance may generate excess noise over that of an ordinary resistor of the same value. This may happen, for example, when the contacts between the conducting particles which make up the conductance are not secure, and the sum of all these contact resistances may fluctuate with time because of the thermal motion of the molecules. Because of this factor and because it is usually not possible to determine an accurate value for T, the amount of noise produced by a given leakage conductance can be estimated only roughly from a measurement of the magnitude of G. With proper care in design and manufacture, it is possible to hold the leakage conductance low enough to make noise from this source negligible.

During the operation of an electron tube, it may happen that a few of the electrons strike the walls of the glass envelope, lodging there and building up a negative charge. As more electrons approach the spot of charge, some are repelled, and others are collected. Since the primary stream of electrons toward the glass fluctuates in time, and the number of electrons reflected by the spot also fluctuates, the result is a fluctuating stream of electrons bouncing off the glass wall and terminating on one or more of the electrodes. In addition, the size of the spot may vary in time, and the spot may drift slowly (perhaps under the influence of external fields), causing additional noise currents. If the magnitude of the charge on the spot is great enough, it may also influence the distribution of current between electrodes.

Another phenomenon which occurs occasionally in pentodes or tetrodes results in the formation of a positively charged spot or area on the glass envelope. This happens when a portion of the stream of electrons misses an electrode, such as the plate, and hits the glass envelope with enough energy to dislodge on the average more than one secondary electron per primary. A positive charge builds up which tends to pull electrons toward itself. In some cases a faint fluorescent spot (greenish blue) can be observed where the electron beam hits the glass. Usually the edges of the fluorescent area are sharply defined and bear a resemblance to the outlines of some of the tube elements. A positively charged spot can produce noise currents by affecting the distribution of current between electrodes and by the exchange of electrons between itself and other electrodes in ways similar to those described above for a negatively charged spot.

These effects have not been studied extensively, and we know of no simple means for estimating the noise to be expected from this source. It is generally assumed that varying wall charges produce a negligible amount of noise under normal conditions.

2. Network Calculations Involving Noise Sources

This section presents the basic principles involved in performing calculations on a network which contains one or more sources of noise. It is pointed out that any four-terminal network containing an arbitrary number of noise sources can be represented by an equivalent network containing only two noise sources. The noise figure of the network is defined and evaluated in terms of these two noise sources, and the effect of feedback on noise figure is discussed briefly.

The particular case of a grounded-cathode triode circuit is used to illustrate the generalizations that are made. A method of taking into account the correlation between grid noise and plate noise is presented, and the manner in which various parameters affect the noise figure is discussed. The section is concluded with a few remarks concerning the noise figures of cascaded networks and of wideband systems.

Circuit Representation of Noise Sources

The noise sources in a network can be represented by noise-current generators, noise-voltage generators, or combinations of the two. A noise-current generator is assumed to have zero internal admittance, and a noise-voltage generator has zero internal impedance. The noise currents and voltages can conveniently be designated by their mean-square values per unit bandwidth $\overline{i^2}$ and $\overline{e^2}$. Two independent noise-current sources, $\overline{i_1^2}$ and $\overline{i_2^2}$, connected in parallel produce a mean-square noise current $\overline{i^2} = \overline{i_1^2} + \overline{i_2^2}$. On the other hand, two completely correlated noise-current sources connected in parallel produce a mean-square noise current $\overline{i^2} = \overline{(i_1 + i_2)^2}$.

Based on the relations

$$\overline{i^2} = 4kTG_{eq}\,\Delta f \tag{4.74}$$

and

$$\overline{e^2} = 4kTR_{eq}\,\Delta f \qquad (4.75)$$

the terms equivalent noise conductance G_{eq} and equivalent noise resistance R_{eq} are often used to specify the magnitudes of the noise sources. (T is taken as the standard noise temperature, 293° K.) The equivalent diode current (I_{eq}) is often used in a similar manner, based on the relationship

$$\overline{i^2} = 2eI_{eq}\,\Delta f \qquad (4.76)$$

The use of such equivalent noise representations will often greatly simplify the final expressions for amplifier noise figures, signal-to-noise ratios, and other noise parameters.

It is shown in network theory that the small-signal electrical properties of any four-terminal linear network can be characterized in terms of admittance parameters by the following set of equations, referring to the sign conventions shown in Fig. 4.6.

$$I_1 = y_{11}V_1 + y_{12}V_2 \qquad (4.77)$$

$$I_2 = y_{21}V_1 + y_{22}V_2 \qquad (4.78)$$

The network may contain any number of passive or active elements. If $y_{12} = y_{21}$, the network is said to be reciprocal. If $y_{12} \neq y_{21}$, the network must contain at least one nonreciprocal element.

Fig. 4.6. Sign conventions for a four-terminal network.

For a linear four-terminal network, it can be shown that the noise characteristics of the network containing an arbitrary number of noise sources can be completely specified in terms of two noise generators,† as indicated in Fig. 4.7.

In general, $\overline{i_1^2}$ and $\overline{i_2^2}$ may be partially correlated. The magnitudes and degree of correlation of these noise generators as well as the admittances in Eqs. 4.77 and 4.78 can be determined by external measurements on the network.

The noise generator $\overline{i_2^2}$, for example, can be determined by short-

† The material that follows is based on a set of notes written in 1947 by J. A Morton and R. M. Ryder for a course in the Communications Development Training Program at the Bell Telephone Laboratories. The development of this point of view was greatly influenced by L. C. Peterson.[62]

circuiting the input terminals and adding noise to the output terminals from a standard source (such as a noise diode) until the available noise power from terminals 2–2 has doubled. The amount of noise added by the standard source is then equal to $\overline{i_2{}^2}$.

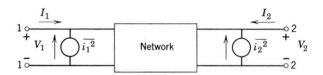

Fig. 4.7. The four-terminal network of Fig. 4.6 modified to take into account noise sources within the network.

For the representation of Fig. 4.7, the network equations become

$$I_1 + i_1 = y_{11}V_1 + y_{12}V_2 \tag{4.79}$$

$$I_2 + i_2 = y_{21}V_1 + y_{22}V_2 \tag{4.80}$$

referring to the Fourier coefficients† of I_1, i_1, V_1, etc.

Instead of dealing with the network of Fig. 4.7 with one noise generator at the input and one at the output, it is often more convenient to make use of a noise specification in which noise generators appear only at the input terminals. Such a representation is shown in Fig. 4.8, making use of one noise-current generator and one noise-voltage generator.‡ Two current generators or two voltage generators obviously would not suffice, since the noise at the output could then be reduced to zero by short-circuiting or open-circuiting the input.

† In calculating the response of a network to periodic signals, we are accustomed to using the Fourier components of currents and voltages. It is convenient to extend this technique and to speak of the Fourier coefficients of fluctuating noise currents and voltages. There is a mathematical difficulty involved in this, however, for the Fourier transforms of noise quantities do not always converge. The difficulty can be resolved by basing our calculations on the mean-square values of the noise quantities in infinitesimal frequency intervals. To be physically meaningful, these means must be averages taken over an ensemble of identical systems rather than time averages. For the mean-square noise currents and voltages which we use in circuit problems, the result of taking a time average over the squared Fourier component of a fluctuating quantity is the same as that obtained by taking an ensemble average in an infinitesimal frequency interval. Consequently it suffices for our purposes to use the familiar Fourier coefficients to represent noise quantities, remembering, however, that, in the strictest sense, our results have no physical meaning unless we take mean-square values. (See also Section 3, Chapter 2.)

‡ This type of equivalent network has also been discussed by Rothe[36,37] and by Rothe and Dahlke.[38]

Referring to the notation of Fig. 4.8, we can write the network equations for this circuit as

$$I_1 + i = y_{11}V'_1 + y_{12}V_2 \tag{4.81}$$

$$I_2 = y_{21}V'_1 + y_{22}V_2 \tag{4.82}$$

where $V'_1 = V_1 + v$.

For the network of Fig. 4.8 to be equivalent to that of Fig. 4.7, there must be a unique relationship between the noise generators i and v and the noise generators i_1 and i_2. The simultaneous solution of Eqs. 4.79 to 4.82 yields the following set of relations between the Fourier coefficients of i, i_1, i_2, and v:

$$i = i_1 - \frac{y_{11}}{y_{21}} i_2 \tag{4.83}$$

$$v = - \frac{i_2}{y_{21}} \tag{4.84}$$

It should be emphasized that there may be partial correlation between i and v as well as between i_1 and i_2. This correlation can be handled conveniently in the analysis of the network in the following manner. We separate the noise current i_1 into two parts

$$i_1 = i'_1 + i''_1 \tag{4.85}$$

such that i'_1 is completely correlated with i_2, and i''_1 is uncorrelated with i_2. Since i_2 is also completely correlated with v (see Eq. 4.84), we can define an admittance

$$y' = - \frac{i'_1}{v} \tag{4.86}$$

Substitution of Eqs. 4.84 to 4.86 into Eq. 4.83 then yields

$$i = i''_1 + (y_{11} - y')v \tag{4.87}$$

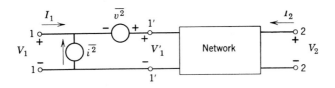

Fig. 4.8. Noise specification of a four-terminal network by means of two noise generators at the input terminals.

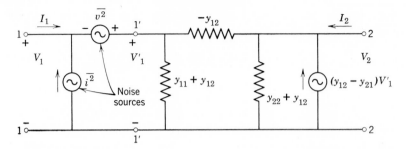

Fig. 4.9.   Equivalent circuit for a four-terminal network containing internal noise sources.

where i''_1 and v are uncorrelated.   The use of this noise representation is illustrated on page 199 in deriving the noise figure of Fig. 4.11.

It is convenient for the purposes of network analysis to substitute an equivalent circuit for the box in Figs. 4.6 to 4.8, choosing the parameters so that the sets of Eqs. 4.77 to 4.82 are satisfied.   Although there are a number of possible equivalent circuits, we shall discuss only the one shown in Fig. 4.9.   The current generator $(y_{12} - y_{21})V'_1$ at the output terminals is required to make the equivalent circuit valid when the network is nonreciprocal.   It will be noted that, if the noise generators are omitted and if $y_{12} = y_{21}$, the circuit reduces to the familiar π-equivalent of a passive reciprocal four-terminal network.

Rothe[37,38] has shown that the effect of the correlation between i and v can be simultated by using a modified form of equivalent circuit, such as that shown in Fig. 4.10.   The noise characteristics of this equivalent circuit are represented by the admittance y' defined by

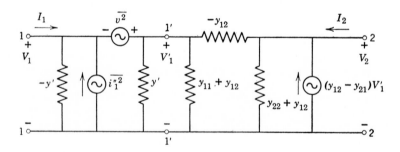

Fig. 4.10.   The equivalent circuit of Fig. 4.9 modified to simulate the effect of correlation between i and v.

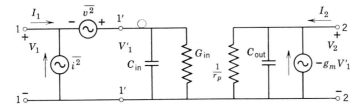

Fig. 4.11. Simplified equivalent circuit for a neutralized triode amplifier.

Eq. 4.86 and the two uncorrelated noise generators $\overline{i''_1}^2$ and $\overline{v^2}$. The results of noise calculations based on this circuit are identical with those obtained by using Fig. 4.9.

We shall illustrate the use of the equivalent circuit of Fig. 4.9 by applying it to the simple case of a single-stage triode amplifier. It is assumed that the triode grid–plate capacitance is neutralized so that $y_{12} = 0$. This also removes the capacitive term from y_{21} and makes $y_{21} = g_m$. Lead-inductance effects are neglected for the sake of simplicity. The equivalent circuit for this particular case is shown in Fig. 4.11, where the admittances y_{11} and y_{22} have each been shown as a parallel combination of capacitance and conductance.

If the input terminals are short-circuited so that $V'_1 = v$, the mean-square noise current appearing in a short circuit placed across the output is given by

$$\overline{(-I_2)^2} = \overline{i_p^2} = g_m^2 \cdot \overline{v^2}$$
$$= g_m^2 \cdot 4kTR_{\text{eq}} \, \Delta f$$

(4.88)

The quantity R_{eq} is the equivalent shot-noise resistance as defined in Section 1, page 161.

If the output terminals are short-circuited so that $V_2 = 0$, the mean-square noise current appearing in a short circuit placed across the input is†

$$\overline{(-I_1)^2} = \overline{(i - y_{11}v)^2}$$

(4.89)

† When the mean-square value of a quantity such as $(i - y_{11}v)$ is to be calculated, the degree of correlation between the two terms must be taken into account. As an example, consider $(a + b)^2$ where a and b are partially correlated. Let

$$a = a' + a''$$
$$b = b' + b''$$

where a' and b' are completely correlated, and a'' and b'' are uncorrelated. Then

$$\overline{(a + b)^2} = \overline{(a' + b')^2} + \overline{a''^2} + \overline{b''^2}$$

Substituting Eqs. 4.83 and 4.84, we obtain

$$\overline{(-I_1)^2} = \overline{(i_1 + y_{11}v - y_{11}v)^2}$$
$$= \overline{i_1^2} \tag{4.90}$$

The result is obvious since this is exactly the way that $\overline{i_1^2}$ was defined in connection with Fig. 4.7.

Referring to Fig. 4.11, we see that

$$y_{11} = Y_{in} = G_{in} + j\omega C_{in} \tag{4.91}$$

In this case Y_{in} can be split into two components: the admittance Y_c of the circuit connected between grid and cathode, and the input admittance Y_g presented by the tube. Correspondingly the noise current $\overline{i_1^2}$ can be written

$$\overline{i_1^2} = \overline{i_c^2} + \overline{i_g^2} \tag{4.92}$$

where $\overline{i_c^2}$ is the thermal noise generated in Y_c, and $\overline{i_g^2}$ is the induced grid noise of the tube. A portion of $\overline{i_g^2}$ is completely correlated with the plate noise.

The correlation between grid noise and plate noise can be taken into account by the following method. If we short-circuit the input and output, and let i'_g represent the component of i_g correlated with the short-circuit plate current i'_p, and i''_g the uncorrelated component, then[†]

$$\overline{i_g^2} = \overline{i'_g^2} + \overline{i''_g^2} \tag{4.93}$$

We let

$$y_{11} = Y_{in} = Y_c + Y_g = Y_c + Y'_g + Y''_g \tag{4.94}$$

where Y'_g is defined as follows

$$Y'_g = g_m \frac{i'_g}{i_p} = -\frac{i'_g}{v} \tag{4.95}$$

This procedure is justified because i'_g and v are completely correlated.

With the use of Eq. 4.83 and Eqs. 4.91 to 4.93, the noise-current generator $\overline{i^2}$ in the equivalent circuit of Fig. 4.11 then becomes

$$\overline{i^2} = \overline{\left(i_1 - \frac{Y_{in}}{g_m} i_p\right)^2}$$
$$= \overline{\left(i_c + i'_g + i''_g - \frac{Y_{in}}{g_m} i_p\right)^2} \tag{4.96}$$

[†] Comparing Eqs. 4.92 and 4.93 with Eq. 4.85, we note that in this example

$$\overline{i''_1^2} = \overline{i_c^2} + \overline{i''_g^2}$$

Since $i'_g = -Y'_g v$ and $i_p = -g_m v$,

$$\overline{i^2} = \overline{i_c{}^2} + \overline{i''_g{}^2} + \overline{v^2}\left|Y_{in} - Y'_g\right|^2$$
$$= \overline{i_c{}^2} + \overline{i''_g{}^2} + \overline{v^2}\left|Y_c + Y''_g\right|^2$$

(4.97)

where $\overline{i_c{}^2}$, $\overline{i''_g{}^2}$, and $\overline{v^2}$ are all independent noise sources. The correlated component of grid noise has been absorbed into the $\overline{v^2}$ term. The advantage of expressing $\overline{i^2}$ in this manner is brought out later in this section in the discussion on noise-figure calculations.

It is sometimes desirable to make use of a correlation coefficient γ to express the degree of correlation between grid noise and plate noise. In general, γ is a complex quantity because of the phase difference that may exist at any given frequency between the Fourier components of the two correlated noise currents. In the notation used above this correlation coefficient is defined as follows:

$$\gamma = \frac{\overline{i_g i_p{}^*}}{\sqrt{\overline{i_g{}^2} \cdot \overline{i_p{}^2}}}$$

(4.98)

where $i_p{}^*$ represents the complex conjugate of i_p. In terms of the correlation coefficient, the correlated component of the mean-square grid-noise current is given by

$$\overline{i'_g{}^2} = |\gamma|^2 \overline{i_g{}^2}$$

(4.99)

and its Fourier coefficient by

$$i'_g = \gamma \sqrt{\overline{i_g{}^2}}$$

(4.100)

Combining Eq. 4.100 with the definition of Y'_g given in Eq. 4.95, we obtain an expression relating the complex correlation coefficient to Y'_g:

$$\gamma = \frac{Y'_g}{g_m} \sqrt{\frac{\overline{i_p{}^2}}{\overline{i_g{}^2}}}$$

(4.101)

The admittance Y'_g is sometimes called the correlation admittance. Although the correlation coefficient has been defined so that it relates to the Fourier components of noise quantities, it is customary to express the degree of correlation between two noise sources as a percentage correlation between their mean-square values. For example, it was stated in Section 1, page 170, that, for a particular tube, the mean-square grid-noise current was found to be 30% correlated with the plate noise ($|\gamma|^2 = 0.30$). The correlation coefficient in this case would be

$$\gamma = \sqrt{0.30} = 0.55$$

Noise Figure†

It is important to have a quantitative measure by which to judge the performance of a system with regard to its noise characteristics. The term noise figure, as defined by Friis,[40] has been almost universally adopted for this purpose. As originally defined, the noise figure of a receiver or amplifier is given by

$$F = \frac{S_i/N_i}{S_o/N_o} \qquad (4.102)$$

where S_i/N_i is the signal-to-noise ratio at the input, and S_o/N_o is the signal-to-noise ratio at the output. Both signal and noise are measured in terms of available power, and the temperature of the source is assumed to be 290° K. The noise-figure ratio of Eq. 4.102 is frequently expressed in decibels. For an ideal device, which adds no noise to the signal, the noise figure is unity, which corresponds to 0 db.

There are several equivalent definitions‡ of noise figure which are often more convenient to use than Eq. 4.102 when noise-figure calculations are to be made. One of these equivalent definitions expresses the noise figure as the ratio of the output noise power of the actual device to the output noise power that would be obtained from an hypothetical ideal device, identical in all respects except that it contains no internal noise sources. The output noise in this case would simply be amplified thermal noise from the signal source connected to the input terminals. Thus

$$F = \frac{N_o}{N_s} = \frac{N_s + N_a}{N_s} = 1 + \frac{N_a}{N_s} \qquad (4.103)$$

where N_o is the noise output of the device, N_s is the output noise component which owes its origin to thermal noise in the signal source, and N_a is the component of output noise that is produced by noise sources within the device.

At this point we propose to restrict our attention to the noise figure of a device at a single frequency—the so-called spot-noise figure. This is done to avoid encumbering the discussion with the concepts of noise spectra and integrated bandwidth and to stick more closely to the

---

† The terms noise figure and noise factor have both been used in the literature. According to the definitions set forth in the IRE Standards on noise,[60] these terms are to be considered equivalent. The noise figure at a single frequency is called the "spot-noise figure," and the term "average noise figure" is used to designate the weighted average of the spot-noise figure over the bandwidth of the system.

‡ See, for example, van der Ziel.[7]

principles involved in simple noise-figure calculations. Most of the formulas can readily be extended to the wide-band case, but the mathematical complications that result tend to obscure the relationship between noise figure and the various noise parameters to be discussed. The subject of wide-band noise figures is treated briefly at the end of this section.

A second equivalent definition of noise figure can be formulated in terms of short-circuit noise currents. Consider the network of Fig. 4.12a where Y_s represents the admittance of the signal source connected to the input terminals, and $\overline{i_s^2}$ represents thermal noise generated in Y_s. The short-circuit output noise current $\overline{i_o^2}$ can be expressed as the sum of two independent components

$$\overline{i_o^2} = \overline{i'_o{}^2} + \overline{i''_o{}^2} \tag{4.104}$$

where $\overline{i'_o{}^2}$ represents the noise current $\overline{i_s^2}$ modified by the transfer characteristics of the network, and $\overline{i''_o{}^2}$ represents the component of

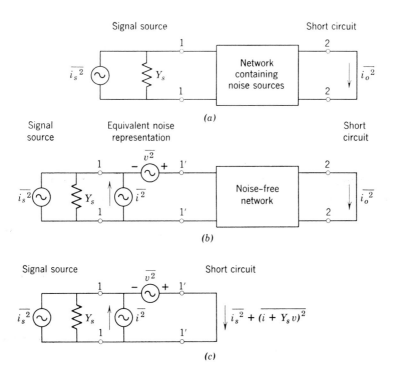

Fig. 4.12. Equivalent circuits for noise-figure calculations,

output noise that is produced by noise sources within the network. The noise figure is given by

$$F = \frac{\overline{i_o^2}}{\overline{i'_o^2}} = 1 + \frac{\overline{i''_o^2}}{\overline{i'_o^2}} \qquad (4.105)$$

The noise figure can be calculated in a similar manner in terms of open-circuit output voltages.

If all the noise sources within the network are referred to the input terminals as equivalent noise generators, the noise figure can be expressed in terms of these equivalent noise sources. One form of equivalent noise representation which is particularly useful in noise-figure calculations is shown in Fig. 4.12b, where one noise-current generator and one noise-voltage generator are used to represent the effects of all noise sources within the network. In calculating short-circuit noise currents and applying the method of Eq. 4.105 to this equivalent circuit, it makes no difference whether we calculate the ratio of short-circuit currents at terminals 1′–1′ or at terminals 2–2, since the network between these pairs of terminals contains no noise sources. Thus the noise figure can also be obtained by using the equivalent circuit shown in Fig. 4.12c. *Although the original network does not appear in Fig. 4.12c, the noise figure does depend on the network configuration since i and v depend on parameters of the network* (see Eqs. 4.83 and 4.84). From Fig. 4.12c we obtain the following expression for the noise figure:

$$F = 1 + \frac{\overline{(i + Y_s v)^2}}{\overline{i_s^2}} \qquad (4.106)$$

An alternate form of this expression can be obtained by combining it with Eq. 4.87, with the result

$$F = 1 + \frac{\overline{i''_1^2}}{\overline{i_s^2}} + \frac{\overline{v^2}}{\overline{i_s^2}} |Y_s + y_{11} - y'|^2 \qquad (4.107)$$

All the noise currents and voltages in this noise-figure expression are uncorrelated.

The single-stage triode amplifier of Fig. 4.11 will be used to illustrate the application of this expression for noise figure. We start with Eq. 4.97, which was derived in this section, page 195, for the equivalent noise-current generator

$$\overline{i^2} = \overline{i_c^2} + \overline{i''_g^2} + \overline{v^2} |Y_c + Y''_g|^2 \qquad (4.97)$$

In terms of Fourier coefficients this becomes

$$i = i_c + i''_g + v(Y_c + Y''_g) \tag{4.108}$$

so that Eq. 4.106 yields

$$F = 1 + \frac{\overline{[i_c + i''_g + v(Y_c + Y''_g) + Y_s v]^2}}{\overline{i_s{}^2}}$$

$$= 1 + \frac{\overline{i_c{}^2} + \overline{i''_g{}^2} + \overline{v^2}|Y_s + Y_c + Y''_g|^2}{\overline{i_s{}^2}} \tag{4.109}$$

Making the substitutions

$$\overline{i_s{}^2} = 4kTG_s\,\Delta f \tag{4.110}$$

$$\overline{i_c{}^2} = 4kTG_c\,\Delta f \tag{4.111}$$

$$\overline{i''_g{}^2} = 4kTG''_{eq}\,\Delta f \tag{4.112}$$

$$\overline{v^2} = 4kTR_{eq}\,\Delta f \tag{4.113}$$

the noise figure can be expressed as

$$F = 1 + \frac{G_c}{G_s} + \frac{G''_{eq}}{G_s} + \frac{R_{eq}}{G_s}|Y_s + Y_c + Y''_g|^2 \tag{4.114}$$

The quantity G''_{eq} is the equivalent noise conductance of the *uncorrelated component* of induced grid noise, and R_{eq} is the conventional equivalent shot-noise resistance.

The noise figure as given by Eq. 4.114 is not in a convenient form for our purposes; to remedy this we set

$$Y_s = G_s + jB_s \tag{4.115}$$

$$Y_c = G_c + jB_c \tag{4.116}$$

$$Y_g = Y'_g + Y''_g = G_g + jB_g \tag{4.117}$$

$$Y'_g = G'_g + jB'_g \tag{4.118}$$

$$Y''_g = G''_g + jB''_g \tag{4.119}$$

Our expression for noise figure then becomes

$$F = 1 + \frac{G_c}{G_s} + \frac{G''_{eq}}{G_s} + \frac{R_{eq}}{G_s} [(G_s + G_c + G''_g)^2 + (B_s + B_c + B''_g)^2]$$

$$= 1 + \frac{G_c}{G_s} + \frac{G''_{eq}}{G_s} + \frac{R_{eq}}{G_s} [(G_s + G_c + G_g - G'_g)^2$$
$$+ (B_s + B_c + B_g - B'_g)^2] \quad (4.120)$$

We next recall the work of Bell[22] who states that to a very good approximation† in triodes

$$Y'_g = j\omega C_e \quad (4.121)$$

where C_e is the space-charge component of input capacitance (see page 169, Section 1).‡ Remembering that

$$B_g = j\omega(C_0 + C_e) \quad (4.122)$$

where C_0 is the "cold" input capacitance of the tube, we obtain for our noise-figure expression

$$F = 1 + \frac{G_c}{G_s} + \frac{G''_{eq}}{G_s} + \frac{R_{eq}}{G_s} [(G_s + G_c + G_g)^2 + \omega^2(C_s + C_c + C_0)^2]$$
$$(4.123)$$

The terms C_s and C_c represent equivalent capacitances of net susceptances; the equivalent capacitance is negative if the inductive susceptance is larger than the capacitive susceptance. It will be noted that the correlated component of induced grid noise does not appear in this expression either explicitly or implicitly in terms of Y'_g.

The dependence of noise figure on input circuit tuning can be predicted by the use of Eq. 4.123. If the susceptance terms alone are allowed to vary, it is seen that the minimum noise figure is obtained

---

† In addition to Bell's work, van der Ziel[7,27] and Stahmann[24] have reported on some triode noise measurements which show that Eq. 4.121 holds very closely up to frequencies on the order of 100 Mc per sec.

‡ In a later work, Bell[39] pointed out the Y'_g should actually be considered a complex quantity rather than a pure imaginary as in Eq. (4.121), since it is known that at any given frequency the phase angle between the plate noise and the correlated component of grid noise is somewhat less than 90°. With this refinement it is necessary to replace G_g in Eq. 4.123 by $G_0 = G_g - G'_g$, where G_0 represents the "cold" input conductance of the tube. Usually G_s and G_c are so large in comparison with G_g that a negligible change in the calculated value of the noise figure results.

for the condition

$$C_s + C_c + C_0 = 0 \qquad (4.124)$$

This corresponds to the input circuit being tuned to resonate with the "cold" capacity C_0 of the tube. This condition is best obtained in practice by tuning the input circuit to resonance with all the elements of the amplifier at operating temperature but with the tube biased beyond "cut-off." (The input circuit will then exhibit a net capacitive susceptance ωC_e when the bias is subsequently changed to adjust the tube for normal operation.)

Tuning the input circuit to resonance with the tube biased for normal operating conditions corresponds to setting

$$C_s + C_c + C_g = 0 \qquad (4.125)$$

The susceptance term in Eq. 4.123 then yields

$$C_s + C_c + C_0 = -C_e \qquad (4.126)$$

and the minimum noise figure is not obtained. In order to achieve the minimum noise figure, it is then necessary to increase C_c by an amount equal to C_e, thus detuning the input circuit. This fact has led to the statement† that a lower noise figure can be obtained in a grounded-cathode stage if the input circuit is detuned slightly to present a net capacitive susceptance. As pointed out by Bell[39] and as shown above, this detuning is effected automatically by the tube itself if the input circuit is initially tuned to satisfy the condition 4.124. Typical values of C_e lie in the range of 1 to 5 $\mu\mu$f.

If a similar analysis is performed on a grounded-grid amplifier, it is found that the minimum noise figure for the grounded-grid connection is obtained with a net inductive susceptance $-\omega C_e$ at the input. This detuning is again provided automatically by the tube if the initial tuning is such as to satisfy Eq. 4.124.

The minimum noise figure that can be obtained by susceptance variation alone is given by

$$F_{\min} = 1 + \frac{G_c}{G_s} + \frac{G''_{\text{eq}}}{G_s} + \frac{R_{\text{eq}}}{G_s} (G_s + G_c + G_g)^2 \qquad (4.127)$$

If this equation is plotted as a function of G_s (see Fig. 4.13), it is found that there is an optimum value of source conductance, $G_{s,\text{opt}}$, for which the lowest noise figure is obtained. This value of source conductance can be calculated by differentiating Eq. 4.127 with respect

† See, for example, Wallman, Macnee, and Gadsden,[42] Strutt and van der Ziel,[41d] and Chapter 13 of reference 5.

to G_s, with the result

$$G^2{}_{s,\text{opt}} = \frac{G_c + G''{}_{\text{eq}}}{R_{\text{eq}}} + (G_c + G_g)^2 \tag{4.128}$$

In most cases $(G_c + G_g) \ll G_{s,\text{opt}}$, so that

$$G_{s,\text{opt}} \cong \sqrt{\frac{G_c + G''{}_{\text{eq}}}{R_{\text{eq}}}} \tag{4.129}$$

If we substitute this into Eq. 4.127, the optimum noise figure is found to be

$$F_{\text{opt}} = 1 + 2\sqrt{R_{\text{eq}}(G_c + G''{}_{\text{eq}})} \tag{4.130}$$

In many cases the Q of the input tuned circuit is high enough so that $G_c \ll G''{}_{\text{eq}}$, and the optimum noise figure is given approximately by

$$F_{\text{opt}} \cong 1 + 2\sqrt{R_{\text{eq}}G''{}_{\text{eq}}} \tag{4.131}$$

This equation is useful in predicting the qualitative effects which various changes in the design parameters of a triode will have on its performance in a low-noise amplifier.

A question often arises concerning the effect of feedback on the noise figure of an amplifier. It is relatively easy by means of feedback to reduce the noise figure of an amplifier as close to unity as desired, but the resulting device may have no practical use. For example, van der Ziel[7] has shown that placing a large capacitor between grid and plate of a grounded-cathode triode stage will reduce the noise figure of that stage. He also points out, however, that the stage gain is reduced so much by this capacitor that the over-all noise figure of a multistage amplifier employing this type of feedback in the first stage is poorer than could be obtained without the feedback capacitor. Thus we are

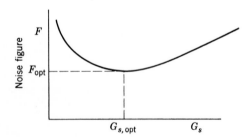

Fig. 4.13. Typical curve of noise figure as a function of source conductance.

cautioned that it is not sufficient to consider merely the effect that feedback will have on the noise figure of a single stage; it is necessary to investigate in addition the effect that the proposed feedback will have on the stage gain and on the over-all noise figure of the system in which the stage is used.

An indication of the manner by which feedback influences noise figure is provided by Eq. 4.100, which expresses the noise figure in terms of general four-terminal-network parameters. For this purpose it is more convenient to deal with the noise figure expressed in terms of the noise specification of Fig. 4.7, using one input and one output noise-current generator. Substituting Eqs. 4.83 and 4.84 into Eq. 4.106, we obtain

$$F = 1 + \frac{\overline{\left(i_1 - \dfrac{Y_s + y_{11}}{y_{21}} i_2 \right)^2}}{\overline{i_s^2}} \tag{4.132}$$

It will be noted that the feedback admittance y_{12} does not appear in this expression. We are thus led to the conclusion that, if feedback is to have any effect on noise figure, it must do so indirectly by means of its effect on the transfer admittance y_{21} and the input admittance y_{11}. In any application using only passive reciprocal elements in the feedback path, y_{21} is always affected by changes in y_{12}, and we learned from the example mentioned above that we need to examine the effects on gain and over-all noise figure before coming to any conclusions about the advisability of using this passive type of feedback. An example of the type of feedback that changes y_{11} as well as y_{21} is furnished by the effect of cathode-lead inductance.

For detailed discussions and for further examples of the relationship between feedback and noise figure, the reader's attention is called to a series of articles by Strutt and van der Ziel[41] and to a discussion by Schremp.[5]

In order to evaluate the noise performance of a proposed system, as opposed to a single-stage amplifier, it is necessary to be able to calculate the over-all noise figure of the system, composed of a cascade of amplifiers or transducers. Friis[40] has shown that the noise figure of a cascade is given by

$$F = F_1 + \frac{F_2 - 1}{G_1} + \frac{F_3 - 1}{G_1 G_2} + \cdots \tag{4.133}$$

where F_1, F_2, etc. are the noise figures of the first stage, second stage, etc., and G_1, G_2, etc. are the available power gains of the individual stages. (The available power gain is defined as the available signal power at the output terminals divided by the available signal power from the source. Available power from a pair of terminals is defined as that power which would flow into a conjugately matched load.)

It is evident from Eq. 4.133 that the gain G_1 should be as large as possible in order to minimize the contribution of second-stage noise to the over-all noise figure. In most cases the third term and all succeeding ones can be neglected. In some cases G_1 may be less than unity, and the term involving F_2 then becomes an important one. This happens, for example, when a crystal mixer is employed as the input stage of the system and a loss in signal power is encountered in the mixer.

Up to this point we have restricted the discussion of noise figures mainly to the subject of spot-noise figures. We shall conclude this section with a few remarks concerning the extension of these calculations to the concept of average noise figure.

The average noise figure of a device is defined as the weighted average of the spot-noise figure over the bandwidth of the system. In general, the spot-noise figure F and the available power gain G may both vary with frequency over the bandwidth of the system. The average noise figure is then given by[60]

$$\overline{F} = \frac{\int_0^\infty F(f) \cdot G(f)\, df}{\int_0^\infty G(f)\, df} \tag{4.134}$$

The calculation of average noise figure by means of this equation is straightforward in principle, although it usually proves to be rather tedious in practice.

In many instances (narrow-band amplifiers, for example) the spot-noise figure $F(f)$ is essentially constant over the frequency band of interest. In these cases the average noise figure is very nearly equal to the spot-noise figure. For some wide-band amplifiers, however, the average noise figure may exceed the spot-noise figure at the center of the band by an appreciable amount. This is especially true for amplifiers that employ a grounded-grid input stage.†

3. The Measurement of Tube Noise

Procedures for the measurement of shot noise, partition noise, induced grid noise, and noise figure are discussed in this section. Although not discussed here, measurements of secondary-emission

† See page 643, reference 5.

noise and cathode noise can be effected by using extensions of these procedures.

Before discussing the measurement procedures themselves, it would be well to mention two basic requirements which must be met by any measurement system that is expected to produce consistently good results. The first of these concerns the amplifier used between the noise source to be measured and the detector. This amplifier should be linear up to quite high input levels, for it must be able to handle peaks in the noise signal without distortion.

The second requirement involves the response of the output detector. A quadratic detector is the most convenient, one whose reading is proportional to the square of the output voltage or directly proportional to output power. Among the devices that fall into this category, one of the simplest commonly used is the thermocouple. Thermocouples are very sensitive to overloads, however, and are apt to burn out rather often if the operator is not extremely cautious. This is not a serious problem if the thermocouple is mounted in such a way as to permit easy replacement.

A silicon crystal rectifier operated at very low levels can also be used as a quadratic detector. If the direct (rectified) current is kept below 1 μa, the crystal performs very well as a square-law device. A galvanometer can be used to measure the rectified current, although readings are then tedious to take owing to the relatively long period of the moving coil. The use of a current amplifier such as a galvanometer–photocell–feedback amplifier combination† is preferable since its response is much more rapid than the galvanometer alone. It is possible to extend the square-law region of a crystal up to about 1 ma by adding the proper amount of resistance in series with the crystal and meter. The value of the resistance has to be determined for each crystal by trial, however, and the accuracy should be checked frequently.

The use of a thermistor in a bridge circuit‡ will also provide a convenient quadratic detector. Some trouble may be experienced with zero drift unless the thermistor is kept in a constant ambient temperature.

The suitability of the amplifier–detector combination for noise-power measurements can be checked by connecting a noise diode to the input of the amplifier. If the amplifier is linear and the detector is quadratic, a plot of the detector output reading as a function of noise diode plate current will be a straight line. The combination can be used if the

† See page 490 of reference 5.
‡ See page 54 of reference 7.

output–input plot is not a straight line, but the operation of such a system is far from convenient and may lead to inaccurate results.

If a calibrated attenuator is available, it can be used in conjunction with any detector (not necessarily quadratic) for taking noise measurements. The procedure requires adjusting the attenuator always to obtain the same detector reading and using the attenuator settings to determine the difference in power levels at the input to the attenuator.

In many instances the noise output of the device being measured may not be sufficient to override the noise generated in the following amplifier. If this is the case, the noise figure of the amplifier must be known and taken into account.

Shot-Noise Measurement

Two methods for measuring shot noise will be outlined here: (*a*) comparison with the noise from a temperature-limited diode, and (*b*) comparison with the noise from a known resistance in the input circuit.

(*a*) *Comparison with a Noise Diode.*[13] In this method a noise diode is connected in parallel with the output of the tube under test (the tube under test is assumed to be a triode here), and the combination is connected to the input of a high-gain linear amplifier. The input of the tube under test is short-circuited for radio frequency so that no amplified input noise can cause errors. A reading of the output power is taken with the tube under test operating alone. The diode is then turned on, and its filament current increased until the output power has doubled. The diode is then producing a mean-square noise current at the input equal to the shot noise of the tube under test. The unknown shot noise is thus given by

$$\overline{i_p{}^2} = 2eI_d\,\Delta f \tag{4.135}$$

where I_d is the temperature-limited diode plate current required to double the power output. In practice, a correction must be applied to Eq. 4.135 to take into account the thermal noise generated in the input circuit of the measuring amplifier (see Eq. 4.139).

Expressing the shot noise of the tube under test as

$$\overline{i_p{}^2} = \Gamma^2 \cdot 2eI_b\,\Delta f \tag{4.136}$$

we see that the space-charge-reduction factor can be calculated from

$$\Gamma^2 = \frac{I_d}{I_b} \tag{4.137}$$

where I_b is the plate current of the tube under test, and I_d is the noise diode plate current as in Eq. 4.135.

The tube under test is left on during the entire procedure so that the shunting effect of its plate resistance on the amplifier input does not change. The diode resistance is so high that it produces negligible damping of the amplifier input circuit. The measurement is usually performed with an amplifier whose pass band is in the region of 100 kc per sec to 1.0 Mc per sec. The bandwidth is not important, provided it does not extend outside this frequency range. A frequency much lower than 100 kc per sec might lead to errors due to cathode flicker noise, and too high a frequency might cause errors owing to induced grid noise and transit-time effects.

It is necessary to make corrections to Eqs. 4.135 and 4.137, owing to the thermal noise of the input circuit of the amplifier. With thermal noise present, doubling the output power by turning on the diode yields

$$2eI_d \, \Delta f = \overline{i_p^2} + \frac{4kT \, \Delta f}{R} \qquad (4.138)$$

where R is the resistance of the input circuit of the amplifier with the tube under test turned off, and T is the temperature of this resistance. Thus the corrected form of Eq. 4.135 becomes

$$\overline{i_p^2} = 2eI_d \, \Delta f - \frac{4KT \, \Delta F}{R} \qquad (4.139)$$

Modification of Eq. 4.137 to take this correction into account yields

$$\Gamma^2 = \frac{I_d}{I_b} - \frac{4kT}{2eI_bR} \qquad (4.140)$$

$$\simeq \frac{I_d}{I_b} - \frac{1}{20I_bR} \qquad (4.141)$$

(b) *Comparison with a Resistance in the Grid Circuit.*[59] The shot noise of a grid-control tube is often measured in terms of its equivalent noise resistance R_{eq}, defined in Section 1, page, 161. In this type of measurement the tube under test serves as the input stage of an amplifier whose pass band lies in the frequency range 100 kc per sec to 1.0 Mc per sec. A known resistance R is connected between grid and cathode of the tube to be tested, and provision is made for short-circuiting this resistance. Readings of the detector output taken with the resistor in the circuit and with the resistor short-circuited are used to determine the value of R_{eq}.

With the resistor R connected between grid and cathode of the tube to be tested, the detector output is given by

$$M_1 = K(4kTR\,\Delta f + 4kTR_{eq}\,\Delta f) \qquad (4.142)$$

where K is a constant of proportionality related to the gain of the amplifier. With the resistor short-circuited, the detector output becomes

$$M_2 = K(4kTR_{eq}\,\Delta f) \qquad (4.143)$$

The simultaneous solution of Eqs. 4.142 and 4.143 yields

$$R_{eq} = \frac{M_2}{M_1 - M_2}\,R \qquad (4.144)$$

A correction (similar to that in Eq. 4.139) may be necessary owing to noise contributed by the amplifier following the tube under test. This correction can be neglected if the detector output falls to a very low value when the tube under test is removed from its socket.

The Measurement of Partition Noise

Partition noise can be measured by an extension of either of the two methods already described in Section 3. In the normal operation of a tetrode or a pentode, the screen grid is by-passed to the cathode, short-circuiting the noise generator labeled $\overline{i_{g2}}^2$ in Fig. 4.1, page 155. The partition-noise generator is thus placed in parallel with the shot-noise generator $\overline{i_p}^2$. Since the two sources of noise are independent, their mean-square noise currents add directly. The resultant equivalent noise-current generator or the equivalent noise resistance (shot noise plus partition noise) can be determined by either of the two methods outlined above for the measurement of shot noise.

Occasionally it may be necessary to determine the magnitude of the partition-noise generator alone. To do this the equivalent noise resistance of the tube is first measured with the tube connected as a triode (screen grid by-passed or connected directly to the plate, depending on the desired operating voltages to be applied to the two electrodes). The tube is then connected normally with the screen grid by-passed to the cathode, and the equivalent noise resistance is again measured with the same value of electrode voltages and currents. The contribution of partition noise to the equivalent noise resistance can be determined from the difference between the two noise resistances (see Eq. 4.65, Section 1, page 180).

The diode comparison method can also be adapted to measure partition noise by a similar procedure, connecting the tube first as a triode and then as a pentode. The partition-noise-current generator in this case is given directly by

$$\overline{i_a{}^2} = 2e \, \Delta f \left[I_{d,\text{pent}} - \left(\frac{I_b}{I_k} \right) I_{d,\text{tri}} \right] \qquad (4.145)$$

The screen–grid shot–noise generator, $\overline{i_{g2}{}^2}$ in Fig. 4.1 can be determined simply by multiplying the shot-noise current $(\overline{i_p{}^2} + \overline{i_{g2}{}^2})$ flowing in the plate circuit with the tube connected as a triode by the ratio of screen current to cathode current. This noise generator is of minor concern, however, since the screen is usually by-passed to the cathode.

The Measurement of Induced Grid Noise

Three methods of measuring induced grid noise will be discussed here: (*a*) comparison with a noise diode, (*b*) calculations based on a curve of noise figure as a function of source conductance, and (*c*) calculations based on measurements of an equivalent noise conductance as a function of input circuit tuning.

(*a*) The most direct method of measuring induced grid noise is that employed by Bakker[20] in the first such measurements recorded in the literature. In this method the grid–cathode terminals of the tube under test are connected to the input of a narrow-band VHF amplifier (center frequency in the range 30 Mc per sec to 100 Mc per sec). A noise diode is also connected to the input terminals. The plate of the tube under test is by-passed to ground. Since this by-pass must be a very good RF short circuit, the use of a capacitor that is in series resonance with its leads and those of the tube is recommended. It is also desirable to neutralize the grid–plate capacitance of the tube under test to avoid any feedback to the grid lead.

The induced grid noise is measured by turning on first the tube under test and then (in addition) the noise diode and comparing the noise outputs for the two conditions. If the tube under test is not left on during the entire procedure, it is necessary to make a correction for the change in input impedance caused by the shunting effect of its input admittance. Loading due to cathode lead inductance is particularly troublesome in this respect. It is also necessary to make a correction for the amplifier noise in this measurement since it is by no means negligible.

(*b*) If the noise figure of a tuned VHF amplifier stage is measured†as a function of source conductance, the induced grid noise of the tube can be calculated from the noise-figure curve which is obtained.

---

† The measurement of noise figure is discussed in Section 3, pages 213 to 215.

Either a grounded-grid or a grounded-cathode circuit can be conveniently used for this purpose; the discussion to follow refers to the grounded-cathode connection.

Eq. 4.123 developed in this section, page 200, gives the noise figure of a grounded-cathode stage as

$$F = 1 + \frac{G_c}{G_s} + \frac{G''_{\mathrm{eq}}}{G_s} + \frac{R_{\mathrm{eq}}}{G_s}[(G_s + G_c + G_g)^2 + \omega^2(C_s + C_c + C_0)^2]$$

$$(4.123)$$

If the input circuit is tuned to resonance with the amplifier in operation (the condition described by Eqs. 4.125 and 4.126), the noise figure becomes

$$F_{\mathrm{tuned}} = 1 + \frac{G_c}{G_s} + \frac{G''_{\mathrm{eq}}}{G_s} + \frac{R_{\mathrm{eq}}}{G_s}[(G_s + G_c + G_g)^2 + \omega^2 C_e^2] \quad (4.146)$$

By making use of Eqs. 4.88, 4.95, and 4.121, it can be shown that

$$4kT\,\Delta f \cdot R_{\mathrm{eq}}\omega^2 C_e^{\,2} = \overline{i_p^{\,2}}\left(\frac{\omega C_e}{g_m}\right)^2 = \overline{i'_g^{\,2}} \quad (4.147)$$

The term $R_{\mathrm{eq}}\omega^2 C_e^{\,2}$ can be thought of as the equivalent noise conductance of the correlated component $\overline{i'_g^{\,2}}$ of induced grid noise. The equivalent noise conductance for the total induced grid noise is designated in Section 1, page 173, and Table 4.1 as G_{eq}. Thus

$$G_{\mathrm{eq}} = R_{\mathrm{eq}}\omega^2 C_e^{\,2} + G''_{\mathrm{eq}} \quad (4.148)$$

By substituting Eq. 4.148 into Eq. 4.146, the noise figure can be expressed as

$$F_{\mathrm{tuned}} = 1 + \frac{G_c}{G_s} + \frac{G_{\mathrm{eq}}}{G_s} + R_{\mathrm{eq}}G_s\left(1 + \frac{G_c + G_g}{G_s}\right)^2 \quad (4.149)$$

In many cases $(G_c + G_g) \ll G_s$, and

$$F_{\mathrm{tuned}} \cong 1 + \frac{G_{\mathrm{eq}}}{G_s} + R_{\mathrm{eq}}G_s \quad (4.150)$$

Summarizing the notation used in these expressions,

G_s = source conductance.
G_c = input circuit conductance.
G_g = input conductance of the tube.
G_{eq} = equivalent-noise conductance for the total induced grid noise.
R_{eq} = equivalent-noise resistance for the plate noise.

A typical curve which might be obtained for the measured values of noise figure as a function of source conductance is shown in Fig. 4.14. The slope of the asymptote to this curve for high values of G_s is equal to R_{eq}, as is evident by inspection of Eq. 4.150. With this value of R_{eq}, the approximate value of G_{eq} can be calculated from Eq. 4.150 and any given point on the noise figure curve. If the magnitudes of G_c and G_g are known from other measurements, or if they can be estimated, a more accurate determination of G_{eq} can be obtained using Eq. 4.149.

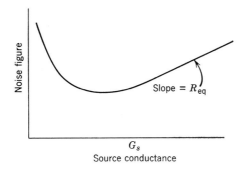

Fig. 4.14. Typical form of the curve obtained for noise figure as a function of source conductance.

There are two major disadvantages to this system of measuring grid noise. In the first place, it is difficult to construct a system in which the source conductance can be changed without slightly affecting the input tuning at the same time. Since noise figure also depends on input circuit tuning, and since it is usually inconvenient to check input resonance at each point, it is difficult to obtain a smooth experimental curve of noise figure as a function of source conductance. Second, the method yields values for the total grid noise only and does not permit separation into correlated and uncorrelated components. Both of these disadvantages can be obviated by means of the method to be described next.

(c) If a temperature-limited diode is placed across the input terminals of our grounded-cathode stage, and if G_s is made equal to zero, then by applying the methods of this section, page 188, it can be shown that the diode plate current required to double the noise output of the stage is given approximately by

$$I_d = \tfrac{1}{20}\{G_c + G''_{eq} + R_{eq}[G_c{}^2 + \omega^2(C_c + C_0)^2]\} \qquad (4.151)$$

It is convenient to convert this (by means of Eqs. 4.74 and 4.76) into an equivalent total noise conductance G_{tot}, which would produce the same noise current as the diode. Thus

$$G_{tot} = G_c + G''_{eq} + R_{eq}[G_c{}^2 + \omega^2(C_c + C_0)^2] \qquad (4.152)$$

By measuring G_{tot} as a function of the susceptance of the input circuit, we can obtain a measure of both the correlated and the uncorrelated components of induced grid noise. A typical plot of G_{tot} as a function of $(C_c + C_0)$ is shown in Fig. 4.15.

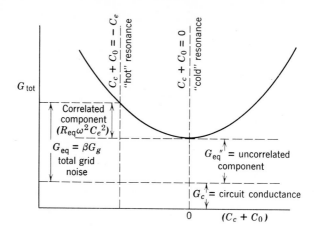

Fig. 4.15.   Typical form of the curve obtained for G_{tot} as a function of input circuit tuning. Measurements discussed in the text are summarized on the curve.

The condition $C_c + C_0 = 0$ corresponds to resonating the input circuit with the "cold" input capacity C_0 of the tube (tube biased beyond cut-off). Inspection of Eq. 4.152 indicates that

$$\text{Cold resonance } G_{tot} = G_c + G''_{eq} + R_{eq}G_c{}^2 \qquad (4.153)$$

The term $R_{eq}G_c{}^2$ is usually negligible so that the equivalent noise conductance G''_{eq} for the uncorrelated component of induced grid noise can be determined provided G_c is known.

Tuning the input circuit for resonance with the "hot" capacity of the tube (tube drawing normal current) corresponds to the condition

$$C_c + C_0 = -C_e$$

and

$$\text{Hot resonance } G_{tot} = G_c + G''_{eq} + R_{eq}(G_c{}^2 + \omega^2C_e{}^2) \qquad (4.154)$$

The term $R_{eq}G_c^2$ is again usually negligible so that the equivalent noise conductance for the total induced grid noise $G_{eq} = \omega^2 C_e^2 + G''_{eq}$ can easily be determined. It is also possible to determine R_{eq} from this latter point. These measurements are summarized in Fig. 4.15. Measurements obtained by the use of this method have been reported by van der Ziel[7,27] and Stahmann.[24]

The Measurement of Noise Figure

Of the various methods[5,7] of measuring noise figure, only two will be mentioned here: the use of a noise diode, and the use of a fluorescent-tube noise source. No attempt will be made to present a complete discussion of the use of these devices. The reader is referred elsewhere[5,7] for detailed descriptions of the procedures involved.

A saturated noise diode is a convenient device to use as an adjustable noise source for noise-figure measurements. The diode load resistance is made equal to the source resistance from which the amplifier is designed to operate, and the diode noise current flowing through the parallel combination of this load resistance and the amplifier input impedance provides an input noise voltage. A detector connected to the output of the amplifier is used to indicate relative changes in output noise power as the noise diode plate current is varied. Usually (but not necessarily) the diode plate current is adjusted to produce a two-to-one increase in output power when the diode is turned on. This can be accomplished conveniently by switching in 3 db of attenuation just ahead of the detector at the same time that the diode is turned on and adjusting the diode current to give the original detector reading. This procedure has the advantage of eliminating errors which might be caused by nonlinearity of the output power detector.

If the diode is connected directly across the input terminals, and if I_d represents the direct diode plate current required to double the power output of the system, then the noise figure is given by

$$F = 20 I_d R_s \qquad (4.155)$$

where R_s is the value of the resistance shunted across the diode to take the place of the source resistance from which the amplifier is intended to be operated. The temperature of R_s should be the standard room temperature, 290° K, or a correction must be applied. If the diode is connected to the amplifier by means of a coaxial cable, or if the lead between the diode noise generator and the input circuit has appreciable inductive reactance, a modified form of Eq. 4.155 must be employed.

The leads for supplying power to the diode must be very well filtered, for the signal level is so low that even a slight amount of regeneration can cause serious errors in the results. The diode filament supply should also be a very stable one, for the temperature-limited plate current of the diode is very sensitive to changes in filament power. The noise diode should have a filament with a very short time constant, made of pure tungsten or thoriated tungsten. The use of a diode developed specifically for noise-measurement work (such as the Sylvania 5722) is desirable.

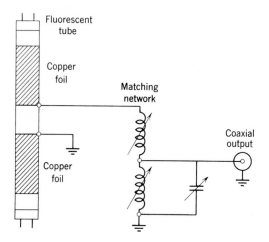

Fig. 4.16. A fluorescent-tube noise source for use in the VHF range.

The noise produced by the gas discharge in a fluorescent tube can also be used for the measurement of noise figure. As originally developed,[61] the fluorescent-tube noise source was intended for microwave noise measurements, although it has since been used successfully at lower frequencies. It was suggested by Hines† that coupling to the fluorescent tube by means of a coil wound directly on the outside of the glass envelope would afford a means of adapting the fluorescent-tube noise source for use in the VHF range (30–300 Mc). A variable capacitor is connected across the coil for tuning purposes. Coupling to the gas discharge can also be effected as shown in Fig. 4.16 by means of two copper foils wrapped around the outside of the glass envelope.‡ A coaxial cable coupled to the fluorescent tube by means of a transformer or a tapped coil provides a convenient output connection. The reactive elements in Fig. 4.16 are adjusted to obtain the proper

† M. E. Hines, Bell Telephone Laboratories, internal memorandum.
‡ B. C. Bellows, Bell Telephone Laboratories, internal memorandum.

impedance match between the fluorescent tube and the coaxial cable. The fluorescent-tube noise source should be well shielded and physically separated from the remainder of the measuring equipment.

The fluorescent tube produces a fixed level of output noise power which cannot be continuously varied as in the noise diode. This represents a slight disadvantage of the fluorescent tube compared with the noise diode. The maximum available noise power is higher for the fluorescent-tube source, however, so that higher values of noise figure can be measured.

The procedure involved in the determination of a noise figure by means of the fluorescent-tube noise source can be summarized as follows. The input of the amplifier to be tested is terminated with the desired value of source conductance and the output noise power noted. The source conductance is then replaced by the noise generator, the noise source being connected through an attenuator to the amplifier input,† and the attenuation is reduced until the output noise power of the amplifier has doubled. The noise figure is then given by

$$F_{\text{db}} = \left(\frac{T_n}{290} - 1 \right)_{\text{db}} - L_{\text{db}} \qquad (4.156)$$

where L_{db} is the amount of attenuation in decibels remaining in the attenuator connected between the noise generator and the amplifier input. For an ordinary fluorescent tube at normal room temperature, the quantity $(T_n/290 - 1)_{\text{db}}$ turns out to be approximately 15.85 db. The noise generator should be calibrated against a noise diode, however, in order to determine an accurate reference for the particular system employed. The first term in Eq. 4.156 also has a slight temperature coefficient which may have to be taken into account in some applications.

When the attenuator is adjustable only in steps, it will not always be possible to obtain exactly a two-to-one power ratio as assumed above. In terms of a general ratio Y between the two values of output power, Eq. 4.156 becomes

$$F_{\text{db}} = \left(\frac{T_n/290 - 1}{Y - 1} \right)_{\text{db}} - L_{\text{db}} \qquad (4.157)$$

If $Y = 2$, this equation reduces to Eq. 4.156.

† The conductance presented to the amplifier input terminals by the noise-generator–attenuator combination must be the same as the desired value of source conductance. It may be necessary to add a resistance in series with or in parallel with the amplifier input terminals to meet this condition. It is then necessary to apply a correction to the measured value of noise figure to take into account the effect of this added resistance.

REFERENCES

Background and Supplementary Material

1. A. H. W. Beck, *Thermionic Valves*, Cambridge University Press, Cambridge (1953).
2. S. Goldman, *Frequency Analysis, Modulation and Noise*, McGraw-Hill Book Co., New York (1948).
3. J. L. Lawson and G. E. Uhlenbeck, *Threshold Signals*, *MIT Rad. Lab. Ser.*, vol. 24, McGraw-Hill Book Co., New York (1950).
4. K. R. Spangenberg, *Vacuum Tubes*, McGraw-Hill Book Co., New York (1948).
5. G. E. Valley, Jr. and H. Wallman, eds., *Vacuum-Tube Amplifiers*, *MIT Rad. Lab. Ser.*, vol. 18, McGraw-Hill Book Co., New York (1948); E. J. Schremp, chap. 12, "Amplifier Sensitivity"; R. Q. Twiss and Y. Beers, chap. 13, "Minimal Noise Circuits"; Y. Beers, chap. 14, "Measurement of Noise Figure."
6. A. van der Ziel, "Fluctuation Phenomena," *Advances in Electronics*, vol. 4, Academic Press, New York (1952).
7. A. van der Ziel, *Noise*, Prentice-Hall, New York (1954).

Shot Noise

8. S. Ballentine "Schrot-Effect in High Frequency Circuits," *J. Franklin Inst.*, **206**, 159 (1928).
9. G. Diemer and K. S. Knol, "The Noise of Electronic Valves at Very High Frequencies," part I, "The Diode," *Philips Tech. Rev.*, **14**, 236 (1952).
10. G. E. Duvall, "The Effects of Transit Angle on Shot Noise in Vacuum Tubes," *MIT Research Lab. Electronics Tech. Rept. 82* (1948).
11. D. B. Fraser, "Noise Spectrum of Temperature-Limited Diodes," *Wireless Engr.*, **26**, 129 (1949).
12. R. Kompfner, J. Hatton, E. E. Schneider, and L. A. G. Dresel, "The Transmission Line Diode as Noise Source at Centimetre Wavelengths," *J. IEE*, **93**, part IIIA, 1436 (1946).
13. D. O. North, "Fluctuations in Space-Charge-Limited Currents at Moderately High Frequencies," part II, "Diodes and Negative Grid Triodes," *RCA Rev.* **4**, 441 (1940); **5**, 106 (1940).
14. A. J. Rack, "Effect of Space Charge and Transit Time on the Shot Noise in Diodes," *Bell System Tech. J.*, **17**, 592 (1938).
15. W. Schottky, "Uber Spontane Stromschwankungen in Verschiedenen Elektrizitätsleitern," *Ann. Physik*, **57**, 541 (1918).
16. W. Schottky and E. Spenke, "Die Raumladungsschwächung, des Schrotefektes," *Wiss. Veröffentl. Siemens-Werken*, **16**, 1 (1937).
17. E. Spenke, "Die Frequenzabhangigkeit der Schroteffektes," *Wiss. Veroffentl. Siemens-Werken*, **16**, 127 (1937).
18. B. J. Thompson, "Fluctuations in Space-Charge-Limited Currents at Moderately High Frequencies," part I, "General Survey," *RCA Rev.*, **4**, 269 (1940).
19. F. C. Williams, "Fluctuation Voltages in Diodes and in Multi-electrode Valves," *J. IEE*, **79**, 349 (1936).

Induced Grid Noise

20. C. J. Bakker, "Fluctuations and Electron Inertia," *Physica*, **8**, 23 (1941).
21. R. L. Bell, "Negative Grid Partition Noise," *Wireless Engr.* **25**, 294 (1948).

22. R. L. Bell, "Induced Grid Noise," *Wireless Engr.*, **27**, 86 (1950).
23. D. O. North and W. R. Ferris, "Fluctuations Induced in Vacuum Tube Grids at High Frequencies," *Proc. IRE*, **29**, 49 (1941).
24. J. R. Stahmann, "Correlation between Induced Grid Noise and Tube Noise," *Trans. IRE*, **ED-2**, 1 (1955).
25. T. E. Talpey and A. B. Macnee, "The Nature of the Uncorrelated Component of Induced Grid Noise," *Proc. IRE*, **43**, 449 (1955).
26. A. van der Ziel "Induced Grid Noise in Triodes," *Wireless Engr.*, **28**, 226 (1951).
27. A. van der Ziel, "Noise Suppression in Triode Amplifiers," *Can. J. Technol.*, **29**, 540 (1951).
28. A. van der Ziel and A. Versnel, "Induced Grid Noise and Total Emission Noise," *Philips Research Repts.*, **3**, 13 (1948).

Partition Noise

29. C. J. Bakker, "Current Distribution Fluctuations in Multi-Electrode Radio Valves," *Physica*, **5**, 581 (1938).
30. D. O. North, "Fluctuations in Space-Charge-Limited Currents at Moderately High Frequencies," part 3, "Multicollectors," *RCA Rev.*, **5**, 244 (1940).
31. W. Schottky, "Zur Theorie des Elektronenrauschens in Mehrgitterröhren," *Ann. Physik*, **32**, 195 (1938).

Secondary-Emission Noise

32. B. Kurrelmeyer and L. J. Hayner, "Shot Effect of Secondary Electrons from Nickel and Beryllium," *Phys. Rev.*, **52**, 952 (1937).
33. W. Shockley and J. R. Pierce, "A Theory of Noise for Electron Multipliers," *Proc IRE*, **26**, 321 (1938).
34. M. Ziegler, "Shot Effect of Secondary Emission," *Physica*, **3**, part I, 1 (1936); part II, 307 (1936).
35. V. K. Zworykin, G. A. Morton, and L. Malter, "The Secondary Emission Multiplier," *Proc. IRE*, **24**, 351 (1936).

Networks and Noise

36. H. Rothe, "Die Grenzempfindlichkeit von Verstärkerröhren," III, "Aquivalenter Rauschleitwert and Geräuschzahl," *Arch. Elektr. Übertragung*, **8**, 201 (1954).
37. H. Rothe, "The Theory of Noisy Four-Poles," *Trans. IRE*, **ED-1**, 258 (Dec. 1954).
38. H. Rothe and W. Dahlke, "Theorie Rauschender Vierpole," *Arch. Elektr. Übertragung*, **9**, 117 (1955).

Noise Figure

39. R. L. Bell, "Induced Grid Noise and Noise Factor," *Proc. IRE*, **39**, 1059 (1951).
40. H. T. Friis, "Noise Figures of Radio Receivers," *Proc. IRE*, **32**, 419 (1944).
41. M. J. O. Strutt and A. van der Ziel, (*a*) "Methoden Zur Kompensierung der Wirkungen Verschiedener Arten von Schroteffekt in Elektronenröhren und Angeschlossenen Stromkreisen," *Physica*, **8**, 1 (1941); (*b*) "Suppression of Spontaneous Fluctuations in Amplifiers and Receivers for Electrical Communication and for Measuring Devices," *Physica*, **9**, 513 (1942); (*c*) "Suppression of Spontaneous Fluctuations in 2n-Terminal Amplifiers and Net-

works," *Physica*, **9**, 528 (1942); (*d*) "Verringerung der Wirkung Spontaner Schwänkungen in Verstärkern fur Meter-und Dezimeterwellen," *Physica*, **9**, 1003 (1942); **10**, 823 (1943).

42. H. Wallman, A. B. Macnee, and C. P. Gadsden, "A Low-noise Amplifier," *Proc. IRE*, **36**, 700 (1948).

Miscellaneous

43. N. R. Campbell, "Discontinuous Phenomena," *Proc. Camb. Phil. Soc.*, **15**, 117 (1909); 310 (1910).

44. H. E. Farnsworth, "Energy Distribution of Secondary Electrons from Copper, Iron, Nickel and Silver," *Phys. Rev.*, **31**, 405 (1928).

45. C. K. Jen, "On the Induced Current and Energy Balance in Electronics," *Proc. IRE*, **29**, 345 (1941).

46. J. B. Johnson, "Thermal Agitation of Electricity in Conductors," *Phys. Rev.*, **32**, 97 (1928).

47. J. L. H. Jonker and A. J. W. M. v. Overbeek, "The Application of Secondary Emission in Amplifying Valves, *Wireless Engr.*, **15**, 150 (1938).

48. A. Lempicki, "Electron Multiplier Valve," *Wireless Engr.*, **30**, 312 (1953).

49. F. B. Llewellyn and L. C. Peterson, "Vacuum Tube Networks," *Proc. IRE*, **32**, 144 (1944).

50. H. Nyquist, "Thermal Agitation of Electric Charge in Conductors," *Phys. Rev.*, **32**, 110 (1928).

51. L. C. Peterson, "Space-Charge and Transit-Time Effects on Signal and Noise in Microwave Tetrodes," *Proc. IRE*, **35**, 1264 (1947).

52. J. R. Pierce, "Noise in Resistances and Electron Streams," *Bell System Tech. J.*, **27**, 158 (1948).

53. S. Ramo, "Currents Induced by Electron Motion," *Proc. IRE*, **27**, 584 (1939).

54. S. O. Rice, "Mathematical Analysis of Random Noise," *Bell System Tech. J.*, **23**, 282 (1944); **24**, 46 (1945).

55. H. Rothe, "Die Grenzempfindlichkeit von Verstärkerröhren, I: Theorie der Triode," *Arch. Elektr. Übertragung*, **6**, 461 (1952).

56. C. N. Smyth, "Total Emission Damping with Space-Charge-Limited Cathodes," *Nature*, **157**, 841 (1946).

57. B. J. Thompson and D. O. North, "Fluctuations in Space-Charge Limited Currents at Moderately High Frequencies," Part 4, "Fluctuations Caused by Collision Ionization," *RCA Rev.*, **5**, 371 (1941).

58. E. C. Titchmarsh; *Introduction to the Theory of Fourier Integrals*, Oxford University Press, New York, (1937).

59. A. van der Ziel, "Direct-Reading Instrument Measures Tube Noise," *Electronics*, **25**, 136 (1952).

60. "Standards on Electron Devices: Methods of Measuring Noise," *Proc. IRE*, **41**, 890 (1953).

61. W. W. Mumford, "A Broad-Band Microwave Noise Source," *Bell System Tech. J.*, **28**, 608 (1949).

62. L. C. Peterson, "The Performance and Measurement of Mixers in Terms of Linear-Network Theory," *Proc. IRE*, **33**, 458 (1945).

CHAPTER 5

Low-Noise Traveling-Wave Tubes

R. W. Peter

Introduction

At present, the designer of a low-noise microwave receiver may choose the input stage from among three major possibilities: triode amplifier, traveling-wave amplifier, or crystal-diode frequency converter. The noise factor of both RF amplifier types increases with frequency, the triode noise factor, however, at a faster rate. The best traveling-wave-tube noise factors at 3000 Mc are somewhat below 4 db, and below 8 db at 10,000 Mc per sec. The shallow noise-factor-versus-frequency curve of the traveling-wave tube is crossed by the steeper curve of the best triodes somewhere below 1000 Mc. The crystal-diode mixer and IF preamplifier combination, on the other hand, has an essentially frequency-independent noise factor of 7 to 8 db in the microwave range of 1 to 10 kMc. The range in which the traveling-wave tube is the leading low-noise amplifier extends, at present, from UHF to X-band frequencies. It is challenged by the triode from below and by the crystal from above as far as noise factor only is concerned. The frequency bandwidths over which these noise factors are available, however, vary by an order of magnitude in these three devices: 10 Mc for the crystal IF amplifier combination, 100 Mc for the triode, 1000 Mc for the traveling-wave tube.

Two new types of low-noise amplifiers, the "molecular" and the "parametric amplifier" are not considered here. They are both still in an early research stage and not yet available to the microwave-receiver designer.

From the large number of presently known beam-type amplifiers we have singled out the traveling-wave tube as the favored low-noise amplifier. This requires some explanation and justification as, according to the theory of Haus and Robinson,[1] all lossless beam-type tubes should have the same minimum noise factor (crossed-field devices are excluded here). A somewhat superficial answer to the question of why the traveling-wave tube and not, say, the klystron has been chosen may be given as follows: The effect of unavoidable circuit loss on the noise factor is smaller for the traveling-wave-tube circuit, and it is easier to produce the low-noise low-current beam needed in the traveling-wave amplifier than the larger current beam required in the klystron.

Transverse-field traveling-wave tubes[2] appear to have very low theoretical noise factors. So far, however, no satisfactory beam-collimating system has been devised, and the best noise factors obtained are much worse than those of the more conventional longitudinal traveling-wave tubes.

The backward-wave amplifier tube appears to have about the same practical limiting noise factors as the forward-wave amplifiers. They are, of course, electrically tunable and have narrow bandwidths at any one setting. The techniques for minimizing the noise in these tubes are similar to those described here for forward-wave amplifiers, although the specific parameters may be somewhat different.

Crossed-field amplifiers, like, for example, the magnetron amplifier, appear to have little chance of competing with devices that operate with a well-defined and strongly focused beam.

1. The Various Causes of Noise in Beam-Type Amplifiers

A brief summary will be presented of the various types of noise that are all present to some extent in a beam-type amplifier. A number of useful methods of reducing the various types of noise are discussed briefly. In Section 4, the mechanism of some of the causes of excess noise and its reduction are treated in more detail.

Five Basic Types of Noise

In every beam-type amplifier five basic types of noise exist and contribute in some degree to the total noise factor of the amplifier. In general, the different noise contributions are uncorrelated such that one can express the total excess noise factor of the amplifier, $F - 1$, as the sum of the individual excess-noise factors $(F_i - 1)$

$$F - 1 = \sum_{i=1}^{m} (F_i - 1) \tag{5.1}$$

The excess-noise factor $(F_i - 1)$ due to a particular type of noise generated in the amplifier and leading to a noise-power output N_i may be defined as

$$F_i - 1 = \frac{N_i/G}{kT \, \Delta f} \tag{5.2}$$

The noise output N_i divided by the power gain of the amplifier is thus referred to the input, and then compared to the available thermal noise power at the input terminals, $kT \, \Delta f$.

Electron-Emission Noise. Electron-emission noise is excited in the beam by the random fluctuations of the electron current and velocity at the emitting cathode. If the beam is finite in its transverse dimension, higher-order radial and circumferential space-charge-wave modes will be excited as the emission is random not only in time but also in space, i.e., across the cathode surface. The mean-square value of the fluctuations at the cathode, $\overline{i_a^2}$, increases with the direct current I_0. The mean-square value of the velocity fluctuation decreases with increasing direct current and is proportional to the cathode temperature; it may be expressed by an equivalent mean-square fluctuation voltage $\overline{V_a^2}$. The product of the two noise excitations $|i_a| \, |V_a|$ is an invariant along an ideal beam and is, therefore, characteristic of the noise in the beam.[3] This product has the dimension of power, and will be called the "noisiness" of the beam. The noise factor of any amplifier is found to be proportional to the beam noisiness and, therefore, under otherwise optimized conditions, is directly proportional to the cathode temperature (see Section 2).

Excess Noise Due to Emission Defects. The noise excitation due to the randomness of current and initial velocity may be increased if the emission-current density and initial velocity across the cathode area are nonuniform. Such nonuniformity may be caused by inhomogeneities of the work function, the surface potential, the surface temperature, the conductivity through the emitting layer to the base, and the physical surface condition (pores, etc.).

Nonuniformity of emission velocity appears to be of primary importance. Adjacent currents of different initial velocity behave like slipping streams if an infinitely strong focusing field holds them apart. Under practical conditions, however, a certain amount of random mixing of electrons from different cathode areas will take place. The mixed current therefore has a broader velocity spectrum, which means a higher effective temperature (see Section 4, page 284).

Excess Noise Due to Beam-Transmission Defects. In this category we include every type of additional noise that may result from defects in the beam or noise-wave transmission. There exists a large number of such transmission defects, which may be grouped in two major categories: those that tend to restore full randomness and those that tend to increase (amplify) noise over its random value.

The group of defects that tend to restore full randomness in a partially ordered beam includes:

(*a*) Nonuniformity of the d-c velocity across the beam due to space-charge depression of the potential.[4]

(*b*) Transverse acceleration of electrons in the beam as a result of lens effects (Section 4, page 278).

(*c*) Nonuniform current density across the beam.

(*d*) Electron interception on gun electrodes or interacting circuit (partition noise), and collisions with residual gas molecules (Section 4, page 271).

The second group of transmission defects includes the mechanisms that tend to amplify any perturbation on the beam beyond complete randomness. Of such "single-beam amplification" mechanisms there are several:

(*e*) Scalloping-beam and rippled-wall amplification.[5,6]

(*f*) Slipping-stream amplification.[7,8]

(*g*) Beam instability in axial magnetic focusing field,[9] and growing-noise mechanisms of other types (Section 4, page 283).

Secondary-Emission Noise. Additional noise may be caused by secondary electrons liberated by primary electrons from intercepting electrodes and, particularly, from the collector. This secondary noise current is random and may have a noise temperature of about two orders of magnitude above the temperature of the primary beam current. Secondary electrons may travel in synchronism with the backward fundamental or some higher space-harmonic modes of the interacting circuit and thereby induce noise.

"Secondary" electrons might conceivably be liberated also by soft X rays, photons, etc. Such noise contributions, however, are negligibly small except under special operating conditions.

Ion-Oscillation Noise. Ion-oscillation noise is caused by plasma oscillations of ions in the electron beam.[10-12] As the ion-plasma frequency usually is about a hundred times smaller than the operating frequency in a microwave amplifier, no direct RF noise is generated by this mechanism. If the ion oscillations are strong enough to modulate the gain of the traveling-wave tube, noisy modulation sidebands of the input signal will be generated which affect the transmission of information in much the same way as the other types of RF noise in the operating frequency band.

Possibilities for Noise Reduction

We shall now briefly discuss possibilities for reducing or eliminating these five types of noise in a beam-type amplifier. All these factors have to be considered in the design of a very-low-noise traveling-wave tube and therefore will be discussed in greater detail later.

1. The *electron-emission noise* can be reduced below the value corresponding to the cathode temperature by reducing either the current or velocity fluctuations. This may be achieved by velocity selection, velocity correction, kinetic-energy extraction, or current smoothing.

"Velocity selection" implies disposing of all but the few electrons in a particular velocity class. Transverse electric or magnetic fields can be used to select axial electron velocities, beam collimation to select transverse velocities. Thus far, experiments have been made only on the collimator used in transverse-field traveling-wave tubes.[2]

"Velocity correction" may be achieved by a servo system which increases the velocity of slow electrons and decreases the velocity of fast electrons in order to produce a monovelocity stream at a particular frequency. Such a scheme has been proposed by R. Kompfner (Clarendon Laboratory, Oxford, 1951). Kinetic-energy extraction might be obtained, for example, by random inelastic or elastic collisions of the electrons with a cool neutral gas. The electron-velocity spread, as a result, can be reduced to a lower value, theoretically to a value corresponding to the cooling-gas temperature. This approach is being investigated at RCA Laboratories.

"Current smoothing" has been provided in a simple form by nature in space-charge-limited emission. This mechanism is discussed in Chapter 1.

2. *Excess noise due to emission defects* should not exist in a beam emitted from a metal cathode made of a properly oriented, single crystal. However, if such a cathode could be constructed from tungsten or thorium, for example, it would not be useful in a practical device owing to its high operating temperature.

In low-temperature oxide and matrix-type cathodes, inhomogeneities play a decisive role in determining the final noise factor of the amplifier.

Unfortunately, a large number of factors seem to be involved which are little understood in the production of good low-noise cathodes. Experience, however, supports the theory that smooth, more uniformly emitting cathodes are better than the fluffy and porous cathodes commonly preferred in radio tubes for their large emission.

3. *Excess noise due to beam-transmission defects* of both types can be avoided almost completely by careful gun and amplifier design. Randomizing defects are essentially eliminated by working with relatively low space-charge densities in the beam, by avoiding abrupt beam-velocity discontinuities (as, for example, "velocity-jump" regions), by eliminating any current interception on electrodes and circuit, and by operating the tube under hard vacuum.

Noise-amplifying effects may be avoided by using a well-aligned and well-focused beam, with small ripple of short period. This may be achieved by extending the magnetic focusing field (if such is used) through the cathode. If a hollow beam is used, it should be operated at a low enough perveance to assure stable flow.

4. *Secondary current* can be prevented from leaving the collector by applying a transverse magnetic field through the collector region or by internally coating the collector with a material of low secondary-emission coefficient. If no part of the circuit or collector is heated appreciably during operation, thermal secondary emission can be neglected.

5. *Modulation noise due to plasma oscillations* does not occur if the ion density is small enough. The formation of an ion trap has to be avoided. It is preferable to drain the ions formed in the circuit region forward to the collector rather than backward toward the cathode to avoid possible damage of the emitting surface. This can be achieved by raising the last gun electrode sufficiently above helix potential while lowering the collector potential slightly below helix potential.

2. Noise Theory of the Traveling-Wave Tube

The theory of noise in the traveling-wave tube has made great advances since the first computations of Kompfner[13,14] and Pierce.[15] The various theories proposed[1,16−30] consider an idealized one-dimensional model of circuit and interacting beam. They all predict essentially the same *lowest* noise factor for a traveling-wave tube. The generalized theories of Haus and Robinson[1,23,28] show that every lossless beam-type microwave amplifier has the same optimum noise factor under these simplifying conditions.

In reality, however, the beam has a finite cross section and propagates an infinite number of modes, not one mode only. The higher-order modes interact less strongly with the circuit. Nevertheless, Beam, Bloom, and Paschke's studies of the multimode interaction in a traveling-wave tube indicate that the higher-order modes can have an important effect on the noise factor.

In this section, first a summary of the results of the one-dimensional theory will be presented to the extent required for practical applications. (For details the reader is referred to Chapter 3.) Then a discussion follows of some aspects of noise-current smoothing in the potential-minimum region near the cathode. This is still a controversial subject, on which theory and experiment have not yet met. Finally, the problem of the effect on noise of higher-order space-charge-wave modes is introduced and briefly discussed.

One-Dimensional Theory of Traveling-Wave-Tube Noise Factor

The minimum noise factor of the traveling-wave tube has been obtained independently by several authors (Robinson,[17] Bloom and Peter,[20] Pierce and Danielson[21]), under the assumption that the noise at the potential minimum is given by uncorrelated shot-noise current and Rack velocity fluctuations. Further, the assumption was made that the gain of the traveling-wave tube is large.

Then, the optimum noise-figure expression of the traveling-wave tube with a lossless helix becomes (see Chapter 3, Eq. 3.194)

$$F_{\min} = 1 + \sqrt{4 - \pi}\,\frac{T_c}{T} \tag{5.3}$$

If the helix of the tube is lossy, the minimum noise figure is higher. In the notation of Chapter 3, Section 4, the noise figure is then

$$F_{\min} = 1 + \sqrt{4 - \pi}\,|D|\,\frac{T_c}{T} \tag{5.4}$$

where the factor D for a traveling-wave tube can be found from Chapter 3, Eqs. 3.183 and 3.185. For a traveling-wave tube with the beam voltage adjusted for maximum gain, the factor simplifies to[1,25,30]

$$|D| = 1 + \frac{d}{x_1}$$

where d is Pierce's loss parameter, and x_1 is his gain parameter. Since x_1 is a solution of a cubic involving b, the velocity parameter; QC, the space-charge parameter; and d, the loss parameter; we conclude that x_1 is, in general, a function of the three parameters b, QC, and d. But, with the beam velocity adjusted for maximum gain, b becomes a dependent variable, and x_1 is a function of QC and d alone. Figure 5.1

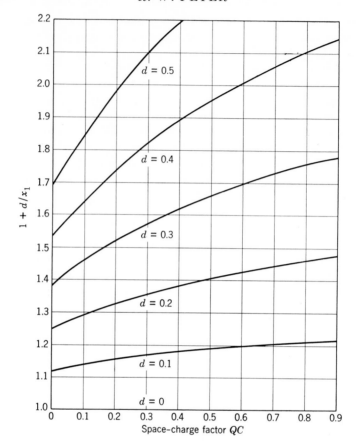

Fig. 5.1. The factor $1 + d/x_1$ that determines the minimum noise factor of a traveling-wave tube (Eq. 5.14) as a function of space-charge factor QC and circuit-loss factor d.[25]

(from reference 25) gives a plot of the function $1 + d/x_1$ as a function of QC and d.

If there exists any correlation between velocity and current fluctuations at the cathode, the noise figure, Eq. 5.4, will be modified by another factor (see Chapter 3, p. 134, and also references 25 and 26).

The standing-wave ratio ρ of noise current in the beam preceding the traveling-wave-tube circuit needed for optimum noise figure can be evaluated using for example Eq. 3.184 of Chapter 3, Section 4. We have

$$\rho^2 = \frac{\Psi_{\text{max}}}{\Psi_{\text{min}}} \tag{5.5}$$

With Eqs. 3.185 introduced into 3.184 we can evaluate ρ^2 in terms of QC and d, provided the beam velocity is optimized for maximum gain. Similarly, we find the angle ψ giving the distance of the current standing-wave minimum from the helix entrance using Eq. 3.179, Chapter 3, where the symbol ϕ_0 is used for ψ. The angle ψ, like ρ, is a function of QC and d only. Closer investigation shows that ρ and ψ are functions of one single parameter[25]

$$\bar{y} = -\frac{y_1}{\sqrt{x_1^2 + y_1^2}} \tag{5.6}$$

where x_1 and y_1 are Pierce's incremental propagation constants of the growing wave. In Figure 5.2 curve $n = 1$ gives the standing-wave ratio ρ and angle ψ required for minimum noise figure as functions of $\bar{y}$. Figure 5.3 gives the parameter $\bar{y}$ as a function of QC and d.

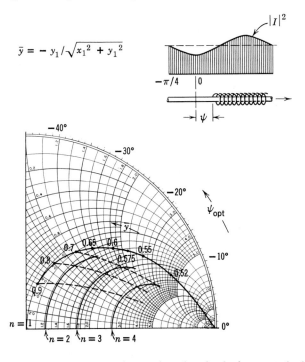

Fig. 5.2. Noise standing-wave ratio ρ and angle ψ in the beam at the helix input required for optimum noise figure as a function of parameter $\bar{y}$ of Fig. 5.3.

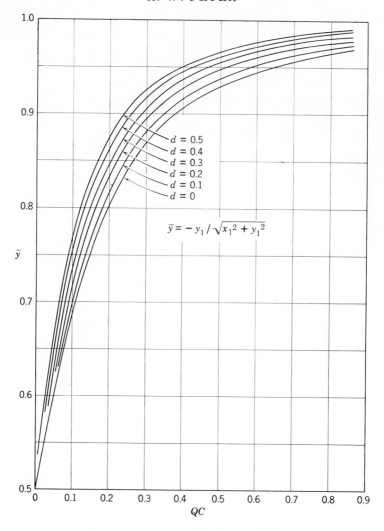

$$\bar{y} = -y_1 / \sqrt{x_1^2 + y_1^2}$$

Fig. 5.3. Parameter $\bar{y}$ versus space-charge factor QC for various circuit-loss factors d.

The extension of the theory for noise optimization to include the effect of noise carried by higher-order beam modes has not yet been accomplished in a rigorous manner because there exists no exact theory for a traveling-wave tube with a thick beam. However, we can get an idea of what happens by applying the above single-mode analysis to each of the higher-order modes separately. We assume that the traveling-wave structure couples only to the noise in the nth mode of the beam. Under this condition there exist optimum ρ_n and ψ_n which lead to a minimum noise figure. Figure 5.2 gives the respective standing-wave ratio ρ_n and angle ψ_n for the modes $n = 1$, 2, 3, 4. These curves were computed under the assumptions that only a single growing wave exists in the traveling-wave tube, and that the values for QC and for the velocity parameter b are the same for all modes. The coupling (Pierce's C) changes from mode to mode because of changes in the field structure.

Noise-Current Smoothing at the Potential Minimum

In our foregoing noise-factor computations, it was assumed that the space-charge noise waves in the beam are excited at the cathode by completely random current i_a and velocity fluctuations V_a. This assumption represents the initial conditions adequately if the cathode operates temperature-limited. If, however, the cathode operates space-charge-limited, a fraction of the emitted electrons returns to the cathode, and a potential minimum is formed in front of the emitter.

From the theories of North[31] or Rack (reference 16, Chapter 1) for the diode noise at lower frequencies, it is well known that the potential minimum acts in a compensating fashion upon the fluctuations of the total current through a diode. It may be surmised that the minimum has still some smoothing effect upon the convection current i_a at higher frequencies where the transit angle between the cathode and anode is very large (see Robinson's early study[17]). An exact treatment of this problem, however, does not exist yet. Several attempts have been made which are based upon simplified models of the rather complex situation (see Chapter 1). Whinnery[22] computed the compensating current produced by a single initial excess-current pulse under simplifying assumptions; Siegman and Watkins[29] improved this theory but assumed similar simplifying conditions. Previously Watkins[24] had suggested a simple circuit to estimate the behavior of an open-circuit space-charge-limited diode. Bloom[32] subsequently proposed an improved circuit analog for the space-charge-limited diode which will be discussed here briefly. Tien[27] has performed a numerical computation for a particular short-circuit diode situation.

None of these treatments represents the actual case adequately. A qualitative idea of the solution, however, can be obtained from any of these studies. In the following, Bloom's circuit analog shall be outlined.

The simplifying assumptions made by Bloom are the following: (1) The beam is infinitely extended in directions transverse to its flow (one-dimensional theory); (2) the total current is equal to zero everywhere in the beam, which means that the RF-convection current equals the capacitive or displacement current; (3) the beam behaves linearly, and its fluctuations are small with respect to the d-c quantities.

Bloom's circuit analog is shown in Fig. 5.4. It replaces the open-circuit retarding-field diode region between cathode and potential minimum by a parallel resonant circuit. The upper, capacitive branch carries the displacement current i_c, which is equal in magnitude to the convection current to be computed, according to assumption 2. The second branch contains, in series with the d-c conductivity of the retarding-field diode G, an inductive reactance L which resonates with the capacity C at the plasma frequency at the potential minimum ω_{pm}. The introduction of this inductance distinguishes Bloom's circuit from

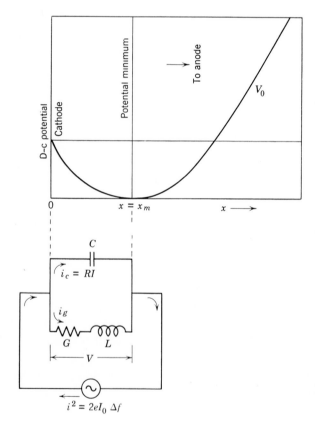

Fig. 5.4.   A circuit analog for action of potential minimum in front of a space-charge-limited cathode.

Watkins'.[24]   The reactance ωL represents the reactive or inertial effect of the space charge at the minimum.

The open-circuit voltage across the diode, V, can be obtained from the short-circuit current, $i^2 = 2eI_0 \Delta f$ (e and I_0 are positive quantities throughout this chapter), and the open-circuit admittance Y of the equivalent diode,

$$Y = j\omega C + \frac{1}{1/G + j\omega L} \qquad (5.7)$$

where $G = eI_0/kT_c$ and $C = \epsilon_0 A/d$, A is the cross-sectional area of the cathode, and d is the cathode-potential minimum distance.   The convection current i_c then becomes

$$i_c = j\omega C V = j\omega C \frac{i}{Y} \qquad (5.8)$$

The absolute value of the ratio between the convection current i_c and the full-shot-noise, short-circuit current ($i = i_c + i_g$) will be called the "noise-current-smoothing" factor as defined by Watkins:

$$R = \left| \frac{i_c}{i_c + i_g} \right| = \left| \frac{j\omega C}{Y} \right| \qquad (5.9)$$

Introducing Eq. 5.7 into Eq. 5.9 leads to

$$R = \left| \frac{j\omega C(1 + j\omega GL)}{G - G\omega^2 LC + j\omega C} \right| \qquad (5.10)$$

The absolute value of the current-smoothing factor (squared) is then

$$R^2 = \frac{(\omega\tau_0)^2 + (\omega/\omega_{\mathrm{pm}})^4}{(\omega\tau_0)^2 + [1 - (\omega/\omega_{\mathrm{pm}})^2]^2} \qquad (5.11)$$

where $\tau_0\text{-}C/G$ is the relaxation time of the diode and $\omega_{\mathrm{pm}}^2 = 1/LC$ represents the plasma frequency at the potential minimum.

According to Fry and Langmuir,[31] the following relation holds for the plane-parallel diode

$$\omega_{\mathrm{pm}}\tau_0 = \sqrt{\pi}/\xi \qquad (5.12)$$

where ξ signifies the ratio of saturation-to-diode current densities J_s/J_0, in parameter form.   With Eq. 5.12, we can write Eq. 5.11 as follows:

$$R^2 = \frac{\pi/\xi^2 + (\omega/\omega_{\mathrm{pm}})^2}{\pi/\xi^2 + (\omega_{\mathrm{pm}}/\omega - \omega/\omega_{\mathrm{pm}})^2} \qquad (5.13)$$

This expression for the current-smoothing factor R^2 is plotted as the heavy curve in Fig. 5.5 for a ratio $J_s/J_0 = 5$ ($\xi = 1.884$) and a plasma frequency at the minimum of $\omega_{pm} = 2\pi \times 3.78 \times 10^9$. These particular values are chosen to compare the result with Tien's[27] curve. A qualitative agreement seems to exist between the two curves: Both show a maximum R value slightly above $\omega/\omega_{pm} = 1$; both become smaller than unity at lower frequencies. The extreme "dip" of Tien's curve and its oscillatory behavior, however, are absent in Bloom's curve. Siegman and Watkins'[29] theory gives a curve similar to Bloom's. Its peak is somewhat toward higher relative frequencies.

Whereas the action of the potential minimum leads to a reduction of the current fluctuations, we may assume that the velocity fluctuations are unaffected and remain at the equivalent Rack value. Arguments in favor of this assumption have been presented by Watkins.[24] If no correlation is introduced between the velocity and current fluctuations, the expression for the minimum obtainable noise figure, Eq. 5.4, becomes

$$F = 1 + R \sqrt{(4 - \pi)} \, \frac{T_c}{T} \left(1 + \frac{d}{x_1} \right) \qquad (5.14)$$

There is a general design conclusion to be drawn from these results: The beam-current density J_0 which determines ω_{pm} should be chosen

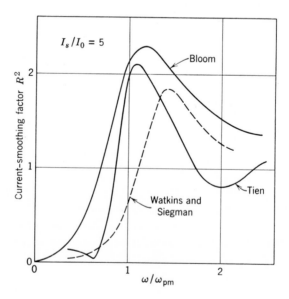

Fig. 5.5. Various predictions for the current-smoothing factor R^2 as a function of normalized frequency.

so as to avoid the peak of R. High current densities may lead to current-smoothing factors considerably below unity.

Finally a physical justification for Bloom's analog circuit will be attempted. The action of the potential minimum at microwave frequencies may be pictured roughly as follows: Assume an excess charge q (per unit area) leaving the cathode. If the anode is very far away, this charge produces a field $E_1 = q/\epsilon$ behind it, but no field ahead of itself. When the charge q passes the minimum, an electric force E_1 is exerted on every electron behind it, accelerating toward the cathode the slow electrons and those stationary at the minimum, and decelerating the fast ones that pass the minimum. The net effect is a deficiency of electron charge (a "hole" charge) traveling toward the anode producing a negative field E_2 at the minimum which grows until it compensates E_1.

Considering now one frequency component ω of the white spectrum of the primary pulse, it is easily seen that the secondary current is lagging the primary field E_1. The space charge, therefore, acts reactive and in the analog has to be represented by an inductance L of such magnitude as to resonate with C at (or near) ω_{pm}. The series connection of L with G seems indicated by the fact that the total voltage across the open-circuit diode is given by the integral of original plus secondary reactive fields.

Finite-Beam Diameter: Higher-Order Space-Charge-Wave Modes and Their Noise Contribution

The electron beam is a medium with finite transverse dimensions which can propagate waves in many different modes. The propagation constants and characteristic impedances are different for the various space-charge modes, just as they are different for the various modes of a wave guide. A few higher-order space-charge modes ($n = 1, 2, 3$), with their radial field distributions and plasma wavelengths λ_{pn}, are sketched in Fig. 5.6.

For a cylindrical beam (radius b) drifting in a concentric tube (radius a), solutions have been obtained by Hahn[33] and Ramo.[34] The propagation constant β_n of the nth mode is given by

$$\beta_n = \beta_e \pm \beta_{qn} = \beta_e \left(1 \pm p_n \frac{\omega_p}{\omega} \right)$$

where ω_p is the usual plasma frequency, and p_n is the plasma-frequency reduction factor. The factor p_n is plotted in Fig. 5.7 for diameter ratios $a/b = 1$ and $a/b = \infty$. For ratios $a/b > 2$, the curves for $a/b = \infty$ may be used with good accuracy.

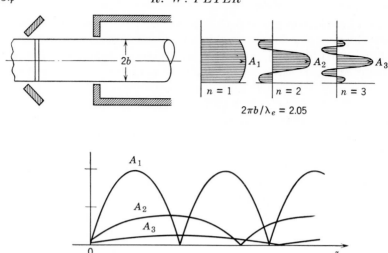

Fig. 5.6.   Transverse and axial variations of the first three radial space-charge-wave modes.

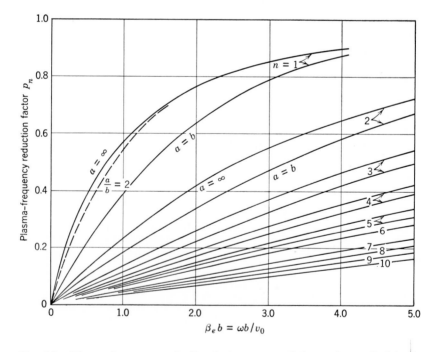

Fig. 5.7.   Plasma-frequency reduction factor versus $\beta_e b$ for various cylindrically symmetric space-charge-wave modes (dominant mode corresponds to $n = 1$).

In an unaccelerated beam, within a drift tube of uniform cross section, the various space-charge modes propagate independently of each other (with no mixing). If the beam is accelerated, or the drift-tube cross section changes, cross-coupling among the modes may occur. If the acceleration is slow, one may argue that the cross-coupling among the modes is negligible. The exponential gun to be described in Section 3, page 259, provides such a slow acceleration.

Owing to the random and uncorrelated electron emission from different points on the cathode surface, higher space-charge modes are excited with relative amplitudes K_e. They propagate with individually different phase constants and beam impedances toward the helix. Associated with each space-charge mode is an amplified noise output of the traveling-wave tube. The nth mode, for example, would contribute a relative noise output M_n which can be expressed by

$$M_n = (K_e K_m K_c)_n \qquad (5.15)$$

where K_e is the mode excitation factor at the cathode, K_m indicates the effect of mode mixing ($K_m = 1$ if no mixing occurs in the gun region), and K_c indicates the degree of coupling of the space-charge-wave mode with the helix wave. The noises in the various space-charge modes are uncorrelated if one assumes no mode mixing in the gun region. In such a case the noise powers contributed to the output by different modes are additive. The sum of these noise contributions constitutes the total "basic noise" in the amplifier.†

For the one-dimensional case, there exists one noise current SWR ρ and one plasma transit angle ψ in the beam at the helix input which will make the noise factor a minimum. Beam[35] has shown that a similar condition holds in the finite beam for each space-charge-wave mode. The problem of designing an optimum gun then reduces to the problem of finding a beam transformer which would transform each mode for the optimum conditions ρ_n and ψ_n at the helix input. An infinite number of adjustable gun parameters would have to be available to satisfy these conditions. However, it can be shown that the noise contributions of higher-order modes decrease rapidly with increasing mode number n. It will be discussed later how an "exponential" gun is able to satisfy the required conditions for at least the fundamental ($n = 1$) and the next higher mode ($n = 2$).

† It has not been definitely determined, so far, whether this sum, under ideal conditions, leads to the same theoretical minimum noise factor as the one-dimensional theory. Preliminary theoretical studies of Beam and Bloom[35] and of Paschke[36] indicate a possibility that this sum may become lower than the one-dimensional value at large values of $\beta_e b$. This means that small beam voltage V_0 and large beam diameter $2b$ would lead to lower noise factor.

3. Low-Noise Electron Guns

In the previous sections, the conditions have been given which must be satisfied by gun and circuit if the lowest noise factor is to be obtained. This section deals with the electron gun. The general requirements are summarized. As they can theoretically be met by an infinite number of different gun arrangements, the fundamental elements, from which every gun is built up, are discussed separately. The relative rate of change of the beam impedance W'/W is found to be the important quantity.

A number of possible low-noise guns are described, of which the "modified exponential gun" seems to have the greatest number of advantages.

Requirements for Low-Noise Guns

According to the results obtained in Chapter 3, a low-noise gun can be considered as a linear beam transducer which transforms the noise between the cathode and the RF-interaction region. In terms of the transmission-line model for the electron beam as a wave-propagating medium,[37] we can describe the low-noise gun as a four-terminal network which transforms an impedance Z_a at the cathode plane a into a desired impedance Z_h at the helix-input plane h (see Fig. 5.9). The problem is to design a transmission-line transformer which transforms the (normalized) impedance at the cathode, $z_a = Z_a/W_a$ into the desired impedance, $z_h = Z_h/W_h$, at the helix input.

The Boundary Conditions. The equivalent impedance Z_a at the potential minimum under the assumptions of full shot noise and random velocity distribution (first proposed by Watkins and extended by Pierce[16]), is (compare to Eq. 3.144 of Chapter 3†)

$$Z_a = \frac{\sqrt{\pi(4 - \pi)}\, kT_c}{2eI_0} \tag{5.16}$$

If it is assumed that (1) the current passing the potential minimum is not full shot noise but is reduced by the action of the potential minimum by a factor R, (2) the velocity fluctuations are unaffected by the potential minimum, and (3) correlation between the current and velocity fluctuations is introduced in the process, producing a phase angle θ in the equivalent impedance, we obtain, for Z_a,

$$Z_a = \frac{\sqrt{\pi(4 - \pi)}\, kT_c}{R2eI_0} e^{j\theta} \tag{5.17}$$

† [Equation 5.16 differs from Eq. 3.144 by $\sqrt{\pi}/2$ because of the use of a different average velocity and the vertical cathode. Editors.]

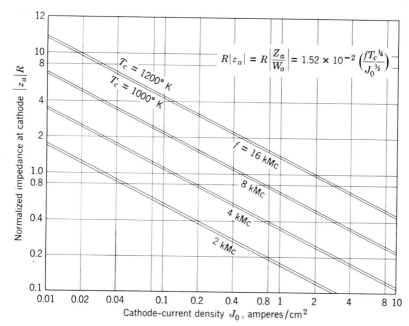

Fig. 5.8. Normalized noise impedance at the cathode versus cathode-current density and temperature.

For most practical purposes, Z_a may be approximated by its absolute value $Z_a \approx |Z_a|$ if the effect of correlation can be neglected. This will be assumed from now on.

Normalization of Z_a with respect to the characteristic beam impedance h/a at the virtual cathode proceeds as follows:

$$W_a = 2 \frac{V_{0a}}{I_0} \frac{\omega_{\mathrm{pm}}}{\omega} \tag{5.18}$$

$$V_{0a} = \frac{\pi}{4} \frac{kT_c}{e} \tag{5.19}$$

$$z_a = \frac{Z_a}{W_a} = 1.52 \times 10^{-2} \left(\frac{fT_c^{1/4}}{RJ_0^{1/2}} \right) \tag{5.20}†$$

where the frequency f is measured in kilomegacycles per second and the cathode-current density J_0 in amperes per square centimeter. This normalized, equivalent impedance presented to the beam by the emitting cathode is plotted in Fig. 5.8.

† Note the use of lower-case z for normalized impedance, and also for distance. Since they never appear together, there should be no cause for confusion.

The desired normalized beam impedance z_h at the helix input may be expressed in terms of the noise-current standing-wave ratio, $\rho = \sqrt{i^2_{max}/i^2_{min}}$, and the plasma angle between noise-current minimum and helix input which give the minimum noise factor of the amplifier. The optimum ρ and ψ can be obtained from Figs. 5.2 and 5.3 for various loss factors d and space-charge parameters QC. The two quantities ρ and ψ determine the normalized impedance z_h as

$$z_h = \frac{Z_h}{W_h} = \frac{1 + \Gamma}{1 - \Gamma} \tag{5.21}$$

where the complex reflection coefficient is

$$\Gamma = \frac{\rho - 1}{\rho + 1} e^{-2_j\psi} \tag{5.22}$$

The noise transformation in the low-noise electron gun will be handled through the differential equations for an accelerated electron beam as developed in Chapter 3, Section 2, Eqs. 3.58 and 3.59. The characteristic admittance as defined there in Eq. 3.51 can be written in an alternate form using definitions, Eqs. 3.48a and 3.18 of Chapter 3. We have for a circular beam of radius b,

$$W = 2 \frac{\left|\dfrac{e}{2m}\right|^{\frac{1}{4}}}{(\pi\epsilon_0)^{\frac{1}{2}}} \frac{V_0^{\frac{3}{4}}p}{\omega I_0^{\frac{1}{2}}b} \tag{5.23}$$

Thus, the ratio of the beam impedances at the cathode W_a and at the helix input W_h is

$$\frac{W_h}{W_a} = p_h \left(\frac{V_{0h}}{V_{0a}}\right)^{\frac{3}{4}} \frac{b_a}{b_h} \tag{5.24}$$

The plasma-frequency reduction factor at the cathode p_a is unity for all practical cathode diameters.

Example

To estimate the magnitude of the impedance transformation in a low-noise gun, we use the following practical values (velocity and current fluctuations are assumed uncorrelated):

$$f = 4000 \text{ Mc/sec}$$
$$V_{0h} = 500 \text{ volts}$$
$$V_{0a} = 0.1 \text{ volt (corresponding to } T_c = 1000°\text{K)}$$
$$2b_a = 0.035 \text{ in. (cathode diameter)}$$
$$b_h/b_a = 1 \text{ (uniform beam diameter)}$$
$$J_0 = 120 \text{ ma/cm}^2$$
$$R = 1 \text{ (no current smoothing)}$$

which yields, with $p_h = 0.44$ and $Z_a/W_a = 1$, $W_h/W_a = 260$.

The Ideal Low-Noise Gun. The ideal low-noise gun should transform the beam impedance at the cathode Z_a over a broad frequency band into the required beam impedance Z_h at the helix input. It is generally not possible to transform exactly an impedance $Z_a(f)$ to another impedance $Z_h(f)$ over a given frequency range Δf. However, it is possible to approximate the exact condition arbitrarily closely. Various solutions of this type have been worked out for conventional networks (e.g., the Tchebycheff approximation). There does not exist a best solution but rather an infinite number of solutions, depending on the maximum tolerable deviation from a perfect transformation, on the allowed complexity of the transformer, and on the effects of unavoidable transformer defects. Similarly, there does not exist a unique optimum low-noise gun. There exists, however, a number of criteria which must be met by any approximation to the ideal gun. Principles of conventional transmission-line theory have to be satisfied for broad-band impedance matching, and, additionally, some particular characteristics of the beam as a wave-propagating medium have to be considered.

The gun may be optimized with respect to various variables. It may be optimized for lowest resulting noise factor of the amplifier under a particular condition and fixed frequency. It may also be optimized for best operation over a large range of operating currents or voltages.

The following four conditions should be met in the design of an ideal gun. They are based on a large amount of empirical knowledge, theoretical evidence, and physical intuition. They should be used, therefore, with a grain of salt.

1. The noise-current SWR at any point in the gun should not exceed either the SWR required at the helix-input plane, ρ_h, or the SWR, ρ_a, caused by mismatch at the cathode, whichever is the larger; i.e.,

$$\rho < \rho_h \text{ or } \rho_a$$

2. The normalized impedance at the cathode should be made equal to unity, $z_a \approx 1$ (by a proper choice of cathode-current density).

3. The gun should optimize noise conditions for a maximum number of space-charge modes.

4. The beam should be accelerated in a smooth fashion, avoiding any discontinuities. The electron flow should deviate as little as possible from rectilinear flow. (The more abrupt any discontinuity and the farther away from the cathode it is located, the larger the resulting increase in noisiness.)

Fundamental Space-Charge-Wave Transformers

The analogy between the differential equations for propagation of space-charge waves along an ideal electron beam and propagation of electromagnetic waves along a lossless transmission line permits us to use the formalism of transmission-line theory.[37] In a well-designed gun, the beam behaves very nearly as if it had a single-velocity, laminar flow. The wave-transducing qualities of a section of an ideal electron beam can be represented by the chain matrix of a purely reactive, lossless network:

$$U_{i+1} = A_i U_i + B_i I_i$$

$$I_{i+1} = C_i U_i + D_i I_i$$

or

$$\begin{bmatrix} U_{i+1} \\ I_{i+1} \end{bmatrix} = \mathbf{M}_i \begin{bmatrix} U_i \\ I_i \end{bmatrix} \tag{5.25}$$

with

$$\mathbf{M}_i = \begin{bmatrix} A_i \ B_i \\ C_i \ D_i \end{bmatrix}$$

As the transducer is lossless, $|\det \mathbf{M}_i| = 1$.

A low-noise gun of the type shown in Fig. 5.9 may be subdivided into a chain of n elementary transformer sections, each with its chain matrix $\mathbf{M}_i$. If the matrices of the elements are known, the matrix of the complete gun can be represented by

$$\begin{bmatrix} U_n \\ I_n \end{bmatrix} = \mathbf{M}_{n-1} \cdots \mathbf{M}_i \cdots \mathbf{M}_1 \begin{bmatrix} U_1 \\ I_1 \end{bmatrix} \tag{5.26}$$

In our problem, the impedance $Z_n = U_n/I_n$ corresponds to the desired Z_h, and Z_1 corresponds to Z_a of Section 3, page 236.

Every transformer $\mathbf{M}_i$ transforms the impedance Z_i at plane i into

$$Z_{i+1} = \frac{A_i Z_i + B_i}{C_i Z_i + D_i} \tag{5.27}$$

at plane $(i + 1)$ according to Eq. 5.25.

In designing a low-noise gun, it is advantageous to use the simplest space-charge-wave transformers as building elements. The following "fundamental" transformers show a particularly simple change of characteristic impedance W:

(a) The parallel-flow diode.
(b) The drift region.
(c) The "Bessel" transformer.
(d) The exponential transformer.

For these space-charge-wave transformers, the equivalent transmission-line equations can be solved explicitly. As these transformers have been discussed in the literature, we will present here only their matrices $\mathbf{M}_i$ in a form useful for our purposes. The "Bessel" transformer[38,39] will not be discussed as no use will be made of it here.

Design Equations for Space-Charge-Wave Transformers. In general, the characteristic impedance W in a low-noise gun is a function of the plasma transit angle ϕ. The designer of such a gun, however, has to know the potential distribution $V_0(z)$ and the beam radius $b(z)$ along the beam axis z, which correspond to a prescribed $W(\phi)$. This relationship is directly obtained from the two Eqs. 3.51a and 3.48 of

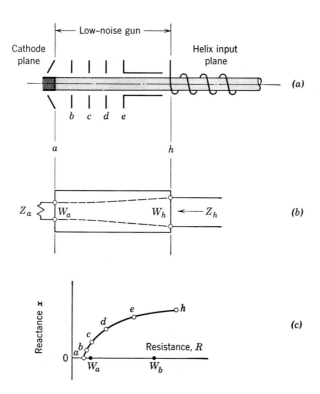

Fig. 5.9. Noise impedance transformation in multi-electrode gun.

Chapter 3, which we write as follows:

$$W = A \frac{p}{b} V_0^{3/4} \tag{5.28}$$

$$\beta_q = \frac{d\phi}{dz} = B \frac{p}{b} V_0^{-3/4} \tag{5.29}$$

where $\qquad p = \dfrac{\omega_q}{\omega_p} = p(V_0, b)$ (see Fig. 5.7).

$$A = 2 \frac{\left| \dfrac{e}{2m} \right|^{1/4}}{(\pi\epsilon_0)^{1/2}} \frac{1}{\omega I_0^{1/2}}$$

$$B = \frac{I_0^{1/2}}{\sqrt{\sqrt{2 \left| \dfrac{e}{m} \right|}} \, 2\epsilon_0 \pi}$$

Equations 5.28 and 5.29 provide two simultaneous equations for $V_0(z)$ and $b(z)$:

$$\frac{dW}{d\phi} = \left(\frac{\partial W}{\partial V_0} \frac{dV_0}{dz} + \frac{\partial W}{\partial b} \frac{db}{dz} \right) \frac{dz}{d\phi} \tag{5.30}$$

$$\frac{d\phi}{dz} = \beta_q(V_0, b) \tag{5.31}$$

If the total differential of W, Eq. 5.30, is carried out according to Eq. 5.28, we obtain

$$\frac{1}{W} \frac{dW}{d\phi} = \frac{1}{B} \frac{b}{p} V_0^{3/4} \left[\left(\frac{1}{p} \frac{\partial p}{\partial b} - \frac{1}{b} \right) \frac{db}{dz} + \left(\frac{1}{p} \frac{\partial p}{\partial V_0} + \frac{3}{4} \frac{1}{V_0} \right) \frac{dV_0}{dz} \right] \tag{5.32}$$

$$\frac{d\phi}{dz} = B \frac{p}{b} V_0^{-3/4} \tag{5.33}$$

If $W(\phi)$ is given, this system yields an infinite number of corresponding solutions, $V_0(z)$ and $b(z)$. In order to obtain a unique solution it is necessary to specify further either the desired beam contour $b(z)$ or the potential distribution $V_0(z)$ or the relation between the two.

In the most common case of a (magnetically confined) *parallel-flow beam*, Eqs. 5.32 and 5.33 with $db/dz = 0$ leads to

$$\left(\frac{1}{W} \frac{dW}{d\phi} \right) dz = \frac{1}{B} \frac{b}{p} V_0^{3/4} \left(\frac{1}{p} \frac{\partial p}{\partial V_0} + \frac{3}{4} \frac{1}{V_0} \right) dV_0 \tag{5.34}$$

$$d\phi = B \frac{p}{b} V_0^{-3/4} \, dz \tag{5.35}$$

For simple functions $W(\phi)$, this system can be solved explicitly, as will be shown for the case of the exponential transformer. This presumes that $p(V_0, b)$, which is a transcendental function, can be approximated over the range of integration by a simple irrational expression as, for example,

$$ p = \left[1 + \left(\frac{2}{\beta_e b} \right)^{3/2} \right]^{-1/2} \tag{5.36} $$

Over the useful range of variables, this approximation is good to about 3%.

The Parallel-Flow Diode. Wave propagation along the parallel electron flow across a planar diode of infinite transverse extent has been treated by Llewellyn and Peterson,[40] Smullin,[41] and others. A simpler and, for our purposes, more convenient form of the diode matrix can be obtained by applying the transmission-line analogy to diode flow.[37]

We define the normalized parameter τ in the region between two planes, 1 and 2, in terms of Llewellyn's space-charge parameter ζ

$$ \tau = \sqrt{\frac{2\zeta}{\zeta - 1}} \tag{5.37} $$

so that the general diode matrix takes the following form†

$$ \mathbf{M}_{\text{diode}} = \begin{bmatrix} 1 - \tau^2 & j\sqrt{2}\,\tau W_1 \\ j\sqrt{2}\,\dfrac{1}{W_1}\dfrac{\tau}{1 + \tau^2} & -\dfrac{1}{1 + \tau^2} \end{bmatrix} \tag{5.38} $$

The space-charge parameter ζ can be defined in terms of the following special model: a space-charge-limited diode in which there may be an average velocity at the plane of the virtual cathode. If electrons are emitted with the velocity spread characteristic of the cathode temperature, there will be an *average potential* V_{01} of electrons at the potential minimum where the acceleration is zero. In terms of the initial average potential and the anode potential, we can write

$$ \zeta = \frac{1 - \sqrt{V_{01}/V_{02}}}{1 + \sqrt{V_{01}/V_{02}}} \tag{5.39} $$

This definition combined with Eqs. 5.37 and 5.38 gives as the value of τ:

$$ \tau = \sqrt{\sqrt{\frac{V_{02}}{V_{01}}} - 1} \tag{5.40} $$

† Equation 29 in reference 37. Note slightly different definition of parameter τ.

The plasma transit angle ϕ between planes 1 and 2 along the beam is found to be

$$\phi = \sqrt{2} \ln (\tau + \sqrt{1 + \tau^2}) \tag{5.41}$$

The relation among ϕ, ζ, τ, and V_{02}/V_{01} is shown in Fig. 5.10.

The ratio of the characteristic beam impedances at planes 2 and 1 is (Eq. 5.24)

$$W_2/W_1 = (V_{02}/V_{01})^{3/4} = (1 + \tau^2)^{3/2} = \cosh^3 (\phi/\sqrt{2}) \tag{5.42}$$

At long plasma transit angles ϕ, the characteristic beam impedance increases exponentially with ϕ, as is evident by introducing the limiting value of Eq. 5.41 into Eq. 5.42

$$\lim_{\tau \to \infty} \phi = \sqrt{2} \ln 2\tau$$

$$\lim_{\tau \to \infty} W_2/W_1 = \tfrac{1}{8} \exp (3\phi/\sqrt{2}) \tag{5.43}$$

W_2/W_1 vs. ϕ is shown in Fig. 5.11. The impedance ratio between planes 1 and 2 increases with increasing plasma phase angle ϕ or space-charge factor ζ. (The logarithmic increment of the asymptotic impedance ratio is $k = 3/\sqrt{2}$. On page 250 it will be shown that an exponential transformer with $k > 2$ represents a transmission line below cut-off i.e., a line which cannot transmit waves and which, if

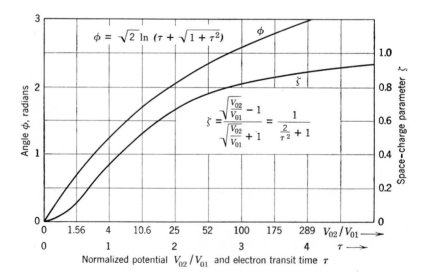

Fig. 5.10. Plasma transit angle ϕ and space-charge parameter ζ versus normalized potential in one-dimensional space-charge-limited diode.

infinitely long, looks like a pure reactance at the input regardless of its termination.) Computing the relative increase of characteristic impedance, $dW/W\,d\phi$, which is the important quantity in these transformers, we find by substituting Eq. 5.41 in 5.42 and differentiating

$$\frac{dW}{W\,d\phi} = \frac{3}{\sqrt{2}}\tanh\frac{\phi}{\sqrt{2}} \qquad (5.44)$$

Example

Let the voltage of the diode space between the cathode (at 1000° K) and the first accelerating electrode in a low-noise gun be 40 volts. In Fig. 5.10 we find the space-charge factor $\zeta = 0.9$ and plasma transit angle $\phi = 3.05$, which

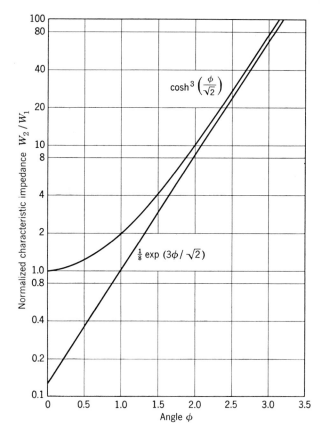

Fig. 5.11. Normalized characteristic beam impedance versus plasma transit angle in one-dimensional space-charge-limited diode.

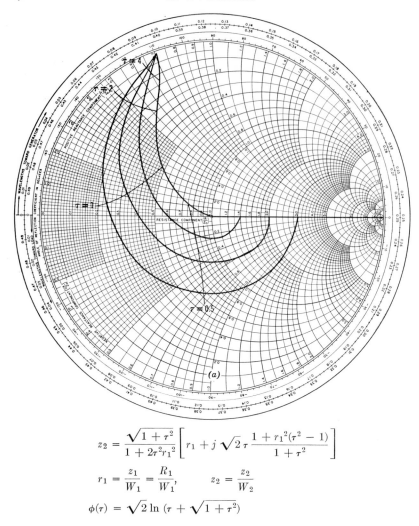

$$z_2 = \frac{\sqrt{1+\tau^2}}{1+2\tau^2 r_1^2} \left[r_1 + j \sqrt{2}\,\tau \frac{1 + r_1^2(\tau^2 - 1)}{1 + \tau^2} \right]$$

$$r_1 = \frac{z_1}{W_1} = \frac{R_1}{W_1}, \qquad z_2 = \frac{z_2}{W_2}$$

$$\phi(\tau) = \sqrt{2}\ln\left(\tau + \sqrt{1+\tau^2}\right)$$

Fig. 5.12.   Beam-impedance transformation in

correspond to a characteristic impedance ratio of about $W_2/W_1 = 83$ according to Fig. 5.11.

The impedance Z_2 of the "diode line" terminated by Z_1 is, with Eqs. 5.27 and 5.38,

$$Z_2 = W_1 \frac{(1 - \tau^4)Z_1/W_1 + j \sqrt{2}\,\tau(1 + \tau^2)}{1 + j \sqrt{2}\,\tau Z_1/W_1} \qquad (5.45)$$

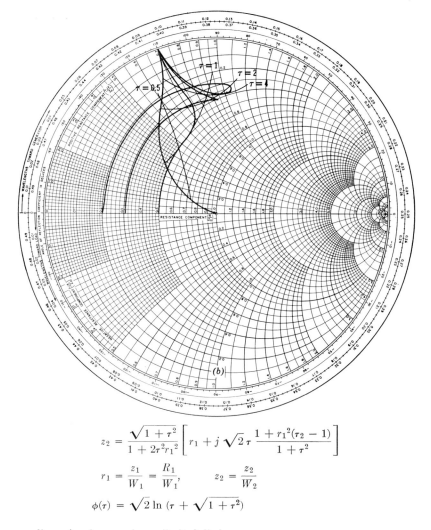

$$z_2 = \frac{\sqrt{1 + \tau^2}}{1 + 2\tau^2 r_1^2} \left[r_1 + j\sqrt{2}\,\tau\, \frac{1 + r_1^2(\tau_2 - 1)}{1 + \tau^2} \right]$$

$$r_1 = \frac{z_1}{W_1} = \frac{R_1}{W_1}, \qquad z_2 = \frac{z_2}{W_2}$$

$$\phi(\tau) = \sqrt{2}\ln(\tau + \sqrt{1 + \tau^2})$$

one-dimensional space-charge-limited diode.

or normalized, with Eq. 5.42,

$$z_2 = \frac{1 + \tau^2}{(1 + \tau^2)^{3/2}} \frac{(1 - \tau^2)z_1 + j\sqrt{2}\,\tau}{1 + j\sqrt{2}\,\tau z_1} \tag{5.46}$$

For the case $z_1 = r_1 =$ real, Eq. 5.46 can be simplified into

$$z_2 = \frac{\sqrt{1 + \tau^2}}{1 + 2\tau^2 r_1^2} \left[r_1 + j\sqrt{2}\,\tau\, \frac{1 + r_1^2(\tau^2 - 1)}{1 + \tau^2} \right] \tag{5.47}$$

This case approximates the actual condition in the diode space of a
low-noise gun where the correlation between current and velocity
fluctuations is small and therefore $z_1 = r_1$ and is real according to
Eq. 5.20.   The asymptotic values of Eq. 5.45 are

$$\lim_{\tau \to 0} z_2 = r_1 \frac{1 + \tau^2/2}{1 + 2\tau^2 r_1{}^2} \to r_1$$

$$\lim_{\tau \to 0} z_2 = \frac{1}{2r_1} \left(\frac{1}{\tau} + j \sqrt{2}\, r_1 \right) \to \frac{j}{\sqrt{2}}$$

(5.48)

A number of diode-impedance curves, Eq. 5.47, are plotted in Fig.
5.12.   It is clearly seen how this transformer in the beginning behaves
essentially like a constant-impedance transmission line (drift region)
which simply shifts the phase angle ϕ but does not change the standing-
wave ratio.   Then, at an angle of about $\phi = 1$ ($\tau = 0.75$), the exponential character of this transformer takes over, and the standing-wave
ratio increases rapidly.

The Drift Region.   Space-charge waves propagating along a uniform beam inside a cylindrical drift tube have been first described by
Hahn[33] and Ramo.[34]   Their original equations were formulated in
terms of a-c velocity.   If instead the kinetic potential is introduced,
the matrix **M** for the drift tube will take the well-known form of the
lossless uniform line of characteristic impedance W (see Chapter 3):

$$\mathbf{M}_{\mathrm{drift}} = \begin{bmatrix} \cos\phi & jW\sin\phi \\ jW^{-1}\sin\phi & \cos\phi \end{bmatrix}$$

(5.49)

where ϕ is the length of the line in electrical degrees.

The Exponential Transformer.   The exponential transformer represents the beam equivalent to the exponential acoustical horn or the
exponential transmission line, both of which are often used for impedance matching.

We stipulate that the impedance be governed by the expression

$$W = W_1 e^{k\phi}$$

(5.50)

where W_1 is the initial characteristic impedance, and ϕ is the plasma
transit angle.   The matrix, Eq. 5.25, for this exponential transformer is[42]

$$\mathbf{M}_{\mathrm{exp}} =$$

$$\begin{bmatrix} \dfrac{\exp(k\phi/2)}{m}\left(m\cos m\phi - \dfrac{k}{2}\sin m\phi \right) & jW_1 \dfrac{\exp(k\phi/2)}{m}\sin m\phi \\ jW_1^{-1} \dfrac{\exp(-k\phi/2)}{m}\sin m\phi & \dfrac{\exp(-k\phi/2)}{m}\left(m\cos m\phi + \dfrac{k}{2}\sin m\phi \right) \end{bmatrix}$$

(5.51)

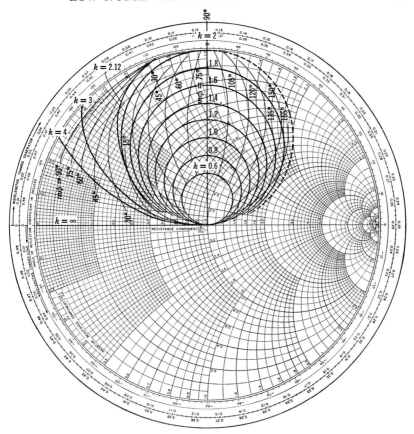

Fig. 5.13. Input impedance of an exponential transmission line terminated in a matched load ($z_1 = 1$).

where
$$m = \sqrt{1 - \left(\frac{k}{2}\right)^2}$$

The exponential line transforms impedances z_1 according to Eqs. 5.27 and 5.51 into

$$z_2 = \frac{Z_2}{W_2} = \frac{(m\cos m\phi - k/2 \sin m\phi)z_1 + j\sin m\phi}{jz_1 \sin m\phi + m\cos m\phi + k/2 \sin m\phi} \quad (5.52)$$

where $z_1 = Z_1/W_1$ at $m\phi = 0$. A number of curves representing the exponential impedance transformation, Eq. 5.52, have been plotted in Figs. 5.13, 5.14, and 5.15.

Figure 5.13 shows the special case of $z_1 = 1$. The larger the log-

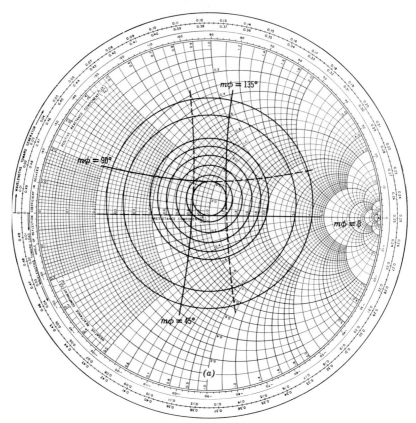

This family of circles is for $k = 0.4$, and values of z_1 from $\frac{1}{4}$ to 4 as determined by
their r intercept.
Use dashed lines for $z_1 < 1$.

Fig. 5.14.   Transformation of various input imped

arithmic increment k, the larger the diameter of the circle which
represents the impedance z_2 for various equivalent line lengths $m\phi$.
The transformation is periodic for real angles $m\phi$, $(k < 2)$, as is evident
from Eq. 5.51.   The maximum reflection coefficient is in this case
equal to $k/2$.   All circles are tangential to the real axis in the center
point $z_1 = 1$.

If the line has a "steepness" larger than $k = 2$, z_2 is no longer
periodic, and the line behaves *for all frequencies* like a transmission line
below cut-off.   If $m\phi$ is infinitely large, the input impedance z_2
becomes purely imaginary (regardless of z_1).   The long, infinite

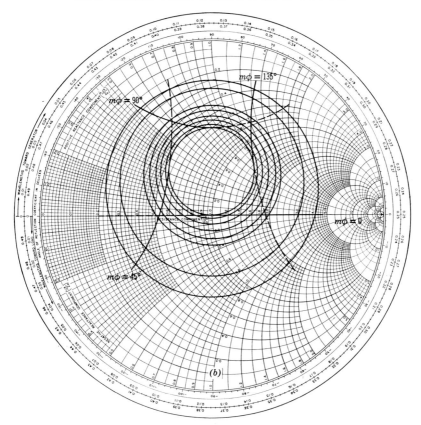

(b)

This family of circles is for $k = 1$, and values of z_1 from $\frac{1}{4}$ to 4 as determined by their r intercept.
Use dashed lines for $z_1 < 1$.

ances by exponential line with $k = 0.4$ and $k = 1$.

plane-parallel diode approximates closely an exponential transformer below cut-off with $k = 3/\sqrt{2} = 2.12$ (see Fig. 5.11). The output impedance z_2 is purely imaginary and frequency-independent.

Figures 5.14a and 5.14b show two families of impedance-transformation circles for an exponential line with $k < 2$, that is, with relatively slow increase of the characteristic impedance. The centers of all circles lie on a line normal to the real axis through the center point of the Smith chart. Figures 5.15a and 5.15b show families of circles for steeper transformers with $k > 2$.

The solid equi-phase-angle curves indicate angles $m\phi$ starting at the

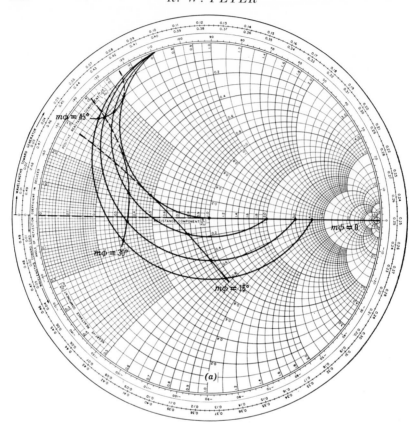

Exponential impedance transformation for $k = 2.12$
Solid $m\phi$ lines for real impedance $r > 1$
Dotted $m\phi$ lines for real impedance $r < 1$

Fig. 5.15.   Transformation of various input imped

right half of the real axis of the Smith chart (impedance Z_1 real and
larger than W_1).   The dotted lines indicate angles $m\phi$ starting at the
left half of the real axis (impedance Z_1 real, but smaller than W_1).

The impedance z_2 of a *uniform* transmission line circles around the
point, $z_1 = 1$, equi-phase lines being straight radial lines through this
point.   In analogy, the input impedance of the *exponential* line rotates
around the *circle* which represents the matched condition.   Lines of
equal phase angle, $m\phi$, now, are circles (or straight lines) tangent to the
matched circle.

We next ask for the potential distribution V_0 along the beam axis
corresponding to the condition

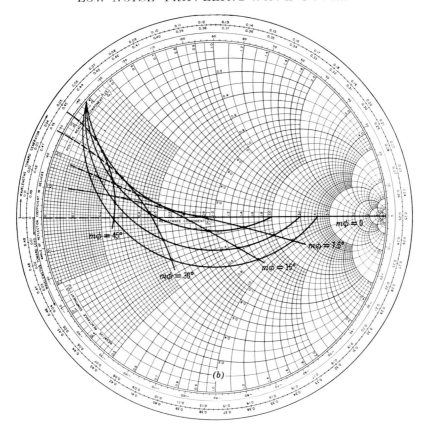

Exponential impedance transformation for $k = 3$
Solid $m\phi$ lines for real impedance $r > 1$
Dotted $m\phi$ lines for real impedance $r < 1$

ances by exponential line with $k = 2.12$ and $k = 3$.

$$W = W_1 e^{k\phi} \qquad (5.53a)$$

or

$$\frac{1}{W}\frac{dW}{d\phi} = k \qquad (5.53b)$$

As a simplifying assumption, we take the beam diameter $b \approx$ constant, the case of a beam in a very large magnetic focusing field. Introducing Eq. 5.53b into Eq. 5.32 yields

$$k\,dz = \frac{1}{B}\frac{b}{p}V_0{}^{3/4}\left(\frac{1}{p}\frac{\partial p}{\partial V_0} + \frac{3}{4}\frac{1}{V_0}\right)dV_0 \qquad (5.54)$$

If the beam is infinitely thick, $p = 1$, the solution for Eq. 5.54 becomes

$$kB(z - z_1) = V_0{}^{3/4} - V_{01}{}^{3/4} \qquad (5.55)$$

or normalized

$$\frac{V_0}{V_{01}} = \left[\frac{kB}{V_{01}{}^{3/4}} (z - z_1) + 1 \right]^{4/3} \qquad (5.56)$$

At large values of the first term (e.g., for large transformer lengths $(z - z_1)$, the potential distribution in the "thick-beam" exponential transformer has the limiting value

$$V_0 = \text{const} \, (z - z_1)^{4/3} \qquad (5.57)$$

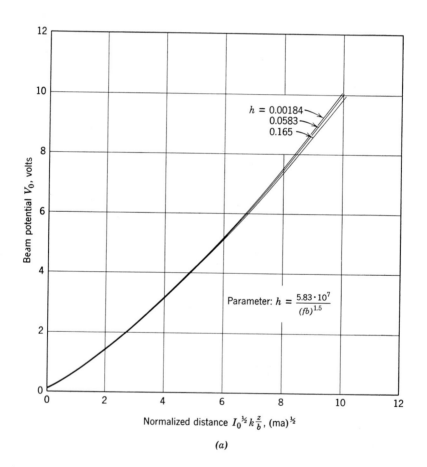

(a)

Fig. 5.16.   Beam voltage versus

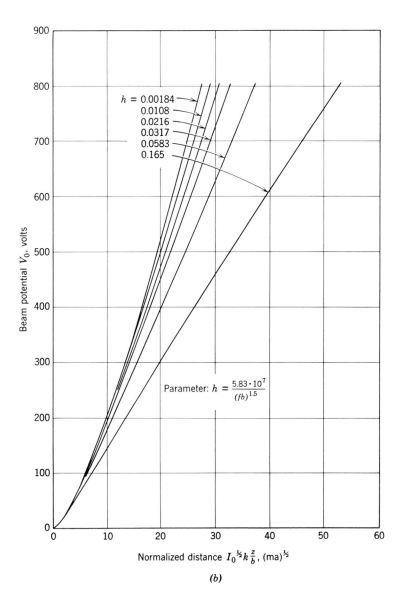

(b)

distance in an exponential gun.

This is the case for infinite space-charge-limited electron flow between parallel-plane electrodes.

In the realistic finite-beam-diameter case with $p < 1$, however, the first term on the right-hand side of Eq. 5.54 adds to the integral of Eq. 5.55, and we obtain, for the exponential transformer with finite-beam diameter,

$$kB \frac{z - z_1}{b} = \frac{3}{4} \int_{V_{01}}^{V_0} \frac{V_0^{-\frac{1}{4}}}{p} \, dV_0 + \int_{V_{01}}^{V_0} \frac{V_0^{\frac{3}{4}}}{p^2} \frac{\partial p}{\partial V_0} \, dV_0 \qquad (5.58)$$

A simple explicit solution of Eq. 5.58 is obtained if we approximate p by Eq. 5.36, and rewrite:

$$p = (1 + hV_0^{\frac{3}{4}})^{-\frac{1}{2}} \qquad (5.59)$$

where

$$h = \frac{5.83 \times 10^7}{(fb)^{1.5}} \qquad (5.59a)$$

with f in cycles per sec and b in meters. The accuracy of this expression is adequate for all design purposes as, on the one hand, the surrounding electrodes modify p slightly, and, on the other hand, the beam diameter usually oscillates about its equilibrium radius, thereby again modifying p.

Introducing Eq. 5.59 into Eq. 5.58 yields, after some simple steps,

$$Bk \frac{z - z_1}{b} = \frac{3}{4} \int_{V_{01}}^{V_0} V_0^{-\frac{1}{4}} (1 + hV_0^{\frac{3}{4}})^{\frac{1}{2}} \, dV_0$$

$$- \frac{3}{8} h \int_{V_{01}}^{V_0} V_0^{\frac{1}{2}} (1 + hV_0^{\frac{3}{4}})^{-\frac{1}{2}} \, dV_0$$

This equation can be integrated, and after some simplifications we obtain

$$Bk \frac{z - z_1}{b} = \frac{1}{3h} [(1 + hV_0^{\frac{3}{4}})^{\frac{3}{2}} - (1 + hV_{01}^{\frac{3}{4}})^{\frac{3}{2}}]$$

$$- \frac{1}{h} [(1 + hV_0^{\frac{3}{4}})^{\frac{1}{2}} - (1 + hV_{01}^{\frac{3}{4}})^{\frac{1}{2}}] \qquad (5.60)$$

This approximate solution of $V_0(z)$ for the exponential transformer is plotted in Figs. 5.16a and 5.16b for a number of values h (or bf) and for an initial potential $V_{01} = 0.1$ volt.

Velocity-Jump Gun

Noise reduction in electron beams was first achieved with velocity-jump guns. It was proposed by Watkins[43] at the time when it was

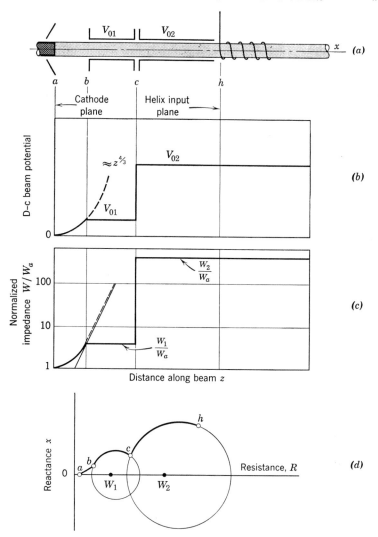

Fig. 5.17. Noise impedance transformation in velocity-jump gun.

still believed that velocity fluctuations constituted the only noise excitation of the beam at the cathode. The velocity-jump gun, Fig. 5.17a, was intended to function as follows: The electron beam, accelerated by the first anode b, was made to drift in the first drift space b-c for such a length that the velocity-fluctuation maximum would fall at point c. At this point, the direct voltage of the beam was suddenly

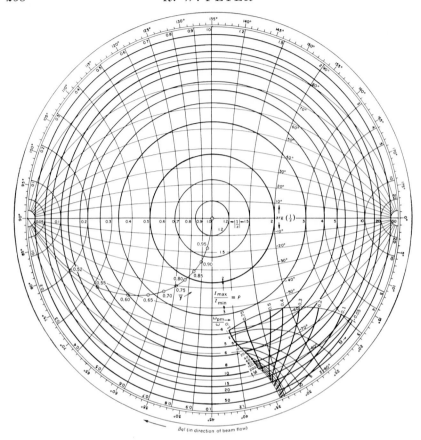

Fig. 5.18.   Transmission-line chart for low-noise tube design.   The chart is used
for calculating the effects of velocity jumps and drifts on the noise-current standing
waves.   The grid of lines of constant M and ω_{pm}/ω give the anode exit conditions.
[M = (first-anode voltage)$(1160/T_c)$, and ω_{pm} is the plasma frequency at the
potential minimum.]   The desired helix entrance conditions for minimum noise
figures are plotted using Bloom's $\bar{y}$ parameter.[44]   The $\omega_{pm}/\omega - M$ nomogram
gives the desired complex normalized admittance $1/z_h$ at the entrance to the helix.

increased to helix potential.   (It is easily shown by considering the
kinetic energy of an electron undergoing this velocity "jump" that
its a-c velocity is reduced in inverse proportion to the ratio of the
direct voltages: $v_2/v_1 = V_{01}/V_{02}$.)   The noise reduction, then was
expected to be the better, the larger this velocity jump V_{02}/V_{01}.   This
argument, logically extended, led Watkins to propose multiple jumps:
i.e., periodically repeated velocity acceleration and deceleration at

points separated by quarter plasma-wave drift regions. Each acceleration was supposed to reduce the velocity fluctuations while the downward jump would not materially affect the process.

The single-velocity jump gun can be considered as consisting of three fundamental transformers, a diode region, and two drift regions of different characteristic impedance; therefore,

$$\mathbf{M} = (\mathbf{M}_{\text{diode}})(\mathbf{M}_{\text{drift}})_1(\mathbf{M}_{\text{drift}})_2 \qquad (5.61)$$

The impedance transformation in the gun is indicated schematically in Fig. 5.17d. It is clearly seen that, if the characteristic beam impedance in the helix region W_2 is given, there is only a limited choice of the first drift-region impedance W_1 due to the requirement that the corresponding circle has to intersect with the second drift-region circle.

Buchmiller[44] et al. have drawn a Smith chart which combines the diode output impedance, Fig. 5.12, and the desired impedance z_h at the helix input, Fig. 5.2, into families of circles which represent drift regions (circles concentric with point 1 on the Smith chart) and impedance changes (circles through the 0 and ∞ points of the Smith chart). Such a chart permits graphical design of velocity-jump guns, Fig. 5.18.

If we compare the single-jump gun with the "ideal" low-noise gun stipulated on page 239, we find that it has two major shortcomings. The major disadvantage lies in the abrupt velocity changes which produce strong lens effects. In Section 4, page 278, it will be shown how lens effects lead to an increase of the noisiness in the beam. A second disadvantage is the limited transformation range of this gun. Once the gun is built, i.e., once the cathode to first anode distance $a-b$ and the drift tube length $b-c$ are chosen, Fig. 5.17, the gun will provide the proper transformation only for one particular beam current. The impedance-jump gun, therefore, is not flexible enough for a traveling-wave-tube amplifier which should have a minimum noise factor over a wide range of beam currents. Both these disadvantages are eliminated in the guns to be discussed in the next section.

The Exponential Gun

Our requirements for the "ideal" low-noise gun on page 239 call essentially for smooth, broad-band matching from a low-impedance line into a high-impedance line at an impedance ratio of several hundred. Cases of impedance matching with similar requirements have been encountered previously in acoustics and in transmission-line theory. The exponential horn and the exponential matching transformer are commonly used for this purpose. In this section, we shall discuss the possibility of applying the same principle to space-charge-wave impedance matching. An electron gun which accelerates the beam in such a way that its impedance between cathode (or potential minimum) and helix input can be characterized by a single exponential rate of change k will be called a "true exponential gun."

A major advantage of the exponential transformer is its analytic simplicity, as was shown on page 248. Of all broad-band matching devices, it is the simplest analytically. It should be pointed out that there exist other somewhat better transformers for particular wide-band matching requirements (see, for example, reference 45). All these solutions, however, are modifications of the basic exponential line.

The Conditions Near the Cathode. From the earlier discussion of the exponential transformer, it is evident that, by applying the proper d-c potential along the axis of a constant-diameter beam, it is possible to produce an exponential impedance increase between two planes 1 and 2. The question arises, however, whether it is possible to put plane 1 at the potential minimum where the impedance $z_a = Z_a/W_a$ is given.

The electron current in a low-noise gun has to be space-charge-limited, in order to derive the benefit of noise-current smoothing ($|R| < 1$). This means that a potential minimum has to exist where, of course, the accelerating field is zero. This leads to the condition

$$\frac{dV_{0a}}{dz} = 0 \tag{5.62}$$

Let us see what this means for a parallel-flow electron beam. Near the potential minimum, $p = 1$, the beam impedance, Eq. 5.28, is essentially proportional to $V_0^{3/4}$. Therefore, at the potential minimum,

$$\frac{dW^{2/3}}{dz} = \frac{dV_0}{dz} = 0 \tag{5.63}$$

Thus, the beam impedance has to be constant. This requirement is obviously in contradiction with the condition of exponential flow where the change of impedance along the beam is always finite. From this result we conclude immediately that initially parallel electron flow will not be tolerable in a "true exponential gun."

To determine the initial flow condition that would be required, we introduce the exponential-line characteristic, Eq. 5.53b,

$$\frac{1}{W}\frac{dW}{d\phi} = k \tag{5.53b}$$

into Eq. 5.32. With the boundary value of Eq. 5.62 at the potential minimum, the initial flow condition for the true exponential gun becomes

$$kB = \frac{b}{p} V_0^{3/4} \left(\frac{1}{p} \frac{\partial p}{\partial b} - \frac{1}{b} \right) \frac{db}{dz} \tag{5.54}$$

If we let $p \approx 1$ for a qualitative estimate, Eq. 5.54 reduces to

$$kB = -V_0^{3/4} \frac{db}{dz} \tag{5.55}$$

This means that the true exponential gun has to have an initially *convergent* beam flow. The condition, Eq. 5.55, could be realized, for example, with a concave cathode surface.

The convergence for the "true exponential gun" condition is required only in the vicinity of the potential minimum. As soon as dV_0/dz takes finite values in Eq. 5.32, db/dz can be reduced, and parallel flow can be established within a short distance from the potential minimum.

The Modified Exponential Gun. A brief initial convergence is technically hard to realize, and, in addition, the importance of an immediately rising characteristic beam impedance is questionable. I am inclined to think that an initially constant impedance is preferable, as this means no transformation in the beam near the potential minimum where the velocity fluctuations are still of the same order of magnitude as the average d-c velocity. If this intuitive reasoning is correct, Nature, for a change, makes the realization of the better solution easier. The characteristic impedance of the infinite-plane-parallel diode in Fig. 5.11 has a zero initial slope. The condition is always present in the cathode region in parallel electron flow. If we confine the electron beam to parallel flow throughout a multi-electrode gun, an almost exponential impedance function may be realized as shown in Fig. 5.19a. This low-noise gun will be called the "modified exponential gun" or briefly the "exponential gun."† For comparison, the infinite-plane-parallel diode carrying the same current density and the true exponential gun with the same steepness, $k = 0.5$, are shown.

In Fig. 5.19b the corresponding potential distribution $V_0(z)$ for the (modified) exponential gun is shown qualitatively. Comparison with the $z^{4/3}$ potential distribution of the parallel-flow gun indicates that the

---

† The (modified) exponential gun, therefore, consists of a very short diode and a true exponential transformer in series. The transition point is given by equality between Eqs. 5.44 and 5.53: i.e., for $\cosh^3 (\phi/\sqrt{2}) = \exp (k\phi)$.

two are identical in the beginning. This is necessary if the same beam-current densities are postulated in both cases. A short distance away from the cathode, however, the potential of the exponential gun flattens out and increases in an approximately linear fashion throughout the gun. The electron beam in the exponential gun, therefore, has the tendency to expand since the potential increase is insufficient to keep it parallel. Magnetic confinement is required to hold the beam diameter constant.

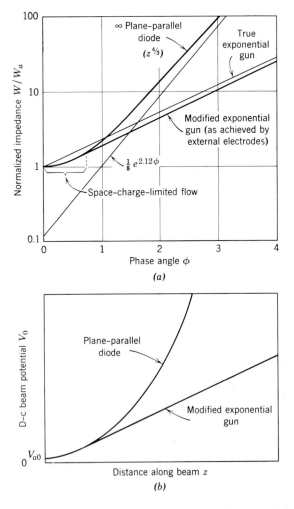

Fig. 5.19.   Comparison of plane-parallel diode and exponential guns.

The Three-Region Gun

The three-region gun was designed with the intention of obtaining as flexible a transformer as possible.[46] A series of three (or four) accelerating electrodes permits a wide range of potential distribution along the beam axis. A typical gun design is shown in the example of Fig. 5.20. The cathode–first-anode region was designed as a parallel-flow Pierce-type gun. However, in order to also test convergent and divergent initial beam flow, a design with a protruding cathode was chosen which allowed great flexibility in the initial electron-flow conditions. Later more detailed studies by Eichenbaum[47] indicated that the electrode geometry shown in Fig. 5.21 can produce an electron beam whose initial angle can be adjusted by about ±5 degrees with little deviation from perfect laminar flow.

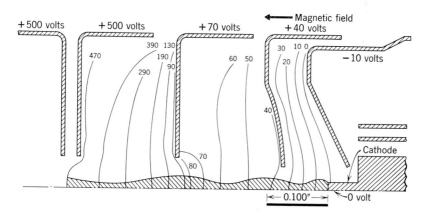

Fig. 5.20. Potential distribution in three-region low-noise gun.

As a result of detailed studies, it was found that an optimum noise factor in a typical 3000 Mc per sec traveling-wave-tube was consistently obtained with conditions closely exemplified by those of Fig. 5.20. An analysis then was made of the impedance transformation along the beam in this gun. The potential $V_0(z)$ along the axis, as obtained from electrolytic tank measurements, is shown in Fig. 5.22. The beam diameter was computed under the assumption of laminar flow in a confining uniform magnetic field of 500 gauss, Fig. 5.20. The plasma phase constant β_q was afterwards determined according to Eq. 5.29 as a function of the beam voltage and diameter and plotted in Fig. 5.23. Finally, the characteristic beam impedance W, Eq.

5.28, was computed as a function of the phase angle. Figure 5.24 shows the impedance plotted in a logarithmic scale from a distance of 0.015 in. ahead of the cathode on to anode 3, the last accelerating electrode.

With this information on the three-region gun, the transfer matrix, Eq. 5.25 was computed in three different ways:

1. By approximating the actual impedance function $W(\phi)$ by a large number of short drift sections and small impedance jumps.

2. By approximating $W(\phi)$ by a small number of exponential transformers with different exponents k (see Eq. 5.53b, page 260).

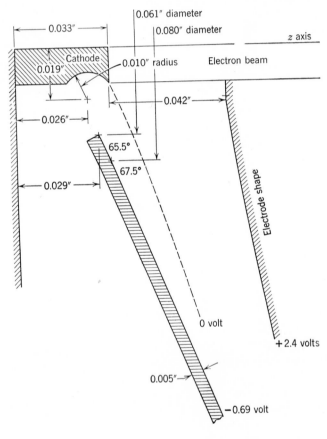

FIG. 5.21.   Design of first region of laminar-flow gun with adjustable initial beam convergence.

3. By approximating the complicated impedance function by a single exponential transformer which was chosen such that the areas above and below the approximating function were the same (with W plotted on a linear scale versus phase angle).

The interesting result of this study was that the simple approximation 3 yielded the same matrix, within a few per cent, as the more exact solutions 1 and 2. It was then concluded that the optimized three-region gun must represent an approximation to an exponential gun of the modified type.

Studies of the transformation characteristics of a three-region gun by Knechtli and Beam[48] confirm this hypothesis. The main result of these studies will be summarized briefly. The impedance transformation of the low-noise gun, schematically shown in Fig. 5.25, was measured at a frequency of 3000 Mc with a noise-current standing-wave detector (Section 5, page 288). The measured noise current SWR ρ and the position of the noise minimum ψ with respect to the reference plane h define a normalized space-charge-wave impedance z_h according to Eq. 5.21. This impedance is plotted in Fig. 5.26 for various electrode potentials V_{0c} and V_{0d}. The drift potential and the beam current were constant, $V_{0h} = 650$ volts and $I_0 = 0.3$ ma. The electrode spacing $a–b$ was 0.070 in., $b–c$ was 0.180 in., and $c–d$ equaled 0.165 in. The cathode diameter was 0.025 in.

As the space charge in the beam considered is relatively small, the potential inside the beam along the gun is very nearly the potential

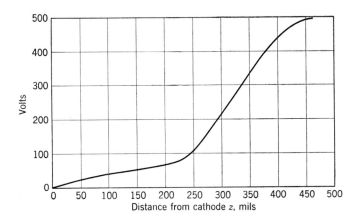

Fig. 5.22. The voltage along the beam axis as a function of distance from the cathode in a low-noise gun.

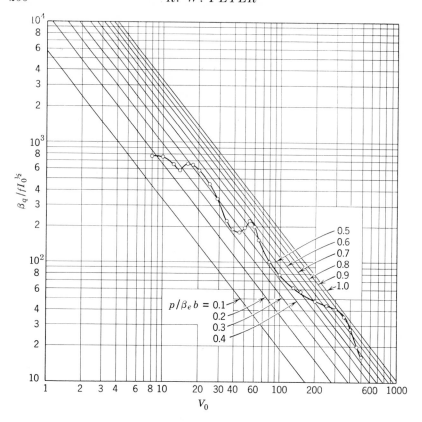

Fig. 5.23. Normalized plasma-propagation constant versus beam voltage V_0 in a low-noise gun.

produced by the electrodes only, with the exception of the region close to the cathode where the electron velocity is small. Between electrodes b, c, and d, the potential is essentially linear; i.e., the gradient is constant. Each section of linear potential increase, therefore, represents approximately an exponential transformer section according to Fig. 5.16, with a particular constant exponent k.

We may, therefore, think of a three-region gun Fig. 5.25, as consisting of three exponential transformer sections a–b, b–c and c–d, of which the first one is of the modified type (Fig. 5.19). It is easily seen that, if the normalized impedance at the cathode $z_a = r_a \approx 1$, a match ($\rho = 1$) at plane h is obtained if approximately

$$k_1 = k_2 = k_3$$

i.e., if the gradient in all three regions is the same.   This is borne out by the experiment.   The condition for constant gradient throughout the gun requires electrode voltages $V_{0c} = 50$ volts and $V_{0d} = 270$ volts if $V_{0h} = 650$ volts is given.   For these potentials indeed, Fig. 5.26 shows a measured impedance close to perfect match.   Any departure of one of the two potentials from the value corresponding to the constant-gradient condition leads to a higher SWR.   According to Fig. 5.8, the normalized impedance at the cathode, r_a/R was 0.8 for the current density used, 0.1 ampere per cm^2 at 3000 Mc.

An increase of potential on any electrode shortens the phase angle through the gun and, therefore (Fig. 5.25), rotates the normalized impedance z_h in a left-handed sense.   Furthermore, a change of V_{0c} has a larger effect on z_h than an equal change in V_{0d} corresponding to the fact that the distance b–c (affected by a V_{0c} change) measured in plasma-wave lengths is longer than the distance c–d (affected by a V_{0d} change).   Figure 5.25 indicates that a considerable range of impedances z_h can be reached without changing the drift space between the gun end d and the reference plane h.   If the reference plane is shifted away from the gun, the pattern of curves rotates in a right-handed sense.

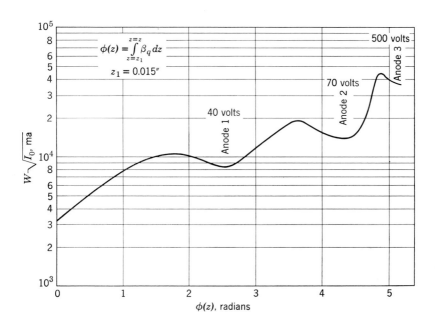

Fig. 5.24.   Characteristic impedance versus distance in a low-noise gun.

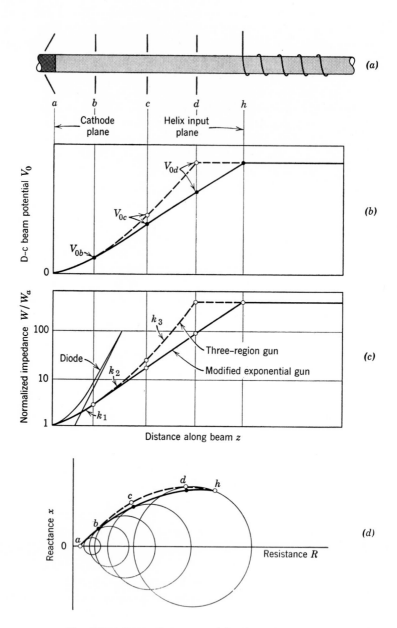

Fig. 5.25.　Low-noise gun used for data of Fig. 5.26.

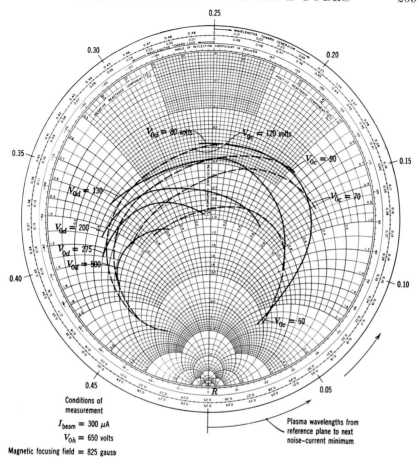

Fig. 5.26.   Measured beam-impedance transformation of a three-region low-noise gun.   The curves indicate the position and amplitude of the noise-current fluctuation in the beam produced by a three-region gun at various electrode potentials.

Matching for Higher-Order Modes

As we discussed in Section 2, page 228, there exists an optimum noise current standing-wave ratio ρ_n and an optimum angle ψ_n for every higher-order space-charge-wave mode at the helix input plane for which the nth mode induces a minimum amount of noise into the growing traveling wave.   Knechtli[49] has observed the second ($n = 2$) radial mode experimentally and demonstrated that the three-region gun is capable of satisfying the condition for both the $n = 1$ and the $n = 2$ mode.

It is possible to design [42] an exponential (or a three-region) gun which provides matching for the first two radial modes. The essence of the method will be briefly described here. From Fig. 5.7, p. 234 we know that the plasma-frequency reduction factor p_n decreases with increasing number n for a given value $\beta_e b$ of the normalized beam diameter. Reference to Eq. 5.28 and Eq. 5.29 shows that both the characteristic impedance W of the beam and the phase constant β_q are proportional to p. The higher the mode number n, therefore, the smaller W and β_q. Consequently the total plasma angle between the cathode and the helix, $\phi_n = \int \beta_{qn} \, dz$, will be smaller for higher mode numbers n.

In general the optimization of the noise in the dominant space-charge mode ($n = 1$) by a given gun does not result in a simultaneous optimization for the modes $n > 1$. Increasing the plasma angle ϕ_1 of the dominant space-charge mode by $m\pi$ (with m an integer) does not affect the optimization of the dominant mode. An increase of ϕ_1 by π in a given structure will produce a smaller increase in ϕ_2. By judiciously adding increments of π to ϕ_1, it is possible to approach closely the optimum phase angle for one or more of the higher-order modes.[42] Figure 5.2 indicates that the optimum phase angle ψ_n increases with mode number n. Consequently, the condition $\phi_3 > \phi_2 > \phi_1$ is required for best adjustment.

For noise optimization the standing-wave ratio ρ has also to be properly adjusted.

Figure 5.2 requires that the noise-current SWR ρ_2 be somewhat larger than ρ_1. Qualitatively this condition is satisfied by an exponential gun, as may be seen by referring to one of the Figs. 5.16. The parameter h which corresponds to the second mode is larger than the h for the fundamental mode (see Eq. 5.59). At constant V_0, z, and b, this means that k for the second mode is larger; i.e. the transformer is steeper, and consequently (Fig. 5.14) the noise current SWR ρ_2 will tend to increase, as required. The optimum condition for the next higher space-charge-wave mode ($n = 3$) can still be matched approximately. Space-charge-wave modes ($n > 3$) will be mismatched more and more as n increases. Their noise contribution fortunately decreases, as discussed in Section 2, page 235, for conventional traveling-wave tubes with relatively small $\gamma a (\leq 2)$.

4. Causes of Additional Noise

In Section 1 we surveyed the various causes of noise in traveling-wave tubes which contribute independently to the total noise factor.

In Section 3, the noise-factor contribution resulting from the basic emission fluctuations was discussed. In this section, we shall list in some detail the mechanisms of major *additional* causes of noise and their contributions to the amplifier noise factor. Most of this excess noise can be avoided by correct construction of the amplifier and by operation of the tube under the necessary optimum conditions as far as vacuum, magnetic fields, and applied voltages are concerned. We shall therefore present only as many details as are required to understand how to avoid the various causes of noise.

Noise Due to Current Interception

If an obstacle, for instance a knife edge or the rim of an aperture, is placed into the stream of electrons, part of the stream will be intercepted. As, however, the number of electrons in each individual partial stream varies randomly, the intercepted current has a random fluctuation. As North[31] has shown, a fluctuation current is added to the stream passing through the intercepting electrode. North's theory, however, holds only for a beam of small transit angle. Beams in traveling-wave tubes, however, are so long that one must consider the effect that interception has on the velocity fluctuation. Furthermore, the position of interception within the noise standing-wave pattern along the beam is of importance.

This is shown schematically in Fig. 5.27. A fine mesh A is placed across the beam and intercepts with uniform probability a portion

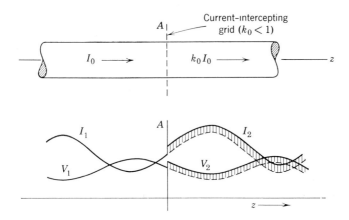

Fig. 5.27. Effect of a current-intercepting grid on the noise in a drifting, laminar-flow electron beam.

$(1 - k_0)$ of the direct current I_0. The mean-square noise current and velocity fluctuation in the beam just beyond the intercepting grid are given by Beam[50] in the following expressions:

$$\overline{i_2{}^2} = k_0{}^2\overline{i_1{}^2} + (1 - k_0)k_0 2eI_0 \,\Delta f \tag{5.64a}$$

$$\overline{v_2{}^2} = \overline{v_1{}^2} + \frac{1 - k_0}{k_0} \frac{2e\,\Delta f}{I_0} \left(\frac{kT_c}{mv_0}\right)^2 \tag{5.64b}$$

By introducing the kinetic potential $V_1 = v_1 v_0 m / e$ instead of the velocity fluctuation v_1, Eqs. 5.64 may be rewritten as follows:

$$\overline{i_2{}^2} = k_0{}^2\overline{i_1{}^2} + (1 - k_0)2ek_0 I_0 \,\Delta f \tag{5.65a}$$

$$\overline{V_2{}^2} = \overline{V_1{}^2} + (1 - k_0)\frac{2(kT_c)^2 \,\Delta f}{ek_0 I_0} \tag{5.65b}$$

These equations are easily understood physically. The mean-square values of current and velocity after interception consist of two components: the (reduced) initial fluctuation term and the new "partition-noise" term. In the case of the noise current i_2, Eq. 5.65a, the original current fluctuation i_1 is reduced to $k_0 i_1$, just as the direct current I_0 is reduced to $k_0 I_0$. To this first term a full shot-noise "partition-noise" current is added. Its value indicates that the intercepted fraction $(1 - k_0)$ of the passing current $k_0 I_0$ is full shot noise.

The mean-square kinetic-potential fluctuations $\overline{V_2{}^2}$, Eq. 5.65b, after interception still conserve the original $\overline{V_1{}^2}$, to which is added a fraction $(1 - k_0)$ times (essentially) the Rack noise value, Eq. 1.74 of Chapter 1. This additional term indicates that the intercepted current has added (by its absence) a fully random velocity component.

If most of the current I_0 is intercepted on the grid electrode, that is, if $k_0 \ll 1$, Eqs. 5.65 reduce to

$$\overline{i_2{}^2} = 2e(k_0 I_0) \,\Delta f \tag{5.66a}$$

$$\overline{V_2{}^2} = \frac{2(kT_c)^2 \,\Delta f}{ek_0 I_0} \tag{5.66b}$$

Equations 5.66 indicate that the small fractional current $k_0 I_0$ which passes the grid is almost completely randomized. A highly intercepting grid can, therefore, be used for noise-current calibration (see Section 5, page 292).

In microwave-tube design, the more important problem is current interception on a round aperture or cylindrical drift tube. In this case the probability of interception changes from zero to unity in the region of the electrode edge. The width of the probability transition region depends upon the strength of the magnetic field constraining the electron paths. The coefficients of the various terms in Eqs. 5.65 are no longer determined by the current interception $(1 - k_0)$ only, but become dependent on the probability distribution across the beam. Equations 5.65 may then be written as follows:

$$\overline{i_2{}^2} = \Omega_1{}^2\overline{i_1{}^2} + \Omega_2 2eI_0 \, \Delta f \tag{5.67a}$$

$$\overline{V_2{}^2} = \overline{V_1{}^2} + \frac{\Omega_2}{\Omega_1{}^2} \frac{2(kT_c)^2 \, \Delta f}{eI_0} \tag{5.67b}$$

with Ω_1 and Ω_2 defined by

$$\Omega_1 = \frac{1}{I_0} \int_A k_0 J_0 \, dA \tag{5.68a}$$

$$\Omega_2 = \frac{1}{I_0} \int_A k_0(1 - k_0) J_0 \, dA \tag{5.68b}$$

and

$$I_0 = \int_A J_0 \, dA \tag{5.68c}$$

The current density J_0, as well as the average fraction of the current intercepted, k_0, is assumed to be a function of the location in the beam cross section.

The coefficients of Eqs. 5.67, for the case of a round beam of constant average current density J_0, partly intercepted by a round concentric aperture, are functions of the magnetic focusing field B (assuming confined flow). The field can be characterized in normalized form by

$$\alpha = B \sqrt{\frac{e}{m} \frac{e}{kT_c}} \tag{5.69}$$

$$\alpha = 4.52 \times 10^3 B_{\text{gauss}} T_c{}^{-\frac{1}{2}} \text{ (meters)}^{-1}$$

Figure 5.28 shows the coefficients for Eqs. 5.67 as functions of the normalized radius difference

$$\alpha(r_b - r_a)$$

(r_b = beam radius, r_a = aperture radius). The field parameter is

$\alpha r_b = 15$; curves for other field values are found in the original paper.[50]

What can be learned from the curves of Fig. 5.28? Increasing interception current is represented by increasing positive values along the abscissa. The fraction Ω_1 of the original noise current $i_1{}^2$, therefore, approaches zero as the interception grows. The partition-noise component Ω_2, created by the interception uncertainty near the aperture edge, however, reaches a maximum, and then decreases.

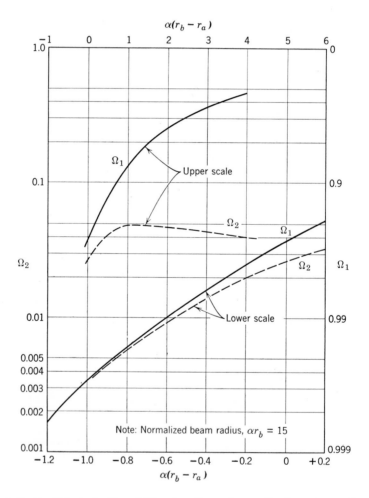

Fig. 5.28. Coefficients for Eqs. 5.67 as functions of normalized aperture radius αr_a at constant normalized beam radius $\alpha r_b = 15$.

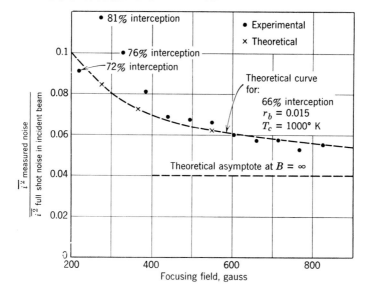

Fig. 5.29. Effect of a magnetic field on beam noise arising from current interception by an aperture.

This occurrence of a maximum partition noise must be expected as the current in the area of uncertainty is zero if the aperture is much larger or much smaller than the beam.

Similarly, the partition-noise component Ω_2 is expected to decrease as the magnetic field B (or the factor α) is increased. This is actually observed experimentally. Figure 5.29 shows the theoretical and measured dependence of relative partition noise on the focusing magnetic field.

The noise created by *collision of beam electrons with gas molecules* in the vacuum tube can be computed on the same basis. Colliding electrons are removed from their velocity classes just as if they had been intercepted on an electrode. They are then reintroduced at random into different velocity classes. Both processes contribute noise.

Both theory and experiment indicate that collision noise becomes negligible in vacua *better* than 10^{-6} mm Hg.

Noise Introduced by Secondary Electrons from Collector

The electrons striking the collector liberate a large number of slow secondary electrons (energies less than about 10 volts), and a few per cent of fast secondary electrons. The latter may be considered as reflected electrons, since they have energies nearly equal to that of the incident beam. Both types of secondaries introduce noise into the circuit.[51]

The returning stream of reflected secondaries is in synchronism with the reverse helix wave and contains noise of the primary beam plus

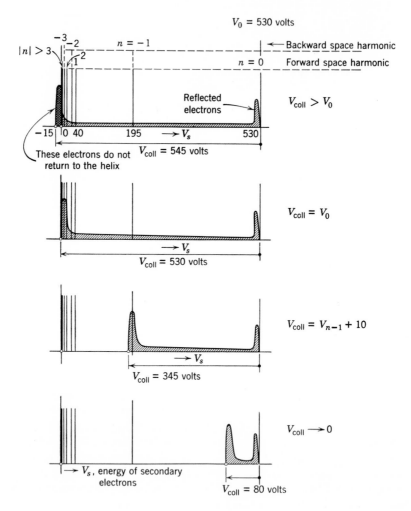

Fig. 5.30. Coupling of secondary electrons to space-harmonic waves of a filter helix. The ordinate of the cross-hatched area corresponds to the number of electrons per unit velocity.

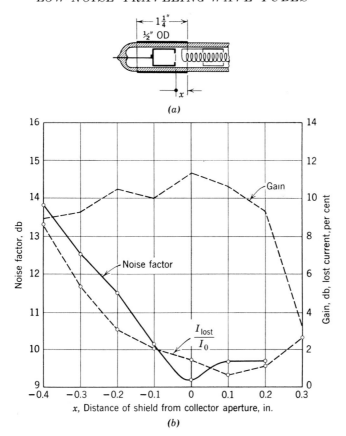

Fig. 5.31. Noise figure of an experimental traveling-wave tube as a function of the collector-shield position x.

shot noise of the secondary-emission process. This noisy secondary beam interacts with the fundamental helix wave traveling in the reverse direction. The noise is amplified, and, at the input coupling, may be partially reflected, thereby contributing to the total noise output. The beam continues on to about the region of the potential minimum, where it may noise modulate the primary beam before itself being reflected, thus increasing the total noise in the forward traveling beam.

The slow secondaries also introduce additional noise by the following mechanism. Every helix (or delay-line circuit) has an infinite number of forward and backward space harmonics due to its periodic nature.[52]

The phase velocities of these harmonics crowd together near very low values. If a random current, like the secondary current from the collector, is traveling back toward the gun at very low speed, say below 10 volts, it will introduce noise into the circuit via all synchronous backward space harmonics. This is shown schematically in Fig. 5.30. In this figure, the electron velocities of the slow secondary electrons as well as the fast reflected electrons entering the helix region are shown for various collector potentials. Also the forward and backward harmonics of a delay-line structure are shown. It can be seen that, under the most common condition of equal collector and circuit potential, most of the slow secondary electrons have velocities near higher space harmonics, and may thus couple noise into them.

In Fig. 5.31, the result of an experiment is shown which was designed to prove the effect of secondary electrons from the collector. A collector arrangement as sketched in Fig. 5.31*a* was used. It consisted of a nonmagnetic collector partially surrounded by a magnetic sleeve outside the tube and adjustable in position. Figure 5.31*b* shows the total noise factor of the amplifier measured at various sleeve positions. It is evident that the noise factor rises steeply in proportion with the intercepted current which is essentially equal to the secondary-emission current from the collector. The further the sleeve was moved away from the collector (larger negative values of x), the higher was the interception current, and the higher was the noise factor. The gain remained approximately constant.

It is, therefore, of utmost importance that secondary electrons be prevented from re-entering the interaction region. The essential factor in the collector design with magnetic sleeve is the strongly divergent magnetic field which has large radial components. Secondary electrons in such a field are deflected in a direction opposite to that of the impinging primary electrons. Any arrangement that produces a large enough transverse magnetic field will serve the purpose.

Noise Increase Due to Lens Effects

An increase of electron-beam noisiness can be caused by the energy-transfer effects of a strong electrostatic (or magnetic) lens through which the beam is passing. This has been long suspected, but only recently Knechtli[49] has provided theoretical and experimental proof. Electrostatic lenses of short focal lengths exist in all low-noise guns operating with potential discontinuities or velocity jumps. The reason for this noise increase is briefly as follows. The axial electron-velocity fluctuations are reduced in the low-noise gun by the increase of the d-c axial electron velocity. The transverse velocities of the electrons, initially at the cathode the same as the axial fluctuations, are not reduced in the process of axial acceleration in the gun. The transverse velocity fluctuations in the beam, therefore, are orders of magnitude larger than the axial velocity fluctuations after the beam has gained some d-c velocity. If, by some process, a fraction of the transverse fluctuation noise is transformed into axial fluctuation noise, the axial noisiness must be increased considerably.

An electrostatic lens has just this effect. An electron entering the lens at an angle to the lens axis leaves the lens at a different angle. In other words, transverse and longitudinal velocities are changed in the lens; therefore, if the transverse velocity fluctuations are initially larger, the axial fluctuations are increased.

The situation will be discussed in the simple example of an "Einzel lens," Fig. 5.32. An Einzel lens is the electron-optical analog of a convergent, optical lens: The d-c potentials on both sides of the lens are equal. The major electron velocities are assumed to be directed at right angles to the principal plane of the lens, and the transverse velocity components are assumed to be small compared to the axial velocities. Following reference 49, it is assumed that the lens equation applies

$$\frac{1}{L} + \frac{1}{L'} = \frac{1}{f} \tag{5.70}$$

where f is the focal length of the lens.

Let v_0 and v'_0 be the d-c velocities in the z direction, and w_0 and w'_0 the transverse d-c components before and after the lens. Let the corresponding velocity fluctuation components be v_1, v'_1, and w_1, w'_1,

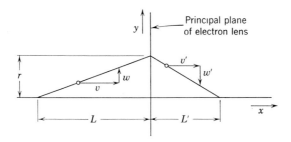

Fig. 5.32. Electron lens diagram used to compute the increase in the beam noisiness caused by a velocity jump.

respectively. Then we have the following total electron velocities before and after the passage through the lens:

	Before the lens:	After the lens:
In axial direction:	$v = v_0 + v_1$	$v' = v'_0 + v'_1$
In transverse direction:	$w = w_0 + w_1$	$w' = w'_0 + w'_1$

$$(5.71)$$

The conservation of the total kinetic energy across the Einzel lens requires that, for each electron,

$$v^2 + w^2 = v'^2 + w'^2 \qquad (5.72)$$

Furthermore, the average kinetic energy is conserved,

$$v_0{}^2 + w_0{}^2 = v'_0{}^2 + w'_0{}^2 \qquad (5.73)$$

From Eq. 5.72 we obtain

$$v_0{}^2 + 2v_0v_1 + v_1{}^2 + w_0{}^2 + 2w_0w_1 + w_1{}^2$$
$$= v'_0{}^2 + 2v_0v' + v'_1{}^2 + w'_0{}^2 + 2w'_0w' + w'_1{}^2 \qquad (5.74)$$

The fluctuation components have no average value,

$$\overline{v_1} = \overline{v'_1} = \overline{w_1} = \overline{w'_1} = 0$$

so that taking an ensemble average of Eq. 5.74 and use of Eq. 5.73 leads to

$$\overline{v_1{}^2} + \overline{w_1{}^2} = \overline{v'_1{}^2} + \overline{w'_1{}^2} \qquad (5.75)$$

The mean-square value of the axial fluctuation is

$$\overline{v'_1{}^2} = \overline{v_1{}^2} + \overline{w_1{}^2} - \overline{w'_1{}^2} \qquad (5.76)$$

The only unknown w'_1 is found with the help of the lens equation, Eq. 5.70. According to Fig. 5.32, where r is the distance from the axis at which an electron crosses the principal plane.

$$\frac{w}{v} = \frac{r}{L} \quad \text{and} \quad -\frac{w'}{v'} = \frac{r}{L'} \qquad (5.77)$$

Hence

$$\frac{w'}{w} = -\frac{L}{L'}\frac{v'}{v} \qquad (5.78)$$

According to Eq. 5.70,

$$\frac{-L}{L'} = 1 - \frac{L}{f} = 1 - \frac{r}{f}\frac{v}{w} \qquad (5.79)$$

so that we may write Eq. 5.78 in the form

$$w' = \frac{v'}{v}\left(w - \frac{r}{f}v\right) \tag{5.80}$$

and its time average is

$$w'_0 = \frac{v'_0}{v_0}\left(w_0 - \frac{r}{f}v_0\right) \tag{5.81}$$

Combination of Eqs. 5.80 and 5.81 leads to the desired expression for the transverse fluctuation after the lens

$$w'_1 = \frac{v'_0}{v_0}\left(w_1 - \frac{r}{f}v_1\right) \tag{5.82}$$

if we make use of the fact that $|v_1| \ll v_0$ and $|v'_1| \ll v_0$. Under the assumption that the transverse and longitudinal velocity fluctuations are uncorrelated, Eq. 5.82 can be written

$$\overline{w'_1{}^2} = \left(\frac{v'_0}{v_0}\right)^2\left[\overline{w_1{}^2} + \left(\frac{r}{f}\right)^2\overline{v_1{}^2}\right] \tag{5.83}$$

The ratio of the axial velocities before and after passage through the lens is obtained by combining Eqs. 5.80 and 5.81. For simplicity we assume that the beam enters the lens without transverse d-c velocity; i.e., $w_0 = 0$. Then Eq. 5.81 becomes

$$w'_0 = -\frac{r}{f}v'_0 \tag{5.84}$$

From Eqs. 5.73 and 5.84,

$$\left(\frac{v'_0}{v_0}\right)^2 = \left[1 + \left(\frac{r}{f}\right)^2\right]^{-1} \tag{5.85}$$

Equation 5.83 therefore may be written after averaging as

$$\overline{w'_1{}^2} = \frac{1}{1 + (r/f)^2}\left[\overline{w_1{}^2} + \left(\frac{r}{f}\right)^2\overline{v_1{}^2}\right] \tag{5.86}$$

and, by use of Eq. 5.86,

$$\overline{v'_1{}^2} = \frac{1}{1 + (r/f)^2}\left[\overline{v_1{}^2} + \left(\frac{r}{f}\right)^2\overline{w_1{}^2}\right] \tag{5.87}$$

Equations 5.86 and 5.87 express the fact that an electrostatic Einzel lens has a tendency to equalize transverse and longitudinal velocity-fluctuation noise: The larger quantity is reduced (in our case the transverse fluctuation $\overline{w_1^2}$), and the smaller quantity is increased (in our case the axial fluctuation $\overline{v_1^2}$). Similar equations hold for the more general case of unequal potentials on the two sides of the lens.

The focal length f in Eqs. 5.86 and 5.87 is affected by space charge and magnetic focusing fields. The space charge has an effect equivalent to that of a divergent lens. Its negative focal length, f_{sc}, partially compensates the converging effect of the positive focal length f of the electrostatic lens:

$$\frac{1}{f_0} = \frac{1}{f} - \frac{1}{f_{sc}} \tag{5.88}$$

The term $1/f_{sc}$ is negligible if the beam is ion-neutralized or has small space-charge density. The effect of a constraining magnetic field depends on the effective axial extent l of the electrostatic fields of the lens measured in radial-oscillation wavelengths λ_s of the electron-beam scalloping in the axial magnetic field B. The scallop wavelength for confined flow is

$$\lambda_s = 2\pi \frac{v_0}{\omega_c} \sqrt{1 + \frac{1}{\sqrt{1 + (\omega_c/\omega_p)^2}}}$$

where the cyclotron frequency $\omega_c = \frac{e}{m} B$ and the plasma frequency

$$\omega_p = \sqrt{\left|\frac{e}{m} \frac{\rho_0}{\epsilon}\right|}.$$

The qualitative behavior of the effective focal length f_L in the presence of a magnetic field—characterized by the scalloping wavelength λ_s— is shown in Fig. 5.33. At very small magnetic field (l/λ_s small), the effective focal length f_L is equal to the space-charge-corrected focal length of the lens f_0. At very large constraining fields (l/λ_s large), the effective focal length f_L goes to infinity, i.e., the electrostatic lens has no effect on the beam. There will be no additional noise due to the lens effect at very large focusing fields.

The theoretical relation, Eq. 5.87, between focal length of an Einzel lens and axial noise increase has been checked experimentally by Knechtli and found to agree well with the measurement. In a specific test where the focal length of the Einzel lens was $f_L = 0.110$ in. and the beam diameter 0.035 in., the measured beam noisiness increased 2.5 db. A similar increase of the tube noise factor must be expected.

The conclusion is: Direct-voltage discontinuities increase the longi-

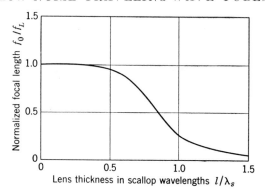

Fig. 5.33. Normalized focal length of electrostatic lens as a function of magnetic field characterized by scalloping wavelength λ_s.

tudinal beam noisiness (as long as the transverse fluctuations are larger than the longitudinal ones); smooth electron acceleration should be used for low-noise guns.

Growing Noise

During measurements of noise current along a beam, workers at MIT, BTL, and RCA found that occasionally the noise distribution, instead of being sinusoidally varying along the beam, suddenly increases very rapidly up to very large saturation values. Figure 5.34 shows an

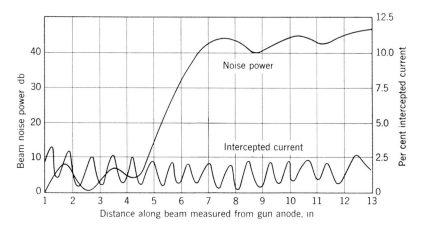

Fig. 5.34. Experimental curve showing noise increase along a drifting electron beam.

example of this peculiar noise behavior.   Extensive measurements at MIT,[53] have indicated that the onset of this increasing noise is moved further away from the gun, the larger the magnetic field in the cathode plane.   The field through the cathode in the RCA experiment reproduced in Fig. 5.34 was close to zero as a shielded gun was used in this experiment.[54]

So far no conclusive theory and explanation have been found to explain satisfactorily the behavior of this growing noise mechanism. One likely hypothesis is the one proposed by Rigrod.[55]   According to his explanation, noise is amplified from the cathode on, owing to scalloped-beam amplification in the various bands up to highest frequencies.   This noise would soon saturate the beam, and consequently nonlinear interference between noise in various frequency bands would occur.   Noise in the 3000 Mc region accordingly might originate by intermodulation between 10,000 and 7000 Mc noise, etc.   This theory may explain why it has not been possible to excite 3000 Mc-signal waves of this type.

Whatever the explanation of this phenomenon, all the available experimental evidence indicates that it is important to have a large enough constraining field at the cathode and through the gun region for lowest noise in the beam.

Noise Increase Caused by Nonuniform Electron Emission

Of all the causes of additional noise, the increase of noise due to imperfect electron emission is probably the most important.   Experience in producing low-noise traveling-wave tubes indicates that seemingly identical tubes may have very different noise factors.   These tubes could have differed only in the quality of their cathodes.   Aging of tubes may either decrease or increase the noise factor considerably. More so than in most other tubes, the production of good and reproducible cathodes is of extreme importance in low-noise traveling-wave tubes.

Large variations of tube noise factor can be caused by the non-uniformities of the emitter.[56]   If the cathode were made of an ideal single crystal, aligned in the proper crystal direction, there could be no difference of emission over its surface.   However, an oxide cathode consists of individual small particles which hang together in a relatively loose way.   The density of an average oxide cathode is around 1.2, the highest density obtained in electrophoretically deposited oxide cathodes is reported to be 2.5, while the crystalline barium carbonate would have a density of 3.5 grams per cc.

Aside from the statistically distributed small nonuniformities due to the grain size, there exist larger nonuniformities across the surface. Electron-optical images of cathodes show large patches emitting at very different current densities. The differences can be very large under nonsaturated conditions.

The matrix-type cathode, whose emitting compound is imbedded in a tungsten or nickel matrix, also shows large nonuniformities on top of the fine-grain structure of the matrix. Figure 5.35 shows electron-optical images of a 0.010 in.-diameter impregnated matrix-type cathode at various temperatures. The individual dots visible in the first three pictures indicate individual emitting pores between the matrix grains. As the temperature, and with it the emission-current density, increases, the space-charge smoothing washes out first the small and then the gross nonuniformities of emission. But it should be noted that, even at 1020° C, a somewhat nonuniform current distribution is seen to exist.

These nonuniformities may be caused by various inhomogeneities of the cathode characteristics which determine the emission: The work function and the surface potential may vary owing to absorption of poisoning molecules or atoms. The surface temperature may be lower at the edge of the cathode than at the center. The resistance drop through the emitting oxide layer may vary by a fraction of a volt at normal operating current densities. Nonuniform pore distribution near the surface, where the maximum potential drop through the oxide layer is concentrated, may also affect the uniformity of the conductance. The physical condition of the cathode surface is important, since emission from pores and crevices may easily lead to very different electron velocities further along the beam.

As long as there are no quantitative data on the nonuniformities of a particular cathode, it is hard to estimate their effect on beam noisiness and amplifier noise factor. However, useful qualitative conclusions may be drawn from a theory, which assumes those quantities to be known. Beam[57] has developed a theory which describes the effect of any emission nonuniformities on mode excitation in the beam. He assumes the velocity and the current density to vary over the emitting surface, and the electrons to mix in the transverse direction, owing to their random transverse initial velocities in a finite axial focusing field. The space-charge-wave excitation is studied at a reference plane some small distance in front of the cathode plane. It is assumed that no space-charge-wave interaction, and therefore no change in spectral distribution of electron velocities, has taken place between the cathode and the reference plane. It is assumed that, from the reference plane on, space-charge-wave concepts will hold.

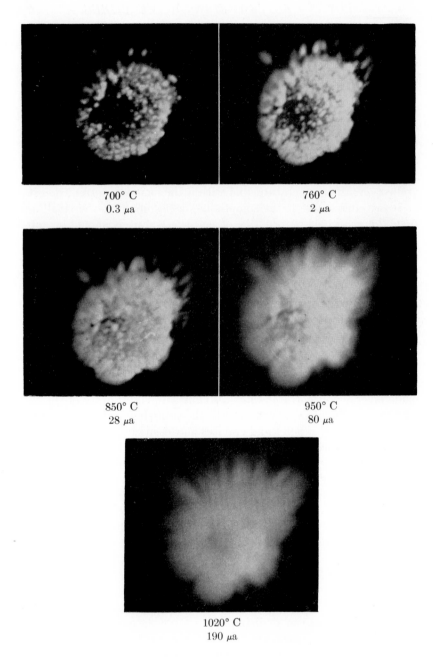

700° C
0.3 μa

760° C
2 μa

850° C
28 μa

950° C
80 μa

1020° C
190 μa

Fig. 5.35.   Electron optical images of a 0.010-in. diameter L-type matrix cathode.

The amplitudes of all higher-order modes excited in the reference plane can be computed if the initial conditions over the cathode surface are known.

The significance of this theory lies mostly in the qualitative conclusions which can be drawn from evaluating the formulas for specific simplified examples. Assume, for example, a *cathode with two different surface potentials* in areas which are close enough together and sufficiently small to allow complete mixing of the electrons originating from the two surfaces. At the reference plane, it will then be impossible to tell whether a specific electron came from the lower- or higher-potential surface. The result turns out to be the same as if the cathode temperature has been raised. A voltage difference of 0.1 volt approximately doubles the noisiness.

As a second qualitative example, consider the effect of *nonuniform current-density distribution across the beam* as a result of nonuniform emission. One can show[57] that higher-order space-charge modes are excited to much larger amplitudes than by a uniformly emitting cathode. The circuit is sensitive to all higher-order (radial) modes, although to a smaller degree than to the fundamental mode. In the three-region gun, it is possible to optimize at least the fundamental and the next higher space-charge mode, as discussed in Section 3, page 269. It is difficult to optimize simultaneously for more than two or three modes. Thus the higher modes, when excited, will increase the over-all noise factor.

From this discussion we can only conclude that the traveling-wave-tube designer has to pay a great deal of attention to producing a cathode that is as uniform in all respects as possible. Electrolytic, electrophoretic, and electrostatic oxide deposition on the cathode have been tried to achieve dense, uniform surfaces. High-pressure spraying of layers of very small (triple carbonate) particles has been used also. Shaving of the oxide cathode after deposition is another method of obtaining a uniform surface. Unfortunately, there is still too much magic and too little understanding of the important quantities in cathode making. At least, we know that the traveling-wave tube is a very sensitive detector of cathode nonuniformities.

5. Methods for Measuring Beam Noisiness and Amplifier Noise Factor

We have shown in previous sections that the noise impedance in the beam has to have a particular value z_h at the helix input if the lowest noise factor is to be obtained in a traveling-wave amplifier. It has also been shown in Chapter 3 that the parameter $(S - \Pi)$ determines the noise factor of the amplifier. In order to optimize a traveling-wave amplifier for lowest noise factor, we have to know the standing-wave ratio ρ and angle ψ of the noise current at the helix input; and, in order to compute the expected noise factor, we have to know the value of $(S - \Pi)$.

Remembering the analogy between space-charge-wave propagation along an electron beam and electromagnetic-wave propagation along a transmission line, we look for analogous measurements on the transmission line. There, impedance measurements are made with the standing-wave detector which has been developed into an instrument of considerable sophistication. In its original form, which is still the most versatile one, the SWR detector consists of a current- or voltage-sensitive probe which is loosely coupled to the transmission-line fields, and which may be moved along the line. A similar method is used to measure the noise-current distribution along an electron beam. Only current-sensitive probes have been used so far. The kinetic voltage, which corresponds to the velocity fluctuation in the stream, is hard to detect directly. In order to measure the noise current, one uses a gridless cavity sliding along the beam, as will be discussed in detail later.

The beam transmits not only the fundamental space-charge-wave mode but also all the higher-order modes which are excited by the noise current at the cathode. The cavity probe is sensitive in various degrees to all higher-order modes, and, therefore, particular care has to be taken to separate out the effects of various modes. This will be discussed later in greater detail.

Noise-Current Measurements

The simplest noise-current measuring apparatus is shown in Fig. 5.36. It consists, as mentioned, of a cavity sliding along the beam. The noise from the cavity is fed into a receiver, detected, and displayed on an indicator (a meter or oscilloscope). The standing-wave ratio is measured by moving the cavity from maximum to minimum of the noise-current wave on the beam.[58,59]

This simple and straightforward system, however, has a number of serious defects. The most obvious one is its limited sensitivity. The noise added by the beam on top of the thermal noise of the cavity and on top of the receiver noise cannot be detected if it is of the same order of magnitude as or smaller than the background noise. By introducing a chopped beam and a synchronously gated (or phase-sensitive) receiver, the sensitivity of this system can be greatly increased, as

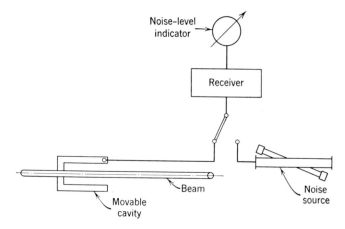

Fig. 5.36. Elementary beam-noise measuring apparatus.

will be described. A second defect of the system of Fig. 5.36 is that it contains no continuous absolute calibration. If the receiver gain changes during the measurement, the measured noise does not correspond to the actual noise. By a continuous comparison of the noise coming out of the cavity with the noise coming from a stable, standard noise source, this defect can be removed. In Fig. 5.37 such a system is depicted.[60] It is in some aspects related to R. H. Dicke's radiometer. The noise introduced into the cavity by a chopped electron

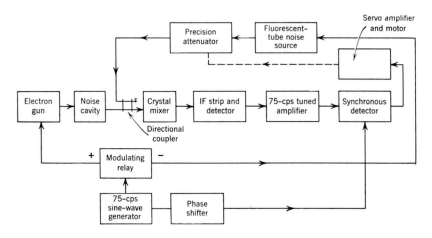

Fig. 5.37. Automatic beam-noise measuring apparatus.

beam (75 cycles), shown on the left side of the diagram, is fed through a crystal mixer, IF amplifier, and detector into a narrow-band amplifier (tuned to 75 cycles per sec) and from there into a synchronous or phase-sensitive detector. Standard noise power from a pulsed (75-cycle) fluorescent-tube noise source is also fed via a precision attenuator into the cavity. This noise, used as a comparison standard, is switched on when the beam from the electron gun is shut off and vice versa. A modulating relay driven by a 75-cycle sine-wave generator is used for the switching-operation. The output of the synchronous detector then activates a servo amplifier which

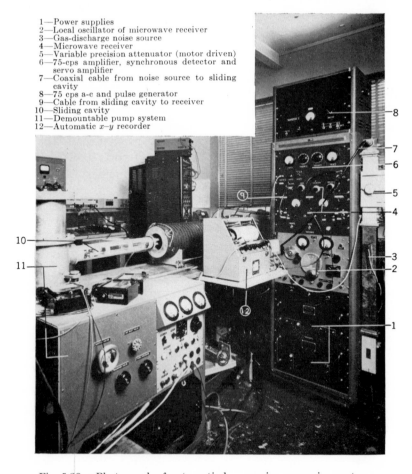

1—Power supplies
2—Local oscillator of microwave receiver
3—Gas-discharge noise source
4—Microwave receiver
5—Variable precision attenuator (motor driven)
6—75-cps amplifier, synchronous detector and servo amplifier
7—Coaxial cable from noise source to sliding cavity
8—75 cps a-c and pulse generator
9—Cable from sliding cavity to receiver
10—Sliding cavity
11—Demountable pump system
12—Automatic x–y recorder

Fig. 5.38.   Photograph of automatic beam-noise measuring system.

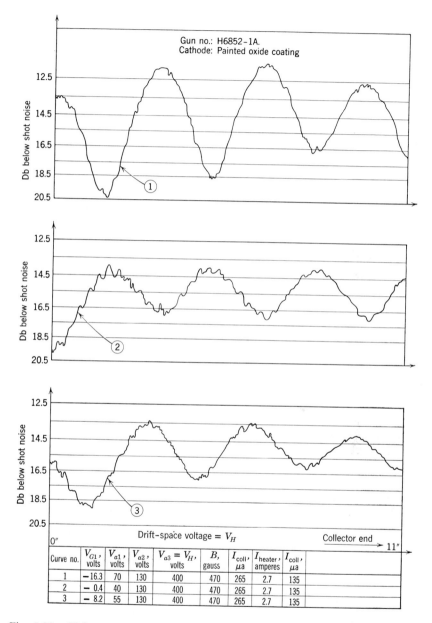

Fig. 5.39. Noise-current distribution measured along a beam as a function of gun potentials.

drives a motor to adjust the precision attenuator to reduce the detector output. At balance, i.e., no output, the standard noise introduced into the cavity equals the noise introduced by the beam into the cavity, and an absolute measurement is obtained if the attenuator calibration is known.

It is of advantage to feed the output of the standard noise source through the cavity in order to obtain identical noise spectra from beam and noise standard. If the standard noise were fed directly into the receiver, white noise would have to be compared with the limited spectral distribution of the beam noise picked up by the cavity.

The position of the precision attenuator is a direct measure of the noise introduced by the electron beam. Its position is transferred via a precision voltage divider onto an x–y recorder. The x movement of the pen is provided by a potential proportional to the position of the cavity which in this particular system is driven by a reversible motor. In Fig. 5.38 a photograph of this automatic noise-measuring system is reproduced.

Three typical recordings of noise from a low-noise gun are shown in Fig. 5.39. The higher-order modes contribute usually only in a small degree to the total measured noise. They affect, therefore, the minima in the fundamental standing-wave patterns much more than the maxima, as is, for instance, seen clearly in the top recording of Fig. 5.39. As the higher-order modes have different wavelengths than the fundamental, analysis of these patterns may help reveal the contributions of the first few higher-order modes to the recorded noise current. By adjusting the voltages of the gun so that approximately unity standing-wave ratio of the fundamental mode is obtained, the higher modes may appear more clearly.

In order to translate the measured receiver output into beam noise current, it is necessary to determine the gap impedance and coupling coefficient of the cavity as seen by the beam. By measuring the receiver output when the cavity is excited by a beam with full shot noise, an absolute calibration can be obtained that includes both of these constants. Full shot noise can be achieved by intercepting the beam with a fine grid of small transmission (see Eq. 5.66a), or by reducing the temperature of the cathode until the beam becomes temperature-limited. (The latter procedure is dangerous because the cathode may be damaged by temperature-limited operation.)

Noise-Factor Measurement

Much has been written on the measurement of amplifier noise factor. This is not the place to discuss in detail the various methods that

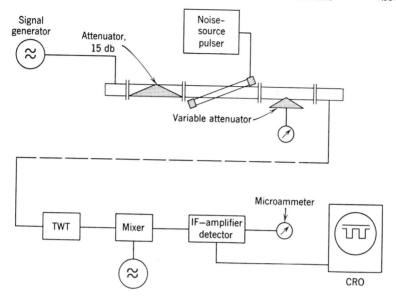

Fig. 5.40. Elementary noise-figure measuring apparatus.

have been employed. We shall describe here only one system which has been used extensively and has proved very satisfactory. Figure 5.40 shows the basic arrangement of this noise-factor measuring method.[61,62]

The noise produced by the amplifier is compared with calibrated noise introduced to the tube from a pulsed noise source through an attenuator. The noise source is usually a gas-discharge tube which has a constant output of about 16 db above thermal noise. The output of the traveling-wave amplifier is fed through a receiver and rectified in a detector. The detector output is displayed on a cathode-ray oscilloscope. If the noise source is pulsed on and off (for instance at a 60-to-80-cycle rate), the display on the cathode-ray oscilloscope permits an easy comparison between amplifier noise and added standard noise. By varying the attenuator, the added standard noise is made equal to the amplifier noise. The noise factor then is obtained directly from the attenuator setting and the temperature of the standard. Figure 5.41 shows noise-figure measurements made to determine the effect of varying the helix input position. One can see that there is an optimum position for each value of magnetic field (which determines the average beam radius).

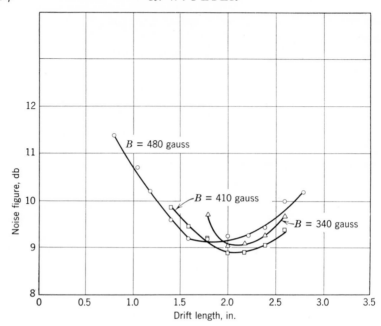

Fig. 5.41. Dependence of noise figure upon helix position as measured in a demountable vacuum system with apparatus shown in Fig. 5.40.

One may combine the noise-factor and gain measuring system of Fig. 5.40 with the automatic noise-current measuring system of Fig. 5.37 to measure and record noise factor, noise current, and gain with the same apparatus. The equipment setup is shown schematically in Fig. 5.42. The beam emitted by an electron gun on the left side of the figure is shot first through the noise-measuring cavity and then through a helix, both of which are connected with the measuring system of Fig. 5.43. In the "noise-figure"-measuring position, the IF-amplifier output is fed for one half cycle through a tap of the second-detector load resistance into the tuned amplifier and synchronous detector. In the other half cycle, the relay connects to the full output of the second detector. If a square-law detector is used, the tap should be at half power in order to give zero output of the synchronous detector after the tuned amplifier 1. If there is some output, the servo voltage actuates the servo motor to adjust the attenuation for correct zero output again.

In order to operate the detector at approximately constant input signal, an automatic volume control is incorporated. The output of the second detector is compared with a reference voltage, and the difference signal is amplified in the tuned amplifier 2 and detected in the synchronous detector, which supplies, via a low-pass filter, the AVC voltage to the IF amplifier.

In the "noisiness" position of the noise-measuring setup, the connections are exactly those described for the noise-current measuring system of Fig. 5.37. In the "gain" position, the amplifier 2 controls the servo system. The attenuator setting is then read with amplifier in, and amplifier by-passed. The difference in setting is the gain.

The measurement of the noise current along the beam allows us to compute directly the following:

$$\text{Standing-wave ratio } \rho = \sqrt{\frac{i^2_{\max}}{i^2_{\min}}}$$

Helix position ψ for optimum noise figure

$$\text{Noise parameter } S = \frac{1}{W} \sqrt{\Psi_{\max} \Psi_{\min}} = \frac{1}{4\pi f W} \sqrt{\left|i^2_{\max}\right| \left|i^2_{\min}\right|}$$

These computations are meaningful only if the standing-wave pattern does not change with distance. Actual measurements usually show a changing pattern (Fig. 5.39). Although some of this change may be attributed to higher radial space-charge modes, there also appear to be other mechanisms at work that are not yet explained. Thus the experimenter must use some judgment in estimating the value of S from the data.

The noise figure has been shown to depend upon $(S - \Pi)$, where Π

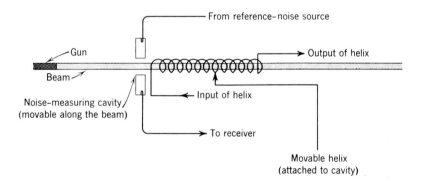

Fig. 5.42.   Combined noise-figure and beam-noise measuring system.

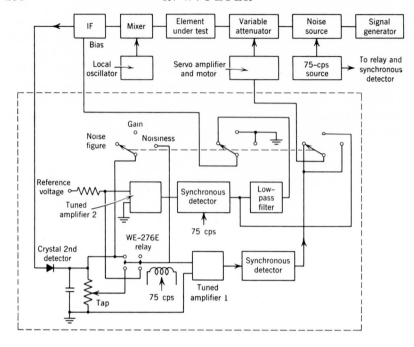

Fig. 5.43. Automatic beam-noise and noise-figure measuring apparatus.

is a measure of the real part of the correlation between current and velocity fluctuations. If $\Pi = 0$, then the standing-wave measurement alone should predict the minimum noise figure.

Measurements of this kind have shown that traveling-wave-tube noise factors computed from the measured noise-current distribution on the beam and, therefore, from $\Psi_{max}\Psi_{min}$ in the beam are in reasonable agreement with the measured noise factor. It should be noted, however, that these experiments in demountable vacuum systems gave noise figures of about 10 db or worse (as compared to the 5-db values found in the better sealed-off tubes).

6. Design, Construction, and Performance of Low-Noise Traveling-Wave Tubes

In this last section, we shall draw design conclusions from the foregoing sections. We will present a few rules which have to be followed to obtain an optimized low-noise traveling-wave amplifier. Unfortunately, these rules do not guarantee, as they stand now, completely optimized performance. There is still a certain amount of experimental know-how involved, particularly in producing a good cathode, which cannot easily be put into words and rules.

The construction of a particular S-band low-noise traveling-wave tube,[46]† will be described, and the measured performance reported. The best noise factors measured on that tube were below 5 db. Other low-noise traveling-wave tubes have been discussed in the literature.[63,64]

Design of Low-Noise Traveling Wave Tubes

Here we present a number of important rules which should be observed in the design of low-noise traveling-wave tubes. These rules are concerned with (1) optimum design of the interaction elements (helix and beam), and (2) optimum design of the low-noise gun. The expression for the minimum attainable excess-noise figure of a given tube, Eq. 5.14 is

$$F_{\min} - 1 = |R| \sqrt{(4 - \pi)} \frac{T_c}{T} \left(1 + \frac{d}{x_1}\right) \tag{5.14}$$

It is immediately obvious that, in order to get the best noise performance, we must

1. *Make electron temperature T_c small.* Substituting an oxide cathode ($T_c = 973°$ K) for an L cathode ($T_c = 1173°$ K) reduces the excess-noise factor by about 1 db.
2. *Make loss coefficient $(1 + d/x_1)$ small.* According to Fig. 5.1, this means reducing the loss parameter[16]

$$d = 0.018L/NC$$

This may be achieved by reducing the distributed loss L by silver-plating the helix (at least up to the lumped attenuator), by using low-loss helix supports, and by choosing a wire-diameter-to-helix-pitch ratio of 0.3 to 0.6. Curves for the attenuation of helices are given elsewhere.[65] According to the definition of the gain factor,

$$C^3 = \frac{KI_0}{4V_0}$$

large C can be obtained by increasing the circuit impedance K. According to Pierce's[16] curve for K and Q, the above conditions are best fulfilled by a beam-to-helix radius ratio b/a close to unity. Dan-

---

† Experimental tube number RCA A1038, commercial tube number RCA 6861.

ger of current interception by the helix limits the optimum ratio to about $b/a = 0.8$.

A different way of increasing the circuit impedance K without reducing $\gamma_0 b$ consists in using filter techniques. By introducing periodic reflections along the delay line, its impedance K is increased but its amplification band width is reduced. The *filter-helix* circuit[66] is most suitable for low-voltage amplifiers. In this circuit, higher impedance is traded for smaller band widths:

$$K = \frac{E^2}{2\beta\epsilon v_g} = \frac{E^2}{2\beta\epsilon v}\left(1 - \frac{\omega}{v}\frac{dv}{d\omega}\right) \qquad (5.89)$$

where v = phase velocity of filter
  v_g = group velocity of filter
  ϵ = stored energy per unit length
  β = propagation constant of circuit wave, $\beta = \omega/v$

As the dispersion $dv/d\omega$ (which is negative) is increased in absolute value, K also rises. As the circuit loss also grows, the maximum permissible dispersion is limited by the attenuation of the specific circuit used. (Compare the proposal in reference 36, page 142.)

Large C may finally be obtained by increasing the current I_0 and decreasing the helix voltage V_0. The maximum current unfortunately is limited by the nonuniformities in electron emission which may increase the beam noise considerably.

3. *Minimize coupling of higher-order space-charge-wave modes to the growing wave.* According to some theories,[35,36] $\gamma_0 b$ should be made large: i.e., small beam voltage V_0 and large beam diameter $2b$ (Section 2, page 224).

4. *Place attenuator far from input.* A lumped attenuator which cuts the circuit effectively in two parts with power gains G_1 and G_2 is commonly used to stabilize traveling-wave amplifiers. In Chapter 3 it is shown that the excess-noise factor, Eq. 3.197 has to be multiplied by the coefficient

$$\left(1 - \frac{1}{G_1 G_2} + \frac{1}{G_1}\right) \qquad (5.90)$$

which requires that G_2/G_1 be made as small as possible. In good practical tubes $G_2/G_1 \approx 0.1$.

The factor, Eq. 5.90, can be made unity by using distributed nonreciprocal-loss "isolators" along the helix.[67] As, however, this factor is very small, this solution is practically never used.

The second part of the low-noise traveling-wave-tube design procedure deals with the *low-noise gun*. The most important factors to be considered are summarized here.

5. *Satisfy required condition for noise waves at circuit input.* The optimum SWR ρ and angle ψ, given in Fig. 5.2, should be satisfied for at least the first two radial space-charge-wave modes $n = 1$ and $n = 2$, by correct design of the gun as a multimode matching transformer (see Section 3, page 269).

6. *Avoid velocity discontinuities.* Abrupt changes of the d-c potential along the beam (as required in the velocity-jump gun) are usually accompanied by strong lens effects which increase the beam noisiness (Section 4, page 278). The noise-wave transformation through the gun should be as smooth as possible. Stated as a general rule,

$$\rho(z) \leq \rho_h \quad \text{(or } \rho_a \text{ whichever is larger)}$$

which means that the noise-current SWR $\rho(z)$ anywhere along the beam should always be smaller than the SWR ρ_h required at the helix input. In particular, ρ_a at the cathode should be chosen as close to unity as possible at the given frequency (see Fig. 5.8). The requirements for wide matching bandwidth go hand in hand with this condition.

7. *Use uniformly emitting cathode.* Every nonuniformity of the emitting surface (porosity, etc.,) and of its electrical qualities (bulk resistance, work function, etc.,) tends to increase the beam noisiness (see Section 4, page 284).

8. *Avoid current interception* (Section 4, page 271).

9. *Use extreme care in the cleaning, evacuation, and processing of the tube* to achieve best cathode emissivity, prevent ion destruction of the cathode and beam noise increase due to gas-electron collisions. The vacuum should be better than 10^{-6} mm Hg.

10. *Prevent re-entry of secondaries* from the collector (Section 4, page 275).

11. *Use a large enough magnetic focusing field in the cathode region.* This will avoid the appearance of growing noise waves (Section 4, page 283).

Construction of an S-band low-noise tube

A detailed description of an S-band low-noise tube has been given earlier,[46] and Fig. 5.44 shows a picture of the tube. This prototype

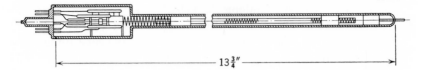

13¾"

Fig. 5.44. Experimental low-noise traveling-wave tube.

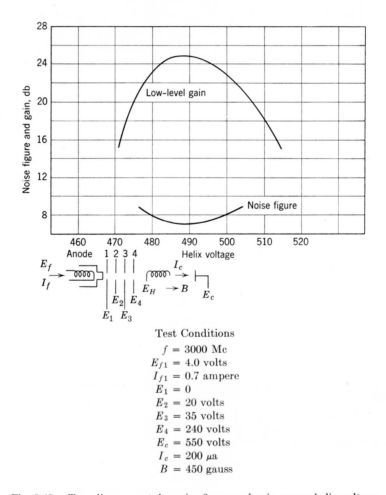

Test Conditions

$f = 3000$ Mc
$E_{f1} = 4.0$ volts
$I_{f1} = 0.7$ ampere
$E_1 = 0$
$E_2 = 20$ volts
$E_3 = 35$ volts
$E_4 = 240$ volts
$E_c = 550$ volts
$I_c = 200$ μa
$B = 450$ gauss

Fig. 5.45. Traveling-wave-tube noise figure and gain versus helix voltage.

has since been developed into a commercial product and has been improved in various ways. However, as far as its electrical character- istics are concerned, they are essentially still the same as in the original tube. We shall briefly discuss here some of the design factors that went into the prototype tube.

The gun-and-helix assembly was aligned concentrically by a pre- cision-bore envelope of Pyrex glass. The envelope was shrunk on a single mandrel with two diameters: one for the bulb, one for the helix. In the commercial tube, the alignment is obtained by three ceramic rods which support the helix and the gun. Gun and helix are now assembled as a single unit and introduced into the vacuum envelope. The three-region gun is operated as a "modified exponential gun." The original gun was assembled with three glass beads and was sup-

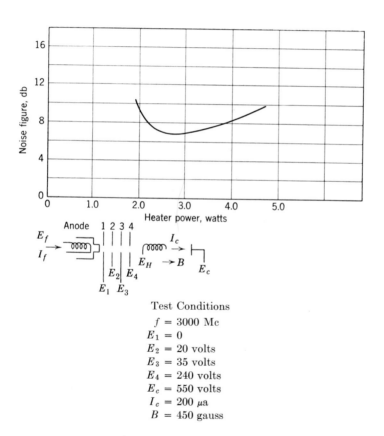

Test Conditions

$f = 3000$ Mc
$E_1 = 0$
$E_2 = 20$ volts
$E_3 = 35$ volts
$E_4 = 240$ volts
$E_c = 550$ volts
$I_c = 200 \ \mu a$
$B = 450$ gauss

Fig. 5.46. Traveling-wave-tube noise figure versus heater power.

ported by three pairs of spacer legs in the precision bulb. The helix was supported by a precision-bore quartz tube, which in turn was suspended by two ceramic rings hidden inside the capacitive-coupling rings. A lumped attenuator was placed two thirds of the helix length away from the gun end. It was made by suspending a very thin alumina film between the turns of the helix. After drying and firing of the film, it was sprayed with Aquadag. Tapering of the film toward both ends produced an adequate match. The helix potential was applied through the choke helix.

The collector was made of magnetic material; e.g., soft iron. It may be coated inside by some material with low secondary-emission ratio (lampblack). This helps to reduce further the secondary emission from the collector, which introduces extra noise as discussed earlier. The spacing of the helix to the gun of about 1 in., and between helix and collector of about $\frac{1}{8}$ in., was maintained by friction on the glass bulb in the prototype tube. The electrical data of helix and gun are the following:

Helix inside diameter	0.095 in.
Wire diameter	0.007 in.
(Silver-plated tungsten wire)	
Helix length	$9\frac{1}{2}$ in.
Helix turns per inch	60
Distributed loss approximately	6 db
Lumped loss approximately	50 db
Cathode diameter	0.035 in.
Electrode aperture diameter	0.100 in.

Measured Performance

The noise and gain characteristics of the RCA low-noise traveling-wave tubes A-1038 and 6861 are presented in this section. The A-1038 is the laboratory prototype tube from which the RCA 6861 was developed.† Although lower noise figures have been achieved in the 6861 than in the A-1038, the general behavior and sensitivity to voltage changes are the same in the two tubes.

In Fig. 5.45, noise factor and gain of the prototype tube are plotted versus helix voltage. It is seen that the maximum gain and minimum noise factor occur at the same helix voltage, as predicted by theory. Previously, lowest noise factor was often obtained at a somewhat lower voltage than that required for optimum gain. This was found

† The data presented here were made available by RCA Microwave Development Group at Harrison, N.J.

to be an indication that the space-charge-wave transformation was not optimized in the gun. It is evident from the curves that the noise factor is only slightly dependent on the helix voltage.

Figure 5.46 shows the noise-factor dependence upon the cathode temperature or heater power. The increase to the right is approximately proportional to the cathode temperature as required by theory. To the left of the minimum, the noise factor increases rapidly as portions of the cathode begin to saturate and the nonuniformities of emission increase.

Figure 5.47 shows that there is very little dependence of noise factor upon the beam current. Up to 800 μa, the noise factor can be optimized at the same value if the gun electrode voltages are adjusted as in the lower figure. The gain, of course, increases with current. As the beam diameter stays nearly constant owing to the constraining magnetic field, increased current means increased current density. Thus, the electrode potentials in the exponential gun have to be raised in order to keep the space-charge-wave phase angles between electrodes constant when the current density is increased.

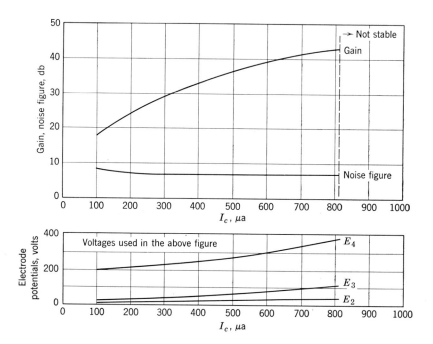

Fig. 5.47. Traveling-wave-tube optimum noise figure, gain, and electrode voltages versus beam current.

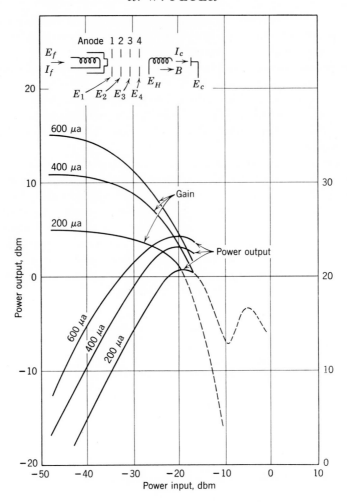

Fig. 5.48.   Power output versus power input for an A-1038 traveling-wave tube.

Power output versus power input and gain curves are plotted in Fig. 5.48 for various beam currents. It can be seen that a power input of 20 db below 1 mw produces saturation.

The following three figures pertain to the commercial tube (RCA 6861) which has a lower operating voltage, $V_H = 375$ volts, and larger $\beta_e b$. The dependence of noise factor upon relative changes of the different electrode potentials and the magnetic field B is shown in

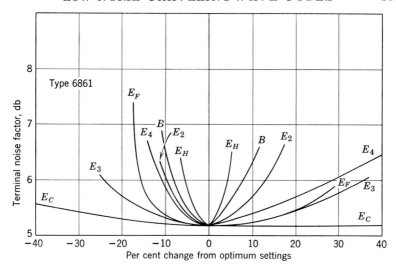

Fig. 5.49.   Effect of operating parameters on traveling-wave-tube noise figure.

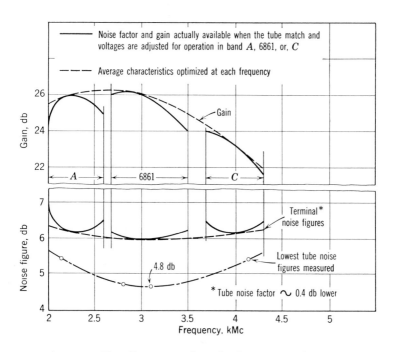

Fig. 5.50.   Traveling-wave-tube noise figure versus frequency.

Conditions for all 4 curves:
Parameters optimized at rated condition for minimum noise
Gain, db
Power output, dbm
Power input, dbm
(a)
Gain of 3.1 kMc signal*
Power input at 3 kMc, dbm
(b)
Power output at 3.5 kMc, dbm
Power input at 1.75 kMc, dbm
(c)
Power output at difference frequency, dbm
(2f1 − f2 = 3 kMc)
Power input at 2.9 kMc, (f)†
(d)
*Fixed power level = − 85 dbm
†In presence of f2 at fixed power level = − 40 dbm

Fig. 5.51. Large-signal behavior of 6861 traveling-wave tube.

Fig. 5.49. The optimum settings refer to 3000 Mc per sec. The terminal noise factor is slightly more than the tube noise factor owing to the loss in the coaxial cable input. (The indices refer to gun electrodes as shown in Fig. 5.48.) Note the dependence upon the magnetic field which probably arises from slight changes of the equilibrium radius of the beam. This effect is also evident from Fig. 5.41.

Figure 5.50 shows typical gain and noise-factor curves versus frequency. The bandwidth of about 1000 Mc around any center frequency between 2 and 4 kMc can be covered with a noise-factor variation of only about 0.5 db at fixed operating voltages, while the gain

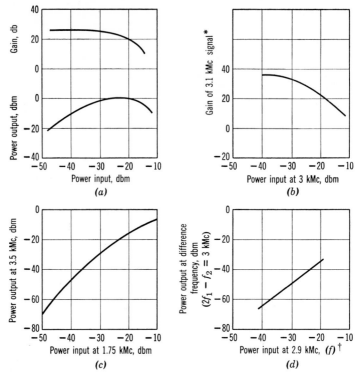

varies by about 2 db. Consider the dashed noise-factor curve which is obtained if the noise factor is optimized at every single frequency. It is seen that the lowest terminal noise factors are the same within 0.3 db. The curve for the best *tube noise factors* measured indicates a total variation of 1 db. This may be the result of mismatch at the input. It may, however, also be due to incomplete transformation in the exponential gun.

The curves in Fig. 5.51 show the nonlinear behavior of the low-noise tube 6861 and are essentially self-explanatory.

REFERENCES

1. H. A. Haus and F. N. H. Robinson, "Minimum Noise Figure of Microwave Amplifiers," *Proc. IRE*, **43**, 981 (1955).
2. R. Adler and O. M. Kromhout, "Transverse-Field Traveling-Wave Tubes with Periodic Electrostatic Focusing," *Proc. IRE*, **44**, 82 (1956).
3. J. R. Pierce, "A Theorem Concerning Noise in Electron Stream," *J. Appl. Phys.*, **25**, 931 (1954).
4. F. N. H. Robinson and R. Kompfner, "Noise in Traveling-Wave Tubes," *Proc. IRE*, **39**, 918 (1954).
5. R. W. Peter, S. Bloom, and J. A. Ruetz, "Space-Charge-Wave Amplification along an Electron Beam by Periodic Change of the Beam Impedance," *RCA Rev.*, **15**, 113 (1954).
6. C. K. Birdsall, "Rippled Wall and Rippled Beam Amplifiers," *Proc. IRE*, **42**, 1628–35 (1954).
7. L. S. Nergaard, "Analysis of a Simple Model of a Two-Beam Growing-Wave Tube," *RCA Rev.*, **9**, 585 (1948).
8. A. B. Haeff, "The Electron-Wave Tube—a Novel Method of Generation and Amplification of Microwave Energy," *Proc. IRE*, **37**, 4 (1949).
9. H. F. Webster, "Breakup of Hollow Electron Beams," *J. Appl. Phys.*, **26**, 1386 (1955).
10. J. R. Pierce, "Possible Fluctuations in Electron Streams Due to Ions," *J. Appl. Phys.*, **19**, 231 (1948).
11. K. G. Hernquist, "Plasma Oscillations in Electron Beams," *J. Appl. Phys.*, **26**, 1029 (1955).
12. T. G. Mihran, "Positive Ion Oscillations in Long Electron Beams," *Trans. IRE*, **ED-3**, 117 (1956).
13. R. Kompfner, "The Traveling-Wave Tube as Amplifier at Microwaves," *Proc. IRE*, **35**, 124 (1947).
14. R. Kompfner, "The Traveling-Wave Tube," *Wireless Engr.*, **24**, 255 (1947).
15. J. R. Pierce, "Theory of the Beam-Type Traveling-Wave Tube," *Proc. IRE*, **35**, 111 (1947).
16. J. R. Pierce, *Traveling-Wave Tubes*, D. Van Nostrand Co., New York (1950).
17. F. N. H. Robinson, "Space-Charge Smoothing of Microwave Shot Noise in Electron Beams," *Phil. Mag.*, **43**, 51 (1952).
18. J. R. Pierce, "A New Method of Calculating Microwave Noise in Electron Streams," *Proc. IRE*, **40**, 1675 (1952).

19. F. N. H. Robinson, "Microwave Shot Noise in Electron Beams and the Minimum Noise Factor of Travelling Wave Tubes and Klystrons," *J. Brit. IRE*, **14**, 79 (1954).
20. S. Bloom and R. W. Peter, "A Minimum Noise Factor for the Traveling-Wave Tube," *RCA Rev.*, **15**, 252 (1954).
21. J. R. Pierce and W. E. Danielson, "Minimum Noise Figure of Traveling-Wave Tubes with Uniform Helices," *J. Appl. Phys.*, **25**, 1163 (1954).
22. J. R. Whinnery, "Noise Phenomena in the Region of the Potential Minimum," Univ. Calif. Berkeley Electronic Research Lab. (Feb. 28, 1955).
23. H. A. Haus, "Noise in One-Dimensional Electron Beams," *J. Appl. Phys.*, **26**, 560 (1955).
24. D. A. Watkins, "Noise at the Potential Minimum in the High-Frequency Diode," *J. Appl. Phys.*, **26**, 622 (1955).
25. S. Bloom, "The Effect of Initial Noise Current and Velocity Correlation on the Noise Figure of Traveling-Wave Tubes," *RCA Rev.*, **16**, 179 (1955).
26. W. R. Beam, "Extension of the Effects of Initial Noise Current and Velocity Correlation on the Noise Figure of Traveling-Wave Tubes," *RCA Rev.*, **16**, 458 (1955).
27. P. K. Tien, "A Dip in the Minimum Noise Figure of Beam-Type Microwave Amplifiers," *Proc. IRE*, **44**, 938 (1956).
28. F. N. H. Robinson, "Microwave Shot Noise and Amplifiers," *Trans. IRE*, **ED-3**, 128 (1956).
29. A. E. Siegman and D. A. Watkins, "Potential Minimum Noise in the Microwave Diode," *Trans. IRE*, **ED-4**, 82 (1957).
30. S. W. Harrison, "On the Minimum Noise Figure of Traveling-Wave Tubes," *Proc. IRE*, **43**, 227 (1955).
31. D. O. North, "Fluctuations in Space-Charge-Limited Currents at Moderately High Frequencies," part 2, *RCA Rev.*, **4**, 441 (1940).
32. S. Bloom, "Noise Convection Current at the Potential Minimum," unpublished RCA Lab. Rept. (1956).
33. W. C. Hahn, "Small Signal Theory of Velocity-Modulated Electron Beams," *GE Rev.*, **42**, 258 (1939).
34. S. Ramo, "Space Charge and Field Waves in an Electron Beam," *Phys. Rev.*, **56**, 276 (1939).
35. W. R. Beam and S. Bloom, "Minimum Noise Figure of Traveling-Wave Tubes, including Higher Space-Charge Wave Modes," paper presented at IRE Electron Tube Research Conf., Boulder, Col. (June 1956).
36. F. E. Paschke, "The Mutuality of Coupling in Traveling-Wave Tubes," *Arch. Elektr. Ubertragung*, **11**, 137 (1957).
37. S. Bloom and R. W. Peter, "Transmission-Line Analog of a Modulated Electron Beam," *RCA Rev.*, **15**, 95 (1954).
38. P. K. Tien and L. M. Field, "Space-Charge-Waves in Accelerated Electron Stream for Amplification of Microwave Signals," *Proc. IRE*, **40**, 688 (1952).
39. R. Müller, "Raumladungswellen in Elektronen Strömungen," *Arch. Elektr. Übertragung*, **9**, 505 (1955).
40. F. B. Llewellyn and L. C. Peterson, "Vacuum-Tube Networks," *Proc. IRE*, **32**, 144 (1944).
41. L. D. Smullin, "Propagation of Disturbances in One-Dimensional Accelerated Electron Streams," *J. Appl. Phys.*, **22**, 1496 (1951).

42. A. Eichenbaum and R. W. Peter, "The Exponential Gun a Low Noise Gun for Traveling-Wave Amplifiers," *RCA Rev.* **19** (1958).

43. D. A. Watkins, "Noise Reduction in Beam Type Amplifiers," *Proc. IRE*, **40**, 65 (1952).

44. L. D. Buchmiller, R. W. de Grasse, and G. Wade, "Design and Calculation Procedures for Low-Noise Traveling-Wave Tubes," *Stanford Univ. Electronics Lab. Rept. AF 33 600-27784 Contract* (Oct. 1955).

45. R. E. Collin, "The Optimum Tapered Transmission-Line Matching Section," *Proc. IRE*, **44**, 539 (1956).

46. R. W. Peter, "Low-Noise Traveling-Wave Amplifier," *RCA Rev.*, **13**, 344 (1952).

47. A. Eichenbaum, "Design of Beam-Type Electron Guns," unpublished RCA Lab. Rept. (Aug. 1956).

48. R. C. Knechtli and W. R. Beam, "Performance and Design of Low-Noise Guns for Traveling-Wave Tubes," *RCA Rev.*, **17**, 410 (1956).

49. "Research and Development on Microwave Generators, Mixing Devices, and Amplifiers," *U.S. Signal Corps. Contract DA 36-039-SC-64443, Quart. Rept.* 6 (Apr. 1956).

50. W. R. Beam, "Interception Noise in Electron Beams at Microwave Frequencies, *RCA Rev.*, **16**, 551 (1955).

51. R. W. Peter and J. A. Ruetz, "Influence of Secondary Electrons on Noise Factor and Stability of Traveling-Wave Tubes," *RCA Rev.*, **14**, 441 (1952).

52. J. C. Slater, *Microwave Electronics*, D. Van Nostrand Co., New York (1950).

53. L. D. Smullin and C. Fried, "Microwave Noise Measurements on Electron Beams," *Trans. IRE* (PGED), **ED-1**, 168 (1954).

54. "Research and Development on Microwave Generators, Mixing Devices, and Amplifiers," *U.S. Signal Corps Contract DA 36-039-SC-5548, Quart. Rept.* 4 (Apr. 1952).

55. W. W. Rigrod, paper presented at the IRE Electron-Tube Research Conf., Boulder, Col., "Noise Spectrum of Electron Beams in Longitudinal Magnetic Field," *Bell System Tech. J.*, **36**, 831 (1957).

56. R. W. Peter, "Problems of Producing Low-Noise Electron Beams," *Proc. Symp. on Modern Advances in Microwave Techniques*, 159 (Nov. 1954).

57. W. R. Beam, "Noise Wave Excitation at the Cathode of a Microwave Beam Amplifier," *Trans. IRE* (PGED), **ED-4** (1957).

58. C. C. Cutler and C. F. Quate, "Experimental Verification of Space-Charge and Transit-Time Reduction of Noise in Electron Beams," *Phys. Rev.*, **80**, 875 (1950).

59. "Research and Development on Microwave Generators, Mixing Devices, and Amplifiers," *U.S. Signal Corps Contract DA 36-039-SC-5548, Quart. Rept.* 3 (Feb. 1952).

60. "Research and Development on Microwave Generators, Mixing Devices, anP Amplifiers," *U.S. Signal Corps Contract DA 36-039-SC-64443, Quart. Rept.* 3 (Aug. 1955).

61. R. W. Peter, "Direct Reading Noise-Factor Measuring Systems," *RCA Rev.*, **12**, 269 (1951).

62. W. R. Beam and R. D. Hughes, "Microwave Noise Source Modulator and Power Supply," *Trans. IRE* (PGED), **ED-4,** 185 (1957).

63. D. C. Rogers, "Traveling-Wave Amplifier for 6 to 8 Centimeters," *Elec. Commun.*, **26**, 144 (1949).
64. A. G. Mungall, "Noise in Traveling-Wave Tubes," *Trans. IRE* (PGED), **ED-2**, 12 (1955).
65. R. W. Peter, J. A. Ruetz, and A. B. Olson, "Attenuation of Wire Helices in Dielectric Supports," *RCA Rev.*, **13**, 558 (1952).
66. W. J. Dodds and R. W. Peter, "Filter-Helix Traveling-Wave Tubes," *RCA Rev.*, **14**, 502 (1953).
67. B. N. Enander, "A New Ferrite Isolator," *Proc. IRE*, **44**, 1421 (1956).

CHAPTER 6

Semiconductor Noise

A. van der Ziel

1. Introduction

The most important noise sources in semiconductors are thermal noise, shot noise, and $1/f$ noise. This chapter will be mainly devoted to a discussion of shot noise. For a discussion of $1/f$ noise the reader is referred to Chapter 2.

This chapter is divided into three sections. The first section deals with the various types of noise sources in semiconductors and also discusses a special class of noise phenomena that can be described by resistance fluctuations. The second section deals with noise in semiconductor devices in which the electric current is caused by carriers drifting under the influence of an electric field. The third section considers noise in semiconductor devices in which the electric current is caused by carrier diffusion due to a concentration gradient of the carriers. This type occurs in junction diodes and junction transistors, and for that reason these devices are discussed more fully. Finally, the results are applied to the calculation of the noise figure of transistor amplifiers.

Discussion of the Noise Sources in Semiconductors[1]

Thermal noise is due to the random motion of current carriers in the semiconductor. It is present whether or not an electric field is applied. The noise spectrum is flat up to the far-infrared frequencies. Its

311

magnitude may be calculated from thermodynamical considerations so that its result is independent of the "model" used in deriving the spectrum.

Shot noise resembles the corresponding effect in vacuum tubes; it only occurs if a voltage is applied to the semiconductor. There are some differences, however. In vacuum tubes all electrons are emitted by the cathode; in semiconductors most of the carriers appear and disappear in the sample itself and not at the electrodes. In vacuum tubes all current flow is due to *drift;* in semiconductors the current flow may be due either to *drift* or to *diffusion.* In vacuum tubes the fluctuating emission at the cathode causes fluctuations in the local electron concentration in the cathode–anode space. In those semiconductor devices in which the current flow is due to *carrier drift*, the local fluctuations in carrier concentration are due to fluctuations in the processes which create new carriers and take existing carriers out of circulation (recombination of electrons and ionized donors and thermal liberation of electrons from neutral donors: hole-electron pair creation and hole-electron recombination, etc.). In those semiconductor devices in which the current flow is due to *carrier diffusion*, the local fluctuations in carrier concentration are partly due to the above processes and partly due to fluctuations in the diffusion process. Both fluctuations are independent and have to be taken into account separately.

The $1/f$ noise also occurs only if an electric field is applied. Its cause is not completely understood; the magnitude of the noise and its characteristic spectrum, which is of the form $f^{-\alpha}$ with α close to unity, distinguish it from shot noise. It is also called flicker noise (because of its resemblance to the flicker noise of vacuum tubes) or excess noise (because it is the noise over and above thermal noise and shot noise).

The resistance of a semiconductor is caused by the random collisions of the carriers with the lattice; the same mechanism also gives rise to the thermal noise of the semiconductor. At first sight one might think that this mechanism should also contribute to the shot noise and the $1/f$ noise, but a calculation shows that this contribution is insignificant.[1]

To understand this, consider a piece of n-type semiconductor (that is, the current is carried by electrons) between plane electrodes at a distance d. The motion of an individual electron gives rise to a current in the external circuit; owing to the collisions of the electron with the lattice, this current consists of a number of short current pulses. Let an electric field be applied in the axial direction, which will be chosen as the x direction, and let it give the electrons an acceleration α_x. Let collisions occur at $t = t_0$, $t = t_1$, $t = t_2$, etc. Between collisions the current in the external circuit is

$$i(t) = \frac{ev_x}{d} \qquad (6.1)$$

where v_x is the velocity of the electron in the x direction. Let

$$v_x = v_{x0} + \alpha_x(t - t_0) \qquad (t_0 < t < t_1) \qquad (6.1a)$$

$$v_x = v_{x1} + \alpha_x(t - t_1) \qquad (t_1 < t < t_2), \quad \text{etc.} \qquad (6.1b)$$

It is usually assumed that v_{x0} and v_{x1} are independent, and that the velocity terms v_{x0}, v_{x1}, etc., are independent of the drift terms $\alpha_x(t - t_0)$, etc.; the drift terms $\alpha_x(t - t_0)$, etc., in Eqs. 6.1a and 6.1b are completely correlated, however, and this correlation will last as long as the electron moves between the electrodes. Because of these correlations it is possible to split the noise due to the individual current pulses into two independent parts:

(a) The contributions of the random velocities v_{x0}, v_{x1}, etc.; we identify these with the thermal noise.
(b) The contributions from the drift terms $\alpha_x(t - t_0)$, $\alpha_x(t - t_1)$, etc.; we identify these with the shot noise and the $1/f$ noise.

This separation of noise terms is valid if the above assumptions hold.

Since the correlation between the drift terms lasts as long as the electron moves between the electrodes, the magnitude of the shot noise and the $1/f$ noise is therefore not determined by the individual short-current pulses caused by a single carrier, but by fluctuations in the carrier concentration in the semiconducting material. The thermal noise, however, is determined by the individual, independent current pulses

$$i(t) = \frac{ev_{x0}}{d} \quad \text{for} \quad t_0 < t < t_1, i(t) = \frac{ev_{x1}}{d} \quad \text{for } t_1 < t < t_2, \quad \text{etc.} \quad (6.1c)$$

Fluctuations in Resistance, Carrier Concentration

Certain types of noise, both of the shot-noise and of the $1/f$ noise varieties, can be represented by resistance fluctuations. In those cases in which the current is due to carrier drift, such resistance fluctuations should be expected because of fluctuations on the carrier concentration. We discuss this type of noise first.

Let a resistance R show fluctuations δR, and let a direct current I flow through it. A noise emf $\delta V = I \, \delta R$ is then developed in the circuit. The noise may also be represented by a current generator

$\delta I = \delta V / R$ in parallel to R. Their mean-square values are

$$\overline{\delta V^2} = I^2\,\overline{\delta R^2}, \qquad \overline{\delta I^2} = I^2\,\frac{\overline{\delta R^2}}{R^2} \tag{6.2}$$

Both quantities are thus proportional to I^2 as long as R is linear and $\overline{\delta R^2}$ is independent of I.

This relation has been verified experimentally in many cases. However, a quadratic dependence on current does not necessarily indicate that the noise is due to resistance fluctuations (example: flicker noise in saturated tubes in which the noise can be described by a fluctuating work function). Neither does the fact that $\overline{\delta V^2}$ and (or) $\overline{\delta I^2}$ are not proportional to I^2 necessarily indicate that the noise cannot be represented by a resistance fluctuation; R may be nonlinear, or $\overline{\delta R^2}$ may depend upon I.

If the current flow is due to *drift*, fluctuations in carrier concentration may cause fluctuations in resistance. Let the sample have a length L, cross-sectional area A, a resistance R, and let N be the number of carriers. Then $R = C/N$, where C depends on the sample geometry and on the semiconducting material. Let δN be the fluctuation in the number of carriers, and let $\overline{\delta N^2} = \alpha N$, where α is independent of N.† The fluctuation in the number of carriers causes a fluctuation δR in resistance:

$$\delta R = -R\,\frac{\delta N}{N}, \qquad \overline{\delta R^2} = R^2\,\frac{\overline{\delta N^2}}{N^2} = \frac{\alpha}{C}\,R^3 \tag{6.3}$$

Substituting into Eq. 6.2, we have

$$\overline{\delta V^2} = \frac{\alpha}{C}\,I^2 R^3, \qquad \overline{\delta I^2} = \frac{\alpha}{C}\,I^2 R \tag{6.4}$$

We now make use of the fact that the resistance R of the sample is[2,3]

$$R = \frac{L/A}{e\mu n_0} = \frac{L^2}{e\mu N} \tag{6.5}$$

where μ is the carrier mobility, and n_0 the carrier concentration, so that $C = L^2/e\mu$. Consequently,

$$\overline{\delta I^2} = \frac{\alpha I^2}{ALn_0} = \frac{\alpha I^2}{N} \tag{6.6}$$

$\overline{\delta I^2}$ is then inversely proportional to the volume AL of the sample.

† This assumption covers practically all the cases discussed in Section 2, pages 315–324.

We also see that $\overline{\delta I^2}$ is inversely proportional to the *total number of carriers* $N = ALn_0$ of the sample.[†] These considerations, which hold for shot noise as well as $1/f$ noise, make it understandable why $1/f$ noise is so important in thin-layer resistors and in point-contact diodes. Actually other physical phenomena play a part and enhance the effect.

Introducing the noise resistance R_n and the equivalent saturated diode current I_{eq} of the sample, we find, therefore,

$$R_n = \text{const } I^2 R^3, \qquad I_{eq} = \text{const } I^2 R \qquad (6.7)$$

Experiments carried out at the University of Minnesota verified these relations for the $1/f$ noise developed in cathode coatings at low current levels.[4] The resistance R was varied by varying the cathode temperature; the corresponding variation of the constant C is very small in comparison with the large variation of N with temperature.

At higher-current levels the coating resistance R became strongly current-dependent, and the noise resistance R_n varied roughly as R^2 instead of varying as R^3. The resistance R increased with increasing current, because the region close to one of the electrodes becomes strongly depleted of donors by electrolytic action so that most resistance becomes concentrated in that region.[‡]

The same relation $R_n = \text{const } I^2 R^2$ was found for the $1/f$ noise generated in the Ba_2SiO_4 interface layer that develops in some cathode coatings during aging.[5,6] That the same relationship is found here is understandable, for the high resistance of the Ba_2SiO_4 layer is caused by a depletion region existing near one of the faces of the Ba_2SiO_4 layer.

2. Shot Noise Due to Drifting Carriers

Formulation of the Basic Problem

Consider a semiconducting sample of length L and cross-sectional area A. If E is the field strength in the sample, then the average drift velocity u_d of the carriers is μE, where μ is the carrier mobility. The average drift time τ_d of the carriers from the one electrode to the other is:

$$\tau_d = \frac{L}{u_d} = \frac{L}{\mu E} \qquad (6.8)$$

[†] This result holds for any noise mechanism in which the current flow is due to drift and in which the noise can be described by fluctuations in the carrier concentration. This pertains both to shot noise and to $1/f$ noise.

[‡] This is an example of a noise mechanism that can be described by a fluctuating resistance, but for which Eqs. 6.3 through 6.7 do not hold. The reason for this difference is the inhomogeneous distribution of carriers in the sample.

The carriers produce current pulses of duration τ_f, where τ_f is the free time of the carriers. If the carriers all appear and disappear at the electrodes, then the free time of the carriers is equal to τ_d; if the carriers nearly all appear and disappear between the electrodes, then the free time of the carriers is smaller than τ_d.

Let N be the number of carriers in the sample. All carriers give the same contribution (e/τ_d) to the direct current I; the fluctuating current $i(t)$ is caused by a fluctuation δN in the number of carriers. Then,

$$I = \frac{eN}{\tau_d} = \frac{eN\mu E}{L}, \qquad i(t) = \frac{e\,\delta N\mu E}{L} \qquad (6.9)$$

(which also shows that the resistance R is equal to $L^2/(e\mu N)$, as used in Section 1, page 314. Representing the noise by a current generator $\sqrt{\overline{i^2}}$ across the sample, we have, according to the Wiener–Khintchine theorem,

$$\overline{i^2} = 4\Delta f \int_0^\infty \overline{i(t)\,i(t+\tau)}\,\cos\omega\tau\,d\tau$$

$$= 4\overline{i^2(t)}\,\Delta f \int_0^\infty c(\tau)\,\cos\omega\tau\,d\tau \qquad (6.10)$$

where $c(\tau) = [\overline{i(t)\,i(t+\tau)}/\overline{i^2(t)}]$ is the *normalized* autocorrelation function of $i(t)$ (normalization means $c(0) = 1$). Substituting Eqs. 6.8 and 6.9 into 6.10 yields

$$\overline{i^2} = 4\Delta f \left(\frac{e\mu E}{L}\right)^2 \overline{\delta N^2} \int_0^\infty c(\tau)\,\cos\omega\tau\,d\tau \qquad (6.11)$$

thus showing that $\overline{i^2}$ is proportional to the square of the field strength or, what amounts usually to the same thing, $\overline{i^2}$ is proportional to the square of the current as expected from Section 1, page 314.

The fundamental problem of shot noise in semiconductors caused by carrier drift consists thus in calculating $\overline{\delta N^2}$ and $c(\tau)$. Here $c(\tau)$ represents that part of the fluctuation δN_0 present at $t = 0$ that is, on the average, still present at $t = \tau$. Earlier theories [2,7–10] gave the right expressions for $c(\tau)$ but erroneous expressions for $\overline{\delta N^2}$.

First, let the average free time of the carrier be small in comparison with the drift time τ_d; this means that most carriers appear and disappear somewhere in the sample and not at the electrodes. If the appearance and disappearance of a carrier is caused by a *single-step*† process, the fluctuation δN in bulk material then satisfies Langevin's

---

† A single-step process is a process in which the recombination is direct; in a multistep process the recombination takes place by means of recombination centers.

differential equation,[11]

$$\frac{d(\delta N)}{dt} = -\frac{\delta N}{\tau_0} + H(t) \tag{6.12}$$

where $H(t)$ is a random source function; the time constant τ_0 is related to the average free time of the carrier. The function $c(\tau)$ is then

$$c(\tau) = \exp(-\tau/\tau_0) \tag{6.13}$$

since this part of the fluctuation δN_0 at $t = 0$ is on the average still present at a time $\tau = \tau_0$. This case holds if $\tau_0 \ll \tau_d$.

Next consider the case in which the time constant τ_0 of Eq. 6.12 is much longer than τ_d. In that event nearly all carriers appear and disappear at the electrodes, just as in a diode, and not inside the material. If a fluctuation δN_0 is present at $t = 0$, then, since the carriers move with constant drift velocity u_d, the part $(\tau_d - \tau)/\tau_d$ of δN_0 will on the average be present at a time $t = \tau$ and the part τ/τ_d has been swept out of the region between the electrodes. Consequently

and
$$c(\tau) = 1 - \frac{\tau}{\tau_d} \quad (\tau < \tau_d)$$
$$\tag{6.14}$$
$$c(\tau) = 0 \quad (\tau > \tau_d)$$

The more general case where τ_0 and τ_d have arbitrary values follows easily. At the time τ all but the part $\exp(-\tau/\tau_0)$ would have disappeared by recombination, and of those left all but the part $(1 - \tau/\tau_d)$ would have been swept out of the region between the electrodes. Consequently

$$c(\tau) = \left(1 - \frac{\tau}{\tau_d}\right) \exp(-\tau/\tau_0) \quad \text{for} \quad \tau < \tau_d,$$
$$\tag{6.15}$$
$$c(\tau) = 0 \quad \text{for} \quad \tau > \tau_d$$

reducing to Eq. 6.13 if $\tau_d \gg \tau_0$ and to Eq. 6.14 if $\tau_d \ll \tau_0$. Formula 6.15 was first used by Davydov and Gurevich.[9]

Substituting Eq. 6.13 or Eq. 6.14 into Eq. 6.11, we obtain

$$\overline{i^2} = 4\Delta f \left(\frac{e\mu E}{L}\right)^2 \overline{\delta N^2} \frac{\tau_0}{1 + \omega^2 \tau_0^2} \quad (\tau_0 \ll \tau_d) \tag{6.16}$$

$$\overline{i^2} = 2\Delta f \left(\frac{e\mu E}{L}\right)^2 \overline{\delta N^2} \tau_d \left(\frac{\sin \frac{1}{2}\omega\tau_d}{\frac{1}{2}\omega\tau_d}\right)^2 \quad (\tau_0 \gg \tau_d) \tag{6.17}$$

For *majority*-carrier density fluctuations Eq. 6.17 cannot occur, since the space-charge neutrality condition forbids density fluctuations to enter or leave the sample at the contacts. For minority-carrier density fluctuations the above theory is correct, provided that the ambipolar drift mobility μ_a is used in τ_d instead of the carrier mobility.[12]

Earlier Applications of the Theory

The earlier calculations of $\overline{\delta N^2}$ showed large errors. For instance, it was assumed by Bernamont,[7] Surdin,[8] Davydov and Gurevich,[9] Gisolf,[10] and van der Ziel[2] that $\overline{\delta N^2} = N$. This assumes a Poisson distribution for δN, which certainly is not the case for n-type or p-type semiconductors at room temperature. At room temperature nearly all donor atoms in n-type material and nearly all acceptor atoms in p-type material are ionized; the number of free carriers in the sample is thus almost constant and shows extremely small fluctuations. Consequently this noise mechanism should not occur at room temperature. No experimental evidence for this noise mechanism, based upon fluctuations in the processes:

Electron + ionized donor $\rightleftarrows$ neutral donor

Hole + ionized acceptor $\rightleftarrows$ neutral acceptor

exists at room temperature. The assumption $\overline{\delta N^2} = N$ must therefore be in error; it will be shown below that $\overline{\delta N^2}$ is usually considerably smaller.

The following method, proposed by van der Ziel,[1,13] avoids this error to a large extent. Let the sample have N_d donors, and let α be the probability that the donor is neutral. Then $N = N_d(1 - \alpha)$, and, according to probability theory,

$$\overline{\delta N^2} = N_d\alpha(1 - \alpha) = N\alpha \qquad (6.18)$$

This assumes a binomial-law distribution for δN and implies that individual carriers can be treated as independent. Though not exact, this method gives reasonably accurate results in many cases.

At normal room temperature, $\alpha \ll 1$ and $\overline{\delta N^2} \ll N$, so that no measurable amount of noise due to fluctuations in carrier concentration would be expected. At sufficiently low temperatures such that $\alpha \simeq 1$, we find $\overline{\delta N^2} \simeq N$; at those temperatures a measurable amount of noise might be expected. Experimental evidence for this effect has been found by Gebbie[14] and Fassett.[15]

We now consider an n-type sample that is not too far from intrinsic. In that case both fluctuations in the electron and in the hole concentration must be considered. Let N be the number of free electrons (mostly from donors which are assumed to be all ionized), and P the number of free holes (all from hole-electron pair creation). If N_i is the number of electrons and (or) holes in intrinsic material,[3] we have $PN = N_i^2$. Fluctuations in the number of carriers may occur due to fluctuations in processes of the type

Free electron + free hole $\rightleftarrows$ electron bound in valence band

Since holes and electrons appear and disappear in pairs, we have $\delta N = \delta P$. If μ_p and μ_n are the hole and electron mobilities, respectively, then Eq. 6.9 must be written

$$I = e(\mu_n N + \mu_p P)E/L, \qquad i(t) = e\, \delta p(\mu_p + \mu_n)E/L \quad (6.19)$$

From Eq. 6.19 we conclude that μ has to be replaced by $(\mu_p + \mu_n)$ in Eqs. 6.16 and 6.17, and that μN in the formula 6.5 for the resistance R has to be replaced by $(\mu_n N + \mu_p P)$.

To calculate $\overline{\delta P^2}$ we may apply Eq. 6.18. Then, by analogy, N_d corresponds to the number of electrons in the valence band, and, unless the gap width between valence band and conduction band is very narrow, we may assume that $P \ll N_d$, so that $\alpha \simeq 1$. (α is the probability that the electron is in the valence band). We would therefore expect $\overline{\delta P^2} = P = N_i^2/N$. The fluctuations are therefore strongest in intrinsic material and decrease rapidly if the sample is more strongly n-type.

Substituting into Eqs. 6.16 and 6.17, we obtain

$$\overline{i^2} = 4\Delta f \left[\frac{e(\mu_p + \mu_n)E}{L}\right]^2 \frac{N_i^2}{N} \left(\frac{\tau_0}{1 + \omega^2 \tau_0^2}\right) \qquad (\tau_0 \ll \tau_d) \quad (6.20)$$

$$\overline{i^2} = 2\Delta f \left[\frac{e(\mu_p + \mu_n)E}{L}\right]^2 \frac{N_i^2}{N} \tau_d \left(\frac{\sin \frac{1}{2}\omega\tau_d}{\frac{1}{2}\omega\tau_d}\right)^2 \qquad (\tau_0 \gg \tau_d) \quad (6.21)$$

Introducing the shot-noise ratio n_s of the sample by the definition

$$\overline{i^2} = n_s 4kT\, \Delta f/R$$

we obtain, from Eqs. 6.20 and 6.21,

$$n_s = \left(\frac{P_d}{N_i kT}\right)\left(\frac{\sigma_i}{\sigma}\right)^2 \left(\frac{N_i}{N}\right)\frac{\tau_0}{1 + \omega^2 \tau_0^2} \qquad (\tau_0 \ll \tau_d) \quad (6.20a)$$

$$n_s = \frac{1}{2}\left(\frac{P_d}{N_i kT}\right)\left(\frac{\sigma_i}{\sigma}\right)^2 \left(\frac{N_i}{N}\right)\tau_d \left(\frac{\sin \frac{1}{2}\omega\tau_d}{\frac{1}{2}\omega\tau_d}\right)^2 \qquad (\tau_0 \gg \tau_d) \quad (6.21a)$$

by setting

$$P_d = \frac{E^2 L^2}{R}, \qquad R = \frac{L^2}{e(\mu_n N + \mu_p P)}, \qquad R_i = \frac{L^2}{e(\mu_n + \mu_p)N_i}, \qquad \frac{\sigma_i}{\sigma} = \frac{R}{R_i}$$

Here P_d is the d-c power dissipated in the sample, L the length of the sample, R_i the intrinsic resistance, σ_i the intrinsic conductivity, and σ the total conductivity.

The noise due to fluctuations in carrier concentration is only important if the n-type sample is nearly intrinsic; otherwise the expression $(\sigma_i{}^2 N_i / \sigma^2 N)$ becomes too small. It was discovered by Herzog and van der Ziel in a Ge sample;[16] Montgomery[17] proved in an interesting experiment that the noise was indeed due to holes. The measurements agreed quantitatively with Eq. 6.20a, which is reasonable since their sample had $L = 0.5$ cm, $I = 3$ ma, and $R = 10{,}000$ ohms, so that $\tau_d \simeq 5$ μsec for the holes, whereas the measured value of τ_0 was 1.1 μsec, so that indeed $\tau_0 < \tau_d$. A slightly larger value of τ_0 would have made a transition from formula 6.20a to 6.21a observable.[12]

The binomial law upon which Eq. 6.18 is based is not an exact expression in all cases; the more detailed study of the probability distribution of δN on page 323 shows that the binomial law should hold if $P \ll N$.

In general, the process of hole-electron recombination is *not* a one-step process; they do not recombine directly but through recombination centers. Such a process may give rise to a frequency dependence differing from Eq. 6.20a.[18,19]

Statistics of Carrier Density Fluctuations

We follow here a discussion given by Burgess[20] to show that significant deviations from the normal law may occur.† These deviations are caused by the fact that individual electrons cannot be treated as independent; this violates the basic assumption of the binomial law.

Let there be N electrons in the conduction band. Let the probability that another electron is generated in the sample during the time dt be $g(N)\, dt$, and the probability that an electron is taken out of the band during the time dt be $r(N)\, dt$. If $P(N)$ is the probability of finding N electrons in the band, then

$$\frac{dP(N)}{dt} = r(N + 1)\, P(N + 1) + g(N - 1)\, P(N - 1)$$
$$- P(N) \left[g(N) + r(N) \right] \quad (6.22)$$

Equation 6.22 may be understood as follows. During the time interval dt, $P(N)$ changes by an amount $dP(N)$, because of transitions

† I am indebted to Professor R. E. Burgess, University of British Columbia, Vancouver, B.C., Canada, for permission to use an earlier unpublished report.

$N \to (N + 1)$, $N \to (N - 1)$, $(N + 1) \to N$, and $(N - 1) \to N$. These transitions give, respectively, the following contributions to $dP(N)$: $-P(N) g(N) dt$, $-P(N) r(N) dt$, $P(N + 1) r(N + 1) dt$, and $P(N - 1) g(N - 1) dt$. Dividing by dt gives Eq. 6.22.

In equilibrium, $dP(N)/dt = 0$. Instead of solving this equation, it is simpler to find the exact solution of Eq. 6.22. As is found by substitution, this turns out to be

$$P(N) = \frac{\prod\limits_{\nu=0}^{N-1} g(\nu)}{\prod\limits_{\nu=1}^{N} r(\nu)} P(0) \tag{6.23}$$

where $P(0)$ is the probability of finding zero electrons in the band. This equation may be used to find the most probable value of N and a Gaussian approximation of $P(N)$.

The derivatives of $\ln P(N)$ are

$$\frac{d \ln P(N)}{dN} \cong \left[\frac{\ln P(N + \Delta N) - \ln P(N)}{\Delta N} \right]_{\Delta N = 1}$$

$$= \ln \left[\frac{g(N)}{r(N + 1)} \right] \frac{d^2 \ln P(N)}{dN^2} = \left[\frac{g'(N)}{g(N)} - \frac{r'(N + 1)}{r(N + 1)} \right], \quad \text{etc.}$$

where g' and r' are derivatives with respect to N. The most probable value N_0 of N is found by equating $d[\ln P(N)]/dN$ to zero. This gives

$$g(N_0) = r(N_0 + 1) \simeq r(N_0) \tag{6.23a}$$

since N_0 is a large number. Making a Taylor expansion of $\ln P(N)$ around $N = N_0$ yields

$$\ln P(N) = \ln P(N_0) + \frac{1}{2} \left[\frac{d^2 \ln P(N)}{dN^2} \right]_{N=N_0} (N - N_0)^2$$

$$\cong -\ln P(N_0) + \frac{1}{2} \left[\frac{g'(N_0) - r'(N_0)}{g(N_0)} \right] (N - N_0)^2 \cdots \tag{6.23b}$$

plus higher-order terms that are negligible if $(N - N_0)$ is not too large; N_0 is again treated as a large quantity.

According to Eq. 6.23b, the function $P(N)$ can be approximated by a Gaussian distribution. Since, for that distribution,

$$\ln P(N) = \ln P(N_0) - \frac{1}{2} \frac{(N - N_0)^2}{(N - N_0)^2} \tag{6.23c}$$

we have, by equating Eqs. 6.23b and 6.23c,

$$\overline{\delta N^2} = \overline{(N - N_0)^2} = \frac{g(N_0)}{r'(N_0) - g'(N_0)} \tag{6.24}$$

where g' and r' again denote derivatives with respect to N.

If $(N - N_0)$ is the excess concentration, it will decay in time as described by Langevin's differential equation[11] 6.12:

$$\frac{d(N - N_0)}{dt} = -(N - N_0)/\tau_0 + H(t) \tag{6.25}$$

where $H(t)$ is again a random source function describing the spontaneous appearance and disappearance of carriers. But we also have

$$\frac{d}{dt}(N - N_0) = g(N) - r(N) + H(t)$$
$$= -(N - N_0)[r'(N_0) - g'(N_0)] + H(t) \tag{6.25a}$$

where g' and r' again denote derivatives with respect to N. The first part of Eq. 6.15a follows directly from the definition of $g(N)$ and $r(N)$; the second part follows from a Taylor expansion of $g(N)$ and $r(N)$ around N_0. Hence:

$$\tau_0 = \frac{1}{r'(N_0) - g'(N_0)} \tag{6.25b}$$

Since $\overline{\delta N^2}$ and τ_0 are now known, the noise problem is hereby solved in principle; we only have to substitute into Eq. 6.11.

In practical cases:

$$g(N) = g_0 - aN, \qquad r(N) = bN + cN^2 \tag{6.26}$$

so that

$$\overline{(N - N_0)^2} = \overline{\delta N^2} = N_0 \frac{b + cN_0}{a + b + 2cN_0} \tag{6.26a}$$

$$\tau_0 = \frac{1}{a + b + 2cN_0} \tag{6.26b}$$

This has to be applied to particular cases.

n-Type Semiconductor. Negligible Amount of Free Holes. In this case $g(N)$ is proportional to $(N_d - N)$, the number of neutral donors, and $r(N)$ is proportional to the product of N, the number of ionized donors, and N, the number of free electrons.   Therefore

$$g(N) = \gamma(N_d - N), \qquad r(N) = \rho N^2 \qquad (6.27)$$

$$\overline{(N - N_0)^2} = \frac{N_0(N_d - N_0)}{2N_d - N_0} \qquad (6.27a)$$

$$\tau_0 = \frac{1}{\gamma + 2\rho N_0} = \frac{N_0}{\gamma(2N_d - N_0)} \qquad (6.27b)$$

Hence $\overline{(N - N_0)^2} = \frac{1}{2}N_0$ if $N_d \gg N_0$ (low temperatures), a factor of 2 different from what would be expected from the binomial law.   If $N_0 \simeq N_d$ (room temperature), $\overline{(N - N_0)^2} \simeq (N_d - N_0)$, as expected from the binomial law (see Eq. 6.18).

n-Type Semiconductor, Close to Intrinsic, at Room Temperature.   In this case all N_d donor levels are ionized.   Let N be the number of free electrons and $P = (N - N_d)$ the number of free holes.   Let their equilibrium values be N_0 and $P_0 = (N_0 - N_d)$.   In this case, since all fluctuations are due to hole-electron pair creation and recombination, g is constant because of the small number of electrons in the conduction band, and r is proportional to the product NP.   Hence

$$g = \text{const}, \qquad r = \rho NP = \rho N(N - N_d) \qquad (6.28)$$

$$\overline{(N - N_0)^2} = \overline{(P - P_0)^2} = \frac{N_0 P_0}{N_0 + P_0} \qquad (6.28a)$$

$$\tau_0 = \frac{1}{\rho(N_0 + P_0)} \qquad (6.28b)$$

If $N_0 \gg P_0$, then $\overline{(P - P_0)^2} = P_0$ as expected from the binomial law.   If $N_0 = P_0$, that is, if the sample is intrinsic, then $\overline{(P - P_0)^2} = \frac{1}{2}P_0$, which differs from the binomial-law result by a factor of $\frac{1}{2}$.

n-Type Semiconductor with N_d Donors and N_t Traps.†   Let all N_d donors be ionized.   Let the traps be neutral when occupied; then $(N_t + N - N_d)$ is the number of empty traps, and $(N_d - N)$ is the number of occupied traps. g is now proportional to the number of filled traps, and r is proportional to the product of N, the number of

† An electron trap is an allowed energy level between the bottom of the conduction band and the top of the filled band, that would normally be empty at $T = 0$. For $T > 0$, however, some of the traps may be occupied because electrons from the donor levels may be excited into the conduction band, or they may wander through the crystal and get trapped in one of those unoccupied energy levels. Since some of the electrons are trapped, we have $N < N_d$.

free electrons, and the number of empty traps.   Hence

$$g = \gamma(N_d - N), \qquad r = \rho N(N_t - N_d + N) \qquad (6.29)$$

Putting $N_{t1} = N_t + N_0 - N_d$ and $N_{t2} = N_d - N_0$, we obtain

$$\overline{(N - N_0)^2} = \left(\frac{1}{N_0} + \frac{1}{N_{t1}} + \frac{1}{N_{t2}}\right)^{-1} \qquad (6.29a)$$

where N_{t1} and N_{t2} are the equilibrium numbers of empty and filled traps.   If $N_0 \gg N_t$, we obtain $\overline{(N - N_0)^2} = N_{t1}N_{t2}/(N_{t1} + N_{t2}) = N_t\alpha(1 - \alpha)$, where $\alpha = N_{t1}/(N_{t1} + N_{t2})$ is the probability that the trap is filled.   If $N_0 \ll N_t$, we obtain $\overline{(N - N_0)^2} = N_0$; both results are the same as for a binomial distribution.   Deviations occur in the intermediate range.

Another Discussion of the Time Constant τ_0.   We finally discuss the time constant τ_0 in a different manner.   We use a model, discussed by Machlup,[21] consisting of a semiconductor with free electrons and traps. Machlup assumed that individual carriers can be treated as independent.   This implies the validity of the normal law; the previous discussion shows that this assumption is not always allowed.

During a time interval dt, let a free electron have the probability dt/τ_1 of being trapped, and let a trapped electron have the probability dt/τ_2 of being liberated.   If we have N_d electrons, and if in equilibrium N of them are free and N_2 are trapped, and if an excess number ΔN of free electrons occur at $t = 0$, then we have Langevin's differential equation,[11]

$$d(\Delta N) = -(N + \Delta N)\,dt/\tau_1 + (N_2 - \Delta N)\,dt/\tau_2 + H(t)\,dt \qquad (6.30)$$

where $H(t)$ is a random source function.   In equilibrium ($\Delta N = 0$) we have on the average $d(\Delta N) = 0$; this gives the equilibrium condition:

$$N/\tau_1 = N_2/\tau_2 \quad \text{or} \quad N = N_d\tau_1/(\tau_1 + \tau_2):$$
$$N_2 = N_d\tau_2/(\tau_1 + \tau_2) \qquad (6.30a)$$

using $N_d = N + N_2$.   The differential equation becomes

$$d(\Delta N)/dt = -\Delta N(1/\tau_1 + 1/\tau_2) + H(t) = -\Delta N/\tau_0 + H(t) \qquad (6.30b)$$

so that

$$1/\tau_0 = 1/\tau_1 + 1/\tau_2 \qquad (6.31)$$

If $\tau_1 \ll \tau_2$ (or $N \ll N_2$), which occurs if τ_2 is sufficiently large, we have $\tau_0 \simeq \tau_1$; otherwise the two quantities differ.

The results of Eq. 6.30a can be translated directly into the binomial form, Eq. 6.18, by introducing $\alpha = \tau_2/(\tau_1 + \tau_2)$.

3. Shot Noise Due to Diffusing Carriers[22]†

Transmission-Line Analogy

The equations for a one-dimensional diffusion of minority carriers, which we shall assume to be holes, in a sample of cross-sectional area A and width w, and in which diffusion predominates over drift, are

$$\frac{\partial p}{\partial t} = -\frac{p - p_n}{\tau_p} - \frac{1}{e}\frac{\partial i_p}{\partial x}$$

$$i_p = -eD_p\frac{\partial p}{\partial x} \tag{6.32}$$

where D_p is the diffusion coefficient and τ_p the lifetime of the holes, e is the electronic charge, i_p the hole current, p and p_n are hole concentrations, p is the actual and p_n the equilibrium hole concentration, both per unit length. We introduce the excess hole concentration $p' = (p - p_n)$ and rewrite Eqs. 6.32 as follows:

$$\frac{\partial p'}{\partial x} = -\frac{1}{eD_p}i_p \tag{6.32a}$$

$$\frac{\partial i_p}{\partial x} = -\frac{e}{\tau_p}p' - e\frac{\partial p'}{\partial t} \tag{6.32b}$$

The equations for a distributed line without inductance are

$$\frac{\partial E}{\partial x} = -RI$$

$$\frac{\partial I}{\partial x} = -GE - C\frac{\partial E}{\partial t} \tag{6.33}$$

where E is the voltage, I the current, and R, G, and C are the resistance, conductance, and capacitance per unit length. $\gamma = \sqrt{R(G + j\omega C)}$ is the propagation constant and $Z_0 = \sqrt{R/(G + j\omega C)}$ is the characteristic impedance. This shows that a complete analogy exists between the one-dimensional diffusion problem and the transmission line:

E corresponds to the excess hole concentration p'.
I corresponds to the hole current i_p.

† The method of attack used in this section was originally developed by D. O. North, RCA Laboratories,[23] and by Petritz.[24]

R corresponds to $1/(eD_p)$.

G corresponds to (e/τ_p).

C corresponds to e.

Z_0 corresponds to $[e^2 D_p(1 + j\omega\tau_p)/\tau_p]^{-\frac{1}{2}}$

$\gamma = a + jb$ corresponds to $[(1 + j\omega\tau_p)/(D_p\tau_p)]^{\frac{1}{2}}$.

We shall call the latter two quantities the *characteristic impedance* Z_0 and the *propagation constant* γ of the semiconductor. For low frequencies,

$$Z_0 = Z_{00} = \left(\frac{e^2 D_p}{\tau_p}\right)^{-\frac{1}{2}}, \qquad \gamma = \gamma_0 = (D_p\tau_p)^{-\frac{1}{2}} \qquad (6.34)$$

so that

$$\frac{1}{Z_0} = \frac{(1 + j\omega\tau_p)^{\frac{1}{2}}}{Z_{00}}, \qquad \gamma = a + jb = \gamma_0(1 + j\omega\tau_p)^{\frac{1}{2}} \qquad (6.34a)$$

To solve the problem of hole flow in n-type material, we solve the corresponding transmission-line problem and then translate back to the diffusion problem by means of the above analogy.

Noise Equivalent Circuit of One-Dimensional Carrier Diffusion

Noise in devices in which the current is carried by diffusion consists of two sources: recombination fluctuations and diffusion fluctuations.

We first discuss recombination fluctuations. For stationary current flow, $\partial p/\delta t = 0$, and hence, according to Eq. 6.32b, a current $\Delta I = ep' \Delta x/\tau_p$ disappears between x and $(x + \Delta x)$. This happens because a small part of the excess holes recombine with electrons in the section Δx. Since the chance of an excess hole to recombine with an electron in the section Δx is quite small, we would expect full shot effect for the recombination current ΔI.† That is, the mean-square value of the Fourier component Δi_{px} of the fluctuation in ΔI should be

$$\overline{\Delta i_{px}^2} = 2e \, \Delta I \, \Delta f = 2e \, \Delta f(ep' \, \Delta x/\tau_p) \qquad (6.35)$$

This gives the contribution to $\overline{\Delta i_{px}^2}$ of that part of the fluctuations in the excess hole concentration p' that is caused by fluctuations in the recombination process. But, since there is no difference between a normal hole and an excess hole, recombination fluctuations in the *normal* hole concentration should also contribute to

† According to the theory of partition noise,[1] we have, if I_p is the hole current entering the section Δx,

$$\overline{\Delta i_{px}^2} = 2e \frac{\Delta I(I_p - \Delta I)}{I_p} \Delta f = 2e \, \Delta I \, \Delta f, \quad \text{since} \quad \Delta I \ll I_p$$

$\overline{\Delta i_{px}}^2$ at the same rate as the other holes. This gives a contribution $[2e\,\Delta f(ep_n\,\Delta x/\tau_p)]$ to $\overline{\Delta i_{px}}^2$, because of the spontaneous recombination fluctuations in the normal hole concentration. This effect can thus be taken into account in Eq. 6.35 by replacing the excess hole concentration p' by the total hole concentration $(p' + p_n)$.

The random process of hole-electron pair creation gives rise to concentration fluctuations which also contribute to $\overline{\Delta i_{px}}^2$. Looking at the normal hole concentration only, we observe that, on the average, as many holes appear by pair creation as disappear by recombination. Hence the mean-square value of the concentration fluctuations due to both processes must be equal; since recombination and pair creation are *independent* processes, these fluctuations must be added quadratically. The process of hole-electron pair creation thus gives also a contribution $[2e\,\Delta f(ep_n\,\Delta x/\tau_p)]$ to $\overline{\Delta i_{px}}^2$. In total we have

$$\overline{\Delta i_{px}}^2 = 2e\,\Delta f(ep'\,\Delta x/\tau_p + ep_n\,\Delta x/\tau_p + ep_n\,\Delta x/\tau_p)$$
$$= 2e^2\,\Delta f(p' + 2p_n)\,\Delta x/\tau_p \cdots \quad (6.36)$$

In our transmission-line analogy, Δi_{px} corresponds to a current generator Δi_x connected across the line at a distance x from the input. Equation 6.36 assumes that the recombination is a one-step process. A different formula may be expected for a two-step process.

We now discuss density fluctuations due to random diffusion fluctuations. A fluctuating quantity $u(t)$ has the following noise intensity[1] at low frequencies (that is, if $\omega\tau_0$ is very small),[†]

$$\overline{u^2} = 4\overline{u^2(t)}\tau_0\,\Delta f \quad (6.37)$$

where τ_0 is the time constant of $u(t)$. We thus have to find $\overline{u^2(t)}$ and τ_0 for the diffusion fluctuations. The number of holes P in the section Δx is $p\,\Delta x$; the fluctuations δP in this number are such that $\overline{\delta P^2} = P = p\,\Delta x$, since p has a Poisson distribution; the concentration fluctuations δp are equal to $\delta P/\Delta x$, and hence

$$\overline{\delta p^2} = \frac{\overline{\delta P^2}}{(\Delta x)^2} = \frac{p}{\Delta x} \quad (6.37a)$$

This gives our $\overline{u^2(t)}$. To obtain the time constant τ_0, we observe that a transmission-line section of length Δx and negligible conductance has a time constant $\tau_0 = (R\,\Delta x)(C\,\Delta x) = RC(\Delta x)^2$. Translating this back to our diffusion problem, this corresponds to a time constant

$$\tau_0 = \frac{(\Delta x)^2}{D_p} \quad (6.37b)$$

[†] This follows from Eq. 6.10 by setting $c(\tau) = \exp(-\tau/\tau_0)$ and $\cos\omega\tau = 1$.

Consequently, since $p = p' + p_n$ is the total hole concentration, we have, for Δp_x, the Fourier coefficient of δp:

$$\overline{\Delta p_x^2} = \frac{4(p' + p_n)\,\Delta x\,\Delta f}{D_p} \tag{6.38}$$

In our transmission-line analogy, Δp_x corresponds to a fluctuating emf Δe_x in series with the line between x and $(x + \Delta x)$.

The noise problem of devices in which the current is carried by diffusion is hereby solved in principle.

Application to Diodes

The Junction Diode. We want to represent the noise of a junction diode by a current generator $\overline{i_p^2}$ in parallel to the junction. To calculate $\overline{i_p^2}$, we assume that *all current is carried by holes*. We short-circuit the external leads; the noise current i_p in the short-circuiting lead is then equal to the noise hole current at the beginning of the n region ($x = 0$). We divide the n region of the junction into sections of length Δx. In our transmission-line analogy the diffusion fluctuations correspond to distributed series noise emf's, and the recombination fluctuations correspond to distributed parallel noise-current generators. The n region corresponds to a transmission line of infinite length short-circuited at one end ($x = 0$); and we must calculate the noise current at $x = 0$. A noise-current generator Δi_x, in parallel to the line at a distance x from the input, gives rise to a current

$$\Delta i = \Delta i_x e^{-\gamma x} \tag{6.39}$$

in the lead short-circuiting the input (Fig. 6.1a). A noise emf Δe_x in series with the line at a distance x from the input gives rise to a current (Fig. 6.1b)

$$\Delta i = (\Delta e_x/Z_0)e^{-\gamma x} \tag{6.40}$$

Applying this to our diffusion problem, adding the contributions of the two noise sources, Eqs. 6.36 and 6.38, in a section Δx and the contribution of the individual sections quadratically, we have (the summation is extended over all Δx and then replaced by an integration):

$$
\begin{aligned}
\overline{i_p^2} &= \sum \left(\overline{\Delta i_{px}^2}\,e^{-2ax} + \frac{\overline{\Delta p_x^2}}{|Z_0|^2}\,e^{-2ax} \right) \\
&= \frac{2e^2\,\Delta f}{\tau_p} \left[\int_0^\infty (p' + 2p_n)e^{-2ax}\,dx \right. \\
&\qquad\qquad \left. + 2\sqrt{1 + \omega^2\tau_p^2} \int_0^\infty (p' + p_n)e^{-2ax}\,dx \right] \tag{6.41}
\end{aligned}
$$

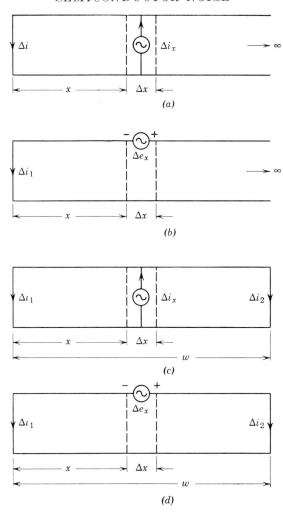

Fig. 6.1. Boundary conditions for noise-transmission-line analog.

We now have to calculate $p'(x)$. If a direct voltage V_0 is applied to the junction, the characteristic is[3]

$$I_p = I_{p0}\left[\exp\left(\frac{eV_0}{kT}\right) - 1\right], \qquad I_{p0} = ep_n\left(\frac{D_p}{\tau_p}\right)^{1/2} \quad (6.42)$$

and the d-c hole concentration in the n region is

$$p'(x) = p'_0 \exp(-\gamma_0 x), \qquad p'_0 = p_n \left[\exp\left(\frac{eV_0}{kT}\right) - 1 \right] \quad (6.42a)$$

We also need the a-c admittance of the junction, its real and imaginary part, and the real part a of γ:

$$Y = G_0(1 + j\omega\tau_p)^{1/2} = G + jB, \qquad G_0 = \frac{e}{kT}(I_p + I_{p0}) \quad (6.42b)$$

$$G = G_0[\tfrac{1}{2}(1 + \omega^2\tau_p^2)^{1/2} + \tfrac{1}{2}]^{1/2} \qquad\qquad (6.42c)$$

$$a = \gamma_0[1(\tfrac{1}{2} + \omega^2\tau_p^2)^{1/2} + \tfrac{1}{2}]^{1/2} = \gamma_0 \frac{G}{G_0} \qquad (6.42d)$$

Formulas 6.42 to 6.42c may also be derived directly with the transmission-line analogy.

Substituting $p'(x)$ into Eq. 6.41 and carrying out the integration, we obtain

$$\overline{i_p^2} = 2e\,\Delta f \left[I_p\left(-1 + 2\frac{G}{G_0}\right) + I_{p0}\left(2\frac{G}{G_0}\right) \right] \quad (6.43)$$

or

$$\overline{i_p^2} = 4kT(G - G_0)\,\Delta f + 2e(I_p + 2I_{p0})\,\Delta f \qquad (6.43a)$$

At low frequencies $G = G_0$, and hence

$$\overline{i_p^2} = 2e(I_p + 2I_{p0})\,\Delta f \qquad\qquad (6.43b)$$

This formula is the same as one derived from the "emission model" of the junction. In that model the current flow is supposed to be due to a simple *emission* of carriers across the junction; a current $I_p + I_{p0} = I_{p0} \exp(eV_0/kT)$ flows in one direction, and a current I_{p0} flows in the opposite direction. Both currents should fluctuate independently and should show full shot effect. Adding these two noise sources gives Eq. 6.43b.[1,25] Actually, the emission model does not apply here; the current flow is by *diffusion*, not by *emission*. It is interesting that this model gives the second term of Eq. 6.43a, though it does not explain the first term; for the explanation of the first term, see van der Ziel and Becking.[26]

We may summarize Eq. 6.43a by saying that *the diode current shows full shot noise at all frequencies, and the conductance $(G - G_0)$ shows thermal noise at all frequencies.*

If we write

$$\overline{i_p^2} = 2eI_{eq}\,\Delta f = 4kTR_n\,\Delta f \cdot |Y|^2 \qquad (6.44)$$

we find that I_{eq} increases with increasing frequency and that R_n decreases. If we equate

$$\overline{i_p^2}_{\mathbf{s}} = n \cdot (4kTG \, \Delta f) \tag{6.45}$$

we find for the noise ratio n of the junction

$$n = 1 - \frac{1}{2}\left(\frac{G_0}{G}\right) \cdot \frac{I_p}{I_p + I_{p0}} \tag{6.46}$$

It is clear that

(a) In forward bias $(I_p > 0)$, n lies in range $\frac{1}{2} \leq n \leq 1$. At very high frequencies $(G \gg G_0)$, $n = 1$.

(b) In back bias $(I_p < 0)$, $n > 1$,

(c) In the absence of bias $(I_p = 0)$, $n = 1$ at all frequencies, as required by Nyquist's theorem for a resistor at the thermodynamic equilibrium.

These theoretical predictions agree very well with the experimental data,[27,28] except for the effect of the series resistance r of the junction. Figure 6.2 shows some of the results obtained by Champlin[28] for a

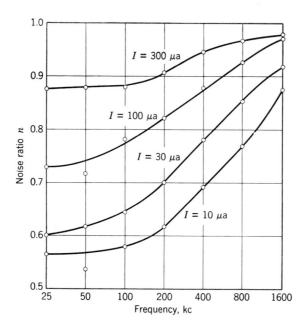

Fig. 6.2. Noise ratio for Raytheon CK-738-diode A.

Raytheon CK-738 silicon diode. In the figure the noise ratio n is plotted against frequency with the direct current as a parameter. The general frequency dependence agrees with Eq. 6.46; owing to the relatively large series resistance, the low-frequency value of n is larger than $\frac{1}{2}$.

The junction may be described by the resistance r in series with the junction impedance $Z = 1/Y$, with Y given by equation 6.42b. One would expect full thermal noise for r, whereas the junction itself should have the noise ratio n given by equation 6.46. The measured noise ratio of the whole junction agrees roughly with these suggestions. More accurate measurements give the same result.

The Point-Contact Diode. The diffusion of minority carriers in n-type or p-type material is a relatively slow process. Microwave point-contact diodes operate successfully at 1-cm wavelength and shorter as mixers, detectors, and generators of higher harmonics.[29] This makes it unlikely that minority carrier injection plays a part in their operation; they operate by majority carrier diffusion across the rectifying barrier, which gives a characteristic similar to the one for a junction. Especially if the barrier region is thin, this may be a very fast process. In that case the "emission model" should apply even at microwave frequencies. *For those devices one would thus expect full shot noise for the diode itself and thermal noise for the series resistance r of the diode.* But, since the current flows through a narrow contact area, considerable local heating may occur; since the resistance r is also concentrated in that region, the noise temperature of r may be considerably above room temperature at current levels of the order of 1 ma. Moreover, a considerable amount of $1/f$ noise is generated, especially if the diode is biased in the back direction.

If the point-contact diode is used as a low-level detector, one may assume that the diode operates so closely to the point $V = 0$ that the diode itself will give thermal noise ($n = 1$ for $I = 0$). One may then design the diode parameters such that optimum sensitivity is obtained. We refer to Torrey and Whitmer's book for details.[29]

If the intermediate frequency in mixer diodes is chosen too small, then the $1/f$ noise predominates. The higher the intermediate frequency, the lower the output noise ratio n_0 of the mixer diode may be. Ultimately one will be restricted by the shot noise of the diode itself and the thermal noise of the series resistance so that output noise ratios of unity or slightly below unity might be expected. The equivalent circuit of the mixer diode consists of the resistance r in series with the diode resistance R and the diode capacitance C connected in parallel. At the highest frequencies the capacitance C may shunt the resistance

R, and most of the microwave input power may be dissipated in the series resistance r instead of in the diode itself. This influences the choice of diode parameters for optimum mixer performance at microwave frequencies. For details we refer again to Torrey and Whitmer's book.[29]

Noise in p-n-p Transistors†

The operation of a junction transistor is based upon the principles of minority-carrier *injection into* the base region by the emitter and minority-carrier *extraction from* the base region by the collector. The current flow in the base region is due to diffusion; for the sake of simplicity it will be assumed that all current is carried by holes. The noise in such a transistor is caused by hole-concentration fluctuations in the base region, for which the theory on page 326 applies.

We want to represent the noise of a transistor first by two current generators, i_{p1} and i_{p2}, across the emitter and the collector junction, respectively. To do this, we introduce the "internal" base b'; the emitter junction extends from the emitter contact e to b', and i_{p1} is in parallel to it; the collector junction extends from the collector contact c to b', and i_{p2} is in parallel to it. It is common practice to introduce in the lead from the internal base b' to the external base b the true base resistance r'_b (see also Fig. 6.4); r'_b represents the influence of the finite base conductivity upon the operation of the transistor. To find i_{p1} and i_{p2}, we simply have to short-circuit the emitter and collector junctions and calculate the noise hole current at $x = 0$ (emitter side of base region) and at $x = w$ (collector side of base region, w being the width of the base region) due to hole-concentration fluctuations in the base region.

The introduction of the internal base b' and of the true base resistance r'_b is illustrated in Fig. 6.3. In this figure, e is the emitter region and c is the collector region, both n-type, which are separated by a narrow strip of p-type material. Consider a small transistor section of cross-sectional area ΔA. If it had a base contact b'', it would act as an ideal transistor with zero base resistance. Actually, there is a resistance r''_b between this base contact b'' and the true base contact b. Since this holds for every small transistor section, and since adding sections means adding the currents, the actual transistor can thus be replaced by an ideal transistor with base contact b' and zero base resistance, and a resistance r'_b connected between b' and the true base contact b, r'_b being found by averaging over all small transistor sec-

† The theory for an n-p-n transistor proceeds along the same lines; in that case one assumes that all current is carried by electrons.

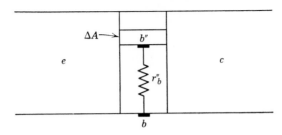

Fig. 6.3.   Internal base b' and true base resistance r'_b of *p-n-p* transistor.

tions.   The introduction of the internal base b' thus takes care of the finite conductivity of the base region, by lumping its effect together in the resistance r'_b.

The current flow from emitter to collector in the actual transistor is thus considered to be the same as for the ideal transistor.   Hence the base region can now be represented by an equivalent transmission line of length w, short-circuited for alternating current at both ends, with distributed series noise emf's and distributed parallel noise-current generators.   Since i_{p1} and i_{p2} come from the same sources, i_{p1} and i_{p2} will be partly correlated; we thus have to calculate $\overline{i_{p1}^2}$, $\overline{i_{p2}^2}$ and $\overline{i_{p1}^*i_{p2}}$.

First we need a few transistor formulas.[3]   We assume that the *equilibrium* (all applied voltages are zero) hole concentration p_n is constant in the base region.   If direct voltages V_e and V_c are connected across the emitter and collector junctions, respectively, there result excess hole concentrations p'_{e0} at $x = 0$ and p'_{c0} at $x = w$:

$$p'_{e0} = p_n[\exp\,(eV_e/kT) - 1], \qquad p'_{c0} = p_n(\exp\,[eV_c/kT] - 1] \quad (6.47)$$

$p'_{c0} \simeq -p_n$ if the collector is properly biased.

The excess hole concentration $p'(x)$ in the base region is

$$p'(x) = p'_{e0}\left[\frac{\sinh \gamma_0(w - x)}{\sinh \gamma_0 w}\right] + p'_{c0}\left(\frac{\sinh \gamma_0 x}{\sinh \gamma_0 w}\right) \quad (6.48)$$

The d-c emitter and collector currents I_e and I_c are

$$I_e = \frac{p'_{e0}}{Z_{00}\tanh \gamma_0 w} - \frac{p'_{c0}}{Z_{00}\sinh \gamma_0 w} \quad (6.49)$$

$$I_c = \frac{p'_{e0}}{Z_{00}\sinh \gamma_0 w} - \frac{p'_{c0}}{Z_{00}\tanh \gamma_0 w} \quad (6.50)$$

If $V_e = 0$, then $p'_{e0} = 0$; if in addition the collector is biased so that it is operating as a saturated diode, we have $I_c = I_{cc}$ and

$$I_{cc} = \frac{p_n}{Z_{00} \tanh \gamma_0 w} \qquad (6.50a)$$

If $V_c = 0$, then $p'_{c0} = 0$, and

$$I_e = I_{ee}[\exp (eV_e/kT) - 1], \qquad I_{ee} = I_{cc} \qquad (6.49a)$$

Defining the saturated collector current I_{c0} as the collector current for zero emitter current, we have

$$I_{c0} = I_{cc}(1 - \alpha_0{}^2) \qquad (6.50b)$$

The collector current I_c may then be expressed as

$$I_c = \alpha_0 I_e + I_{c0} \qquad (6.50c)$$

This discussion neglects leakage across the emitter and collector junctions.

If a small alternating voltage v_e is applied to the emitter junction and the collector junction is a-c short-circuited, an alternating emitter current i_e will flow and give rise to an alternating collector current αi_e, where

$$\alpha = \frac{1}{\cosh \gamma w}; \qquad \alpha_0 = \frac{1}{\cosh \gamma_0 w} \qquad (6.51)$$

α is the *current-amplification factor* of the transistor. Its low-frequency value α_0 may be close to unity for properly designed transistors. The emitter junction admittance under that condition is

$$Y_e = Y_{e0}(1 + j\omega\tau_p)^{1/2} \frac{\tanh \gamma_0 w}{\tanh \gamma w} = G_e + jB_e \qquad (6.52)$$

$$Y_{e0} = G_{e0} = \frac{e}{kT}[I_e + I_{cc}(1 - \alpha_0)] \qquad (6.52a)$$

The formulas 6.48 through 6.52a may also be derived directly with the help of the transmission-line analogy.

We now complete the equivalent circuit of the transistor. If a small a-c collector emf v_c is applied, it modulates the width of the base region. This gives rise to several effects; an additional collector admittance Y_c will be connected in parallel to the capacitance C_c of the collector space-charge region,† and feedback emf's $\mu_{ec}v_c$ and $\mu_{bc}v_c$ have to be inserted in series with the emitter admittance Y_e

† Y_c represents the collector admittance due to base-width modulation effects (Early effect); C_c is the capacitance of the collector transition region.

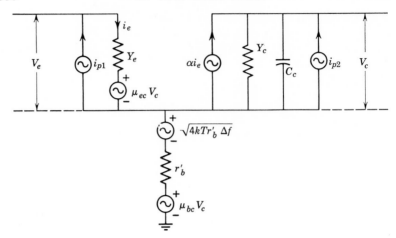

Fig. 6.4.   Equivalent circuit of p-n-p transistor with internal noise.

and the base lead, respectively.[30]   In addition, we have to add the thermal noise emf of the true base resistance r'_b (Fig. 6.4).

To calculate i_{p1} and i_{p2} we observe that in our transmission line a current Δi_x connected in parallel to the line between x and $(x + \Delta x)$ gives currents Δi_1 and Δi_2 in the short-circuited input and output (Fig. 6.1c)

$$\Delta i_1 = \Delta i_x \frac{\sinh \gamma(w - x)}{\sinh \gamma w}, \qquad \Delta i_2 = \Delta i_x \frac{\sinh \gamma x}{\sinh \gamma w} \qquad (6.53)$$

whereas an emf Δe_x connected in series with the line between x and $(x + \Delta x)$ gives currents $\Delta i'_1$ and $\Delta i'_2$ in the short-circuited input and output (Fig. 6.1d).

$$\Delta i'_1 = -\frac{\Delta e_x}{Z_0} \frac{\cosh \gamma(w - x)}{\sinh \gamma w}, \qquad \Delta i'_2 = \frac{\Delta e_x}{Z_0} \frac{\cosh \gamma x}{\sinh \gamma w} \qquad (6.54)$$

We thus have that our noise sources Δi_{px} and Δp_x between x and $(x + \Delta x)$ give currents Δi_{p1} and Δi_{p2} in input and output:

$$\Delta i_{p1} = \Delta i_{px} \frac{\sinh \gamma(w - x)}{\sinh \gamma w} - \frac{\Delta p_x}{Z_0} \frac{\cosh \gamma(w - x)}{\sinh \gamma w} \qquad (6.55)$$

$$\Delta i_{p2} = \Delta i_{px} \frac{\sinh \gamma x}{\sinh \gamma w} + \frac{\Delta p_x}{Z_0} \frac{\cosh \gamma x}{\sinh \gamma w} \qquad (6.56)$$

Consequently, since the sections Δx are independent,

$$\overline{i_{p1}{}^2} = \Sigma\overline{\Delta i_{p1}{}^2}, \qquad \overline{i_{p2}{}^2} = \Sigma\overline{\Delta i_{p2}{}^2}, \qquad \overline{i^*{}_{p1}i_{p2}} = \Sigma\overline{\Delta i^*{}_{p1}\Delta i_{p2}} \quad (6.57)$$

where the summation is extended over all sections Δx. We take into account that Δp_x and Δi_{px} are uncorrelated, substitute for $\overline{\Delta p_x{}^2}$ and $\overline{\Delta i_{px}{}^2}$, Eqs. 6.36 and 6.38, and replace all summations by integrations. This yields after some mathematical manipulations:

$$\overline{i_{p1}{}^2} = 4kTG_e\,\Delta f - 2eI_e\,\Delta f \qquad (6.58)$$

$$\overline{i_{p2}{}^2} = 2eI_c\,\Delta f \qquad (6.59)$$

$$\overline{i^*{}_{p1}i_{p2}} = -2kT\alpha Y_e\,\Delta f \qquad (6.60)$$

For low frequencies, $\alpha = \alpha_0$, $G_e = G_{e0}$, and $Y_e = Y_{e0} = G_{e0}$. Neglecting a few small terms of the order $I_{cc}(1 - \alpha_0)$, these equations may be written:

$$\overline{i_{p1}{}^2} = 2eI_e\,\Delta f \qquad (6.61)$$

$$\overline{i_{p2}{}^2} = 2eI_c\,\Delta f \qquad (6.62)$$

$$\overline{i^*{}_{p1}i_{p2}} = -2eI_c\,\Delta f \qquad (6.63)$$

In this approximation the emitter and collector current thus show full shot noise, and the two noise currents are almost completely correlated. The correlation coefficient c is

$$-c = \frac{\overline{i^*{}_{p1}i_{p2}}}{(\overline{i_{p1}{}^2} \cdot \overline{i_{p2}{}^2})^{1/2}} \simeq \left(\frac{I_c}{I_e}\right)^{1/2} \simeq \alpha_0{}^{1/2} \qquad (6.64)$$

According to Eq. 6.59, the collector junction shows full shot noise at all frequencies. Equation 6.58 may be written

$$\overline{i_{p1}{}^2} = 2e[I_e + 2I_{cc}(1 - \alpha_0)]\,\Delta f + 4kT(G_e - G_{e0})\,\Delta f \quad (6.58a)$$

This can be interpreted as follows. The emitter current can be considered as consisting of a current $[I_e + I_{cc}(1 - \alpha_0)]$ and a current $I_{cc}(1 - \alpha_0)$ flowing in opposite directions. Both show full shot noise at all frequencies. In addition, the conductance $(G_e - G_{e0})$ shows thermal noise.

We thus have determined the noise equivalent circuit of a transistor. It is formally equivalent to another circuit, where the noise is represented by a noise emf $e_s = i_{p1}/Y_e$ in series with the emitter junction and a noise-current generator $i_p = i_{p1} + \alpha i_{p2}$ in parallel with the

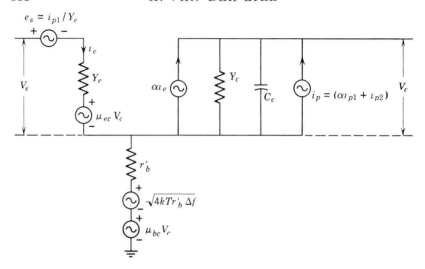

Fig. 6.5.   Alternate equivalent circuit of *p-n-p* transistor with internal noise.

collector junction (Fig. 6.5).   We then find, from Eqs. 6.58 and 6.60 in a simple equivalent-circuit transformation:

$$\overline{e_s^2} = \frac{\overline{i_{p1}^2}}{\left| Y_e \right|^2} \tag{6.65}$$

$$\overline{i_p^2} = 2e(\alpha_0 - |\alpha|^2) I_e \,\Delta f + 2e I_{c0}\,\Delta f \tag{6.66}$$

$$\overline{e^*_s i_p} = \frac{\alpha}{Y^*_e} \left[2kT Y^*_e \,\Delta f - 2e I_e\,\Delta f \right] \tag{6.67}$$

At low frequencies these equations reduce to

$$\overline{e_s^2} \simeq \frac{2kT\,\Delta F}{G_{e0}} \tag{6.65a}$$

$$\overline{i_p^2} = 2e\alpha_0(1 - \alpha_0) I_e\,\Delta f + 2e I_{c0}\,\Delta f \tag{6.66a}$$

$$\overline{e^*_s i p} \simeq 0 \tag{6.67a}$$

so that e_s and i_p are practically uncorrelated.   The emitter resistance $1/G_{e0}$ then shows approximately half thermal noise.   The first term in Eq. 6.66a can be interpreted as partition noise of the emitter current I_e, and the second term as shot noise of the collector saturated current I_{c0}.   This is in full agreement with earlier suggestions.[31]

If the noise generator i_p is represented by its equivalent saturated diode current I_{eq}, we have

$$I_{eq} = \left(1 - \frac{|\alpha|^2}{\alpha_0}\right) I_c + \frac{|\alpha|^2}{\alpha_0} I_{c0} \qquad (6.68)$$

This is equal to $[I_c(1 - \alpha_0) + \alpha_0 I_{c0}]$ at low frequencies, equal to $\simeq \frac{1}{2} I_c$ at the cut-off frequency (where $|\alpha|^2 \simeq \frac{1}{2}$), and equal to I_c at very high frequencies. This quantity is easily measured; if a high impedance is inserted in series with the emitter junction, the equivalent saturated diode current of the device as measured at the collector side is equal to I_{eq}. The emf e_s and the correlation between e_s and i_p are more difficult to measure.

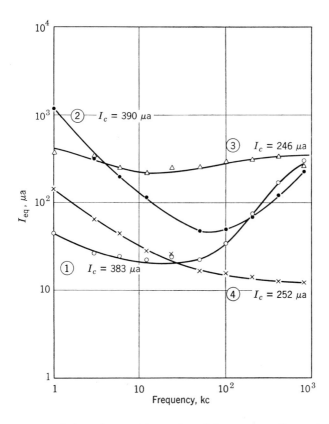

Fig. 6.6. Measured dependance of I_{eq} on I_c and frequency. Curves 1, 2, and 3 refer to transistors with $f_c \simeq 400$ kc; curve 4 refers to a surface-barrier transistor with $f_c > 10$ Mc.

This rise in I_{eq} at high frequencies has been observed by Hanson;[32] Fig. 6.6 shows some of his results. Curves 1 and 2 refer to two transistors having a cut-off frequency of 400 kc; the rise in I_{eq} at high frequencies is clearly shown. Curve 3 refers to the transistor of curve 2 but with emitter and collector interchanged; this results in a very low α_0 (and hence a much higher value of I_{eq} at intermediate frequencies) and a lower cut-off frequency. Curve 4 refers to a surface-barrier transistor with a cut-off frequency above 10 Mc; no rise in I_{eq} is observed even at 800 kc. The rise in I_{eq} at low frequencies for all transistors is due to $1/f$ noise.

Since $\gamma_0 w$ is a small quantity for a good transistor, so that $\sinh \gamma_0(w - x) \ll \cosh \gamma_0(w - x)$, we see from Eq. 6.55 that the term with Δp_x in Δi_{p1} predominates. This means that i_{p1}, and hence e_s, is mainly caused by diffusion fluctuations at low frequencies and that noise due to recombination can be ignored. In the same way we see that, in the expression for i_p obtained from Eqs. 6.55 and 6.56,

$$i_p = \sum (\Delta i_{p2} + \alpha \Delta i_{p1})$$
$$= \alpha \sum \left[\Delta i_{px} \cosh \gamma x + \left(\frac{\Delta p_x}{Z_0} \right) \sinh \gamma x \right] \quad (6.69)$$

the term with Δi_{px} predominates at low frequencies, so that i_p is mainly caused by recombination fluctuations at low frequencies. At higher frequencies both noise sources contribute to e_s and i_p.

The equivalent circuit thus developed shows complete agreement with an earlier circuit proposed by Montgomery, Clark, and van der Ziel.[31] The equivalent circuit with two noise-current generators shows considerable agreement with an earlier circuit proposed by Giacoletto.[33]

Calculation of the Noise Figure of Transistors

Now that the equivalent circuit has been established, it is a simple matter to calculate the noise figure F of a transistor amplifier. Let a signal source (emf e_s, internal impedance $Z_s = R_s + jX_s$ be connected to the emitter input. The thermal noise of the source is then represented by an emf $\sqrt{4kTR_s \, \Delta f}$ in series with the source, and the noise figure F is defined as

$$F = \frac{\text{total output noise power of network}}{\text{output noise power due to thermal noise of source}}$$

Assuming the noise voltage between b and b' to be small in comparison with the voltage developed across Y_c (this condition is usually satisfied), we obtain, if v_{ct} is the noise voltage across Y_c due to all transistor noise sources and v_{cs} the noise voltage across Y_c due to the thermal noise of the source:

$$F - 1 = \frac{\overline{v_{ct}^2}}{\overline{v_{cs}^2}} = \frac{r'_b}{R_s} + \frac{\overline{|-e_s + i_p(Z_s + Z_e + r'_b)/\alpha|^2}}{4kTR_s\,\Delta f} \quad (6.70)$$

where Z_s is the signal-source impedance and $Z_e = R_e + jX_e$ is the impedance of the emitter junction. At low frequencies, $R_e = R_{e0}$ and $X_e = 0$; neglecting the shot noise of I_{c0}, the minimum noise figure becomes

$$F_{\min} = \frac{1}{\alpha_0}\left\{\left[1 + \frac{(1-\alpha_0)r'_b}{R_{e0}}\right] + \sqrt{\left[1 + \frac{(1-\alpha_0)r'_b}{R_{e0}}\right]^2 - \alpha_0}\right\} \quad (6.71)$$

for

$$R_s = \left[(R_{e0} + r'_b)^2 + \frac{\alpha_0}{1-\alpha_0}R_{e0}(2r'_b + R_{e0})\right]^{1/2} \quad (6.71a)$$

The noise figure is thus unity if $\alpha_0 = 1$; this result is achieved for $R_s = \infty$. The result is not surprising, for in that case the transistor does not give any noise (if I_{c0} can be neglected), and the thermal noise of r'_b has no influence if R_s is sufficiently large.

The quantity $(1 - \alpha_0)r'_b/r_{e0}$ is thus the determining factor in the low-frequency noise figure of a transistor; for a low-noise transistor it should be made as small as possible. This condition is best satisfied for small emitter currents. If the shot noise of I_{c0} is taken into account, one finds that $F_{\min}$, considered as a function of I_e, has a minimum value for relatively small I_e.[34]

At low frequencies the noise figure increases as soon as $1/f$ noise becomes important; for well-built transistors this occurs below a few kilocycles. At high frequencies F increases for *two reasons:*

(a) $\overline{i_p^2}$ increases by an important factor. Since most of this increase occurs below the cut-off frequency, this effect is important for the initial rise in noise figure.

(b) Above the cut-off frequency, $\overline{i_p^2}$ remains roughly constant, but $|\alpha|^2$ continues to decrease with frequency, and this causes a corresponding increase in noise figure.

Final Discussion of Shot Noise in Diodes and Transistors

The theory presented here is a one-dimensional approximation, and the recombination is assumed to be volume recombination. Actually, recombination occurs at the surface, and hence the carrier flow is not a one-dimensional process. Nevertheless the agreement between theory and experiment is gratifying.

The basic equations, 6.43a and 6.58 through 6.60, do not contain any specific reference to the model used in their derivation. For that reason it might be assumed that these equations should be true for all models and all geometries. A more detailed investigation has verified this.[26,34] The equations also hold if the current is partly carried by electrons and partly by holes. In addition the assumption that p_n is constant in the base region of a p-n-p transistor may be dropped. Equations 6.43a and 6.58 through 6.60 are thus of general validity.

REFERENCES

1. A. van der Ziel, *Noise*, Prentice-Hall, New York (1954).
2. A. van der Ziel, *Physica*, **16**, 359 (1950); **19**, 742 (1953).
3. W. Shockley, *Electrons and Holes in Semiconductors*, D. Van Nostrand Co., New York (1950).
4. H. J. Hannam, Ph.D. thesis, University of Minnesota (1956).
5. H. J. Hannam and A. van der Ziel, *J. Appl. Phys.*, **25**, 1136 (1954).
6. L. S. Nergaard, *RCA Rev.*, **13**, 464 (1952).
7. J. Bernamont, *Compt. rend.*, **198**, 1755 and 2144 (1934); *Ann. phys.* **7**, 71 (1937); *Proc. Phys. Soc.*, **49**, 138 (1937).
8. M. Surdin, *Rev. gén. élec.* **47**, 97 (1940).
9. B. Davydov and B. Gurevich, *J. Phys. USSR*, **7**, 138 (1943).
10. J. H. Gisolf, *Physica*, **15**, 825 (1949).
11. F. Zernike, *Handbuch der Physik*, vol. 3, 419, Springer, Berlin (1928).
12. J. E. Hill and K. M. van Vliet, *J. Appl. Phys.*, **29**, 177 (1958); K. M. van Vliet and J. E. Hill, *Physica*, to be published.
13. A. van der Ziel, *J. Appl. Phys.*, **24**, 222 (1953); **24**, 1063 (1953).
14. H. A. Gebbie, *Phys. Rev.*, **98**, 1757 (1955).
15. J. R. Fassett, M.Sc. thesis, University of Minnesota, (1958) unpublished.
16. G. B. Herzog and A. van der Ziel, *Phys. Rev.*, **84**, 1249 (1951); R. H. Mattson and A. van der Ziel, *J. Appl. Phys.*, **24**, 222 (1953).
17. H. C. Montgomery, *Bell Syst. Tech. J.*, **31**, 950 (1952).
18. K. M. van Vliet, *Phys. Rev.*, **110**, 50 (1958).
19. K. M. van Vliet, *Proc. I.R.E.*, **46**, 1004 (1958).
20. R. E. Burgess, *Physica*, **20**, 1007 (1954); *Proc. Phys. Soc. B.*, **68**, 661 (1955); *Brit. J. Appl. Phys.*, **6**, 185 (1955).
21. S. Machlup, *J. Appl. Phys.*, **25**, 341 (1954).
22. A. van der Ziel, *Proc. IRE*, **43**, 1639 (1955).
23. D. O. North, IRE—AIEE Conf. on Semiconductor Device Research, University of Pennsylvania (1955).
24. R. L. Petritz, *Proc. I.R.E.*, **40**, 1440 (1950); *Phys. Rev.*, **91**, 204, 231 (1953) (see especially the correction shown on page 204; no detailed account published).
25. W. F. Weisskopf, "On the Theory of Noise in Conductors, Semi-conductors, and Crystal Rectifiers, *NDRC*, *14–33*, May 15, 1943.

26. A. van der Ziel and A. G. Th. Becking, *Proc. I.R.E.*, **46**, 589 (1958).

27. R. L. Anderson, M.S., thesis, University of Minnesota (1952); R. L. Anderson and A. van der Ziel, *Trans. IRE PGED*-1, 20 (1952).

28. K. S. Champlin, M.S. thesis, University of Minnesota (1955).

29. H. C. Torrey and C. A. Whitmer, *Crystal Rectifiers*, McGraw-Hill Book Co., New York (1948).

30. J. M. Early, *Proc. IRE*, **40**, 1401 (1952).

31. H. G. Montgomery and M. A. Clark, *J. Appl. Phys.*, **24**, 1397 (1953); A. van der Ziel, *J. Appl. Phys.*, **25**, 815 (1954).

32. G. H. Hanson, *J. Appl. Phys.*, **26**, 1388 (1955).

33. L. J. Giacoletto, IRE—AIEE Conf. on Semiconductor Device Research, University of Minnesota (1954).

34. A. van der Ziel, *Proc. I.R.E.*, **46**, 1019 (1958).

CHAPTER 7

Noise in Transistors[†]

W. H. Fonger

Introduction

The noise observed in transistors may be classified as diffusion—recombination noise, thermal noise, $1/f$ surface noise, $1/f$ leakage noise, and irradiation noise.

The processes of electron-hole-pair creation, motion, and recombination form the basis of transistor action. Noise unavoidably accompanies these processes. The noise may be calculated to first approximation by assuming that fluctuations in the creations, motions, and recombinations of pairs are completely random. Noise thus idealized will be called diffusion–recombination noise. At frequencies smaller than the reciprocal of an average pair transit time, diffusion–recombination noise is described by $2eI\,\Delta f$ formulas reminiscent of full shot noise in vacuum tubes. This parallelism results because vacuum-tube full-shot-noise formulas are based upon idealizations equivalent to those cited above and because the electron charge e is the quantum of charge for both devices.

† This chapter is intended to substitute for lectures originally presented at the noise course by Dr. D. O. North. The method of presentation and the emphasis on particular topics have been changed somewhat, though the basic understanding of transistor noise is, of course, not changed. A description of North's model for diffusion–recombination noise is not included. This important topic, including some interesting results at high frequencies, is discussed in the preceding chapter by Professor van der Ziel.

Thermal noise in the transistor base-lead resistance is described by Nyquist's $4kTR\,\Delta f$ formula. Base-lead resistance is not essential to transistor action. However, insofar as this resistance is unavoidable practically, its associated noise is unavoidable.

Diffusion–recombination noise and thermal noise in the base-lead resistance account to a good approximation for the white noise observed in transistors. Together these two noises establish a lower limit for the signals which can be usefully amplified. This limit is comparable to the corresponding limit for vacuum tubes. Indeed, noise factors of 1 to 2 db are realized with present-day transistors.

The (diffusion–recombination-noise) approximation that fluctuations in the generations and recombinations of electron-hole pairs are uncorrelated breaks down at low frequencies. The details of these correlations depend on the detailed characteristics of the charge-carrier traffic across the semiconductor forbidden band. These details are not known accurately. However, calculations based on a phenomenological model which assumes a slow $(1/f)$ variation in the probability of electron-hole-pair surface generation and recombination account to a good approximation for the modest $1/f$ noise observed in present-day transistors. Noise accounted for by this model will be called $1/f$ surface noise. This noise is proportional to the square of the emitter current and, for small currents, is small in the audio region. It is possible that future advances in the control of semiconductor surfaces may appreciably decrease the magnitude of $1/f$ surface noise.

$1/f$ leakage noise is associated with electrical breakdown in the high-field region at the perimeter of the collector junction. This noise was responsible for the large noise factors of early transistors, but has been largely eliminated nowadays by proper fabrication techniques. Leakage noise in present-day transistors is negligibly small in the audio region if the collector voltage is not large.

Irradiation noise is associated with the bursts of electron-hole pairs generated when high-energy particles penetrate a transistor. This noise is described by $\overline{M^2}2eI\,\Delta f$ formulas reminiscent of noise in multiplication by secondary emission. The multiplication M is analogous in the two cases. Irradiation noise is negligibly small unless the transistor is exposed to strong artificial sources of radiations.

The program of this chapter on transistor noise is as follows: In Section 1 Shockley's[1][†] theory of the junction transistor is outlined and stated in a form readily generalized to describe noise. Diffusion–

---

† Compare Eqs. 15 on p. 313, 8 on p. 312, and 11 on p. 321 of reference 1b with Eqs. 7.1, 7.2, and 7.3, respectively below.

recombination noise and $1/f$ surface noise are treated in Sections 2 and 3, respectively. Experimental data concerning both are combined in Section 3. The application of these topics to the design of low-noise transistor amplifiers is summarized in Section 4. $1/f$ leakage noise, because of waning interests, is mentioned only briefly at appropriate points in Sections 3 and 4. Irradiation noise is treated in Section 5, in a framework very similar to that used for $1/f$ surface noise. The chapter concludes with a two-page summary of important results.

1. Diffusion Theory of the Junction Transistor

Validity of the Theory

The theory of the junction transistor has been developed by Shockley[1] in the approximation that

    (*a*) Pairs move only via diffusion.

    (*b*) The base thickness W is independent of the junction voltages.

Shockley's theory will be followed here with the additional approximation that

    (*c*) All time-varying quantities are quasistationary.

    (*d*) The conductivities of the emitter and collector electrodes are large.

The validity of the theory is therefore formally restricted to transistors that have uniformly doped, constant-thickness base regions and heavily doped emitter and collector regions and that are operated at moderately low emitter currents and moderately low frequencies. In practice, the results of the theory can be applied with some success under broader conditions.

Statement of the Boundary-Value Problem

For definiteness, the discussion will be phrased in terms of a *p-n-p* transistor. The extension to an *n-p-n* transistor is straightforward.

Let $P(t, \mathbf{r})$ be the density distribution of pairs in the base region, where t is the time and $\mathbf{r}$ is the spatial radius vector. P satisfies the equation of continuity

$$\frac{\partial P}{\partial t} = D \, \nabla^2 P - \frac{P - P_n}{\tau} \tag{7.1}$$

and the boundary conditions

$$
\begin{aligned}
P &= P_n \exp \left(V_{eb'}/V_T \right) \quad \text{at the emitter} \\
&= P_n \exp \left(V_{cb'}/V_T \right) \quad \text{at the collector}
\end{aligned} \tag{7.2}
$$

and

$$R = -D \frac{\partial P}{\partial n} = S(P - P_n) \quad \text{at the base recombination surface} \quad (7.3)$$

where D is the pair-diffusion constant, ∇^2 is the Laplacian, P_n is the base-region thermal-equilibrium hole density, τ is the pair volume lifetime, b' is the approximately equipotential portion of the base region near the junctions where pairs move only via diffusion, $V_{eb'}$ and $V_{cb'}$ are the voltages of the heavily doped emitter and collector electrodes, respectively, relative to the b' region, V_T is the thermal voltage, R is the net pair surface recombination rate per unit area, $\partial P/\partial n$ is the space derivative of P in the direction of the outward normal, and S is the surface recombination velocity. P/τ and SP give the rates of pair recombination per unit volume and area, respectively, while P_n/τ and SP_n give the corresponding rates of pair thermal generation. $V_T = kT/e$, where k is the Boltzmann constant, T the temperature in degrees Kelvin, and e the magnitude of the charge on the electron; at room temperature $V_T = 0.025$ volt. The restriction that time-varying quantities be quasistationary is defined to mean that $\partial P/\partial t$ may be taken equal to zero in Eq. 7.1.

The b' region occurring in the formulation of the boundary-value problem is separated from the external base electrode b by a portion of the base semiconductor. In circuit schematics, this separating element is represented by a base-lead resistance $r_{bb'}$.

The theory is readily applied to a device consisting of $(N - 1)$ heavily doped p-type electrodes attached to an approximately equipotential n-type b' region. This approach will be followed here because it is simple and describes simultaneously the semiconductor diode and triode. The b' electrode, differing in a fundamental way from the junction electrodes, will sometimes require special comment. For the N-electrode device, Eqs. 7.1 and 7.3 are unchanged, and Eqs. 7.2 are replaced by

$$P = P_n \exp \left(V_{jb'}/V_T \right) \quad \text{at the } j\text{th junction} \quad (7.4)$$

where $V_{jb'}$ is the voltage of the jth-junction electrode relative to the b' region.

Subdivision of the Boundary-Value Problem

Because the boundary-value problem above is linear, that is, because D, τ, and S are taken to be independent of P, the problem may be divided into N separate boundary-value problems. Eqs. 7.1, 7.3, and

7.4, subject to the quasistationary proviso, are formally satisfied if P is written as the sum of N pair-density distributions,

$$P = \sum_j P_j, \tag{7.5}$$

where the P_j's satisfy the separate equations of continuity,

$$D \nabla^2 P_j - \frac{P_j}{\tau} = 0 \quad \text{for} \quad j \neq b'$$

$$D \nabla^2 P_{b'} - \frac{P_{b'} - P_n}{\tau} = 0 \quad \text{for} \quad j = b' \tag{7.6}$$

and the separate boundary conditions,

$$\left.\begin{array}{l} P_j = P_n \exp\left(V_{jb'}/V_T\right) \quad \text{at the } j\text{th junction} \\[4pt] P_j = 0 \quad \text{at all other junctions} \\[4pt] -D\dfrac{\partial P_j}{\partial n} = SP_j \quad \text{at the base recombination sur-} \\[4pt] \qquad\qquad\qquad\text{face} \end{array}\right\} \text{for } j \neq b'$$

$$\left.\begin{array}{l} P_{b'} = 0 \quad \text{at all junctions} \\[4pt] -D\dfrac{\partial P_{b'}}{\partial n} = S(P_{b'} - P_n) \quad \text{at the base recombina-} \\[4pt] \qquad\qquad\qquad\text{tion surface} \end{array}\right\} \text{for } j = b' \tag{7.7}$$

This is readily verified by adding the separate equations.

P_j may be interpreted as the density distribution of pairs "injected at the jth junction by the voltage $V_{jb'}$." P_j pairs are subject to recombination in the base volume and at the base surface, and are extracted perfectly at the other junctions. (If the currents associated with these extractions increase the biases $V_{kb'}$ of the extracting electrodes, pairs are reinjected into the base. However, these returned pairs are accounted for by the pair-density distributions P_k, $k \neq j$.) P_j is proportional to $\exp\left(V_{jb'}/V_T\right)$ but is otherwise independent of the junction voltages. For $j = b'$, the phrase "injected at the jth junction by the voltage $V_{jb'}$" must be replaced by "generated thermally in the base region." In order to make the notation uniform, it is convenient to consider $P_{b'}$ proportional to $\exp\left(V_{b'b'}/V_T\right)$, even though $V_{b'b'}$ is always zero,

The Transfer Currents I_{jk}

It is hereafter assumed that the boundary-value problem has been solved. The solutions P_j are related to the device terminal voltages and currents and will presently be used to compute the device current–voltage relationship. The computation will amount essentially to the evaluation of transfer admittances.

The processes of pair injection at the jth junction, pair extraction at the other junctions, pair recombination in the base, and pair diffusion as represented by the pair-density distribution P_j describe $(N - 1)$ electric currents flowing *from* the electrode j *to* the other $(N - 1)$ electrodes. The processes of pair generation and recombination in the base, pair extraction at the junctions, and pair diffusion as represented by the pair-density distribution $P_{b'}$ describe $(N - 1)$ electric currents flowing *from* the b' electrode *to* the $(N - 1)$ junction electrodes. In all, $N (N - 1)$ currents I_{jk} $(k \neq j)$ may be identified, where I_{jk}, the current associated with the diffusion of pairs *from* the electrode j *to* the electrode k, is given by

$$I_{jk} = -e \int_K D \frac{\partial P_j}{\partial n} \, da \qquad (k \neq b')$$

$$I_{jb'} = e \int_{BV} \frac{P_j}{\tau} \, dv + e \int_{BS} SP_j \, da \qquad (k = b')$$

(7.8)

where K is the area of the kth junction, da the element of area, $\partial P_j / \partial n$ the space derivative of P_j in the direction of the outward normal, dv the element of volume, and BV and BS are the base volume and recombination surface, respectively. The I_{jk} will be called transfer currents.

Like P_j, I_{jk} is proportional to $\exp (V_{jb'}/V_T)$ but is otherwise independent of the junction voltages. Evidently the reciprocity relations

$$\frac{I_{jk}}{\exp (V_{jb'}/V_T)} = \frac{I_{kj}}{\exp (V_{kb'}/V_T)} \tag{7.9}$$

hold, since no net current must flow between pairs of electrodes at the same voltage.

The Current–Voltage Characteristic

The net current I_j passing *into* the device through the electrode j is given by

$$I_j = \sum_k (I_{jk} - I_{kj}) \qquad (k \neq j) \tag{7.10}$$

The I_j will be called electrode currents.† If the boundary-value problem for the P_j's has been solved, and if the integrations indicated in Eqs. 7.8 have been carried out, the transfer currents I_{jk} are known functions of the voltages $V_{jb'}$. Thus Eqs. 7.10 constitute the current–voltage characteristic of the intrinsic device (the total device minus the base-lead resistance). The small-signal characteristic is obtained by taking differentials.

The notation

$$A(t) = A_0 + a(t) \tag{7.11}$$

will be used for separating any time-dependent quantity $A(t)$ into d-c and small-signal a-c components A_0 and $a(t)$, respectively. Since the transfer current I_{jk} is proportional to exp $(V_{jb'}/V_T)$ but independent of the other junction voltages, its differential is given by

$$i_{jk} = \frac{I_{jk,0}}{V_T} v_{jb'} + i_{jk}\big|_V \tag{7.12}$$

where $v_{jb'}$ is the small-signal variation in $V_{jb'}$, and $i_{jk}\big|_V$ is the small-signal variation in the transfer current I_{jk} when the voltage $V_{jb'}$ is held constant; that is, when the electrode j is a-c short-circuited to the b' region. The variations $i_{jk}\big|_V$ are caused by stimuli internal to the device and will represent noise sources. The small-signal characteristic corresponding to Eqs. 7.10 is therefore

$$i_j = \sum_k \left(\frac{I_{jk,0}}{V_T} v_{jb'} + i_{jk}\big|_V \right) - \sum_k \left(\frac{I_{kj,0}}{V_T} v_{kb'} + i_{kj}\big|_V \right) \qquad (k \neq j) \tag{7.13}$$

In the signal theory of the device, the transfer currents I_{jk} do not vary when the voltages $V_{jb'}$ are held constant. Thus the $i_{jk}\big|_V$'s are all zero, and Eqs. 7.13 simplify to

$$i_j = v_{jb'} \sum_k \frac{I_{jk,0}}{V_T} - \sum_k \frac{I_{kj,0}}{V_T} v_{kb'} \qquad (k \neq j) \tag{7.14}$$

In the language of circuit theory, Eqs. 7.14 state that every junction electrode j is connected to the b' electrode by the conductance $\dfrac{I_{jb',0}}{V_T}$ and to every other junction electrode k by the transconductance generator

† The sign convention chosen for the electrode currents is depicted in Fig. 7.1a for the particular case of the three-electrode device (transistor). The currents I_b and $I_{b'}$ flowing through the external and internal base electrodes, respectively, are identical.

$\left(\dfrac{I_{jk,0}}{V_T} v_{jb'} - \dfrac{I_{kj,0}}{V_T} v_{kb'} \right).$ Conductance and transconductance elements in these forms should hardly come as a surprise, since only voltages $V_{jb'}$ and not V_{jk} were admitted in the statement of the boundary-value problem.

Equations 7.14, for the particular case of the triode, are represented

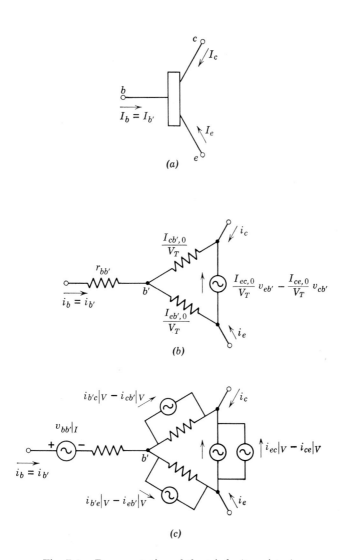

Fig. 7.1.   Representation of the triode (transistor).

by the small-signal schematic shown in Fig. 7.1b. The base-lead resistance $r_{bb'}$ has been included in the schematic. The currents i_b and $i_{b'}$ are identical.

In the signal theory, small-signal currents and voltages are excited by signal sources external to the device. In the noise theory, the transfer currents I_{jk} undergo, in addition, inherent fluctuations not prompted by junction voltages, and the full characteristic given by Eqs. 7.13 must be used. In the language of circuit theory, Eqs. 7.13 state that, in addition to the conductance and transconductance elements cited in connection with the signal theory, every electrode j is connected to every other electrode k by the noise-current generator $i_{jk}|v - i_{kj}|v$. These noise generators, together with the noise generators which characterize the circuitry external to the device, drive such noise currents and voltages as are observed.

Equations 7.13, for the particular case of the triode, are represented by the small-signal schematic shown in Fig. 7.1c. A noise-voltage generator $v_{bb'}|_I$ has been included in the base-lead arm. This generator will be interpreted subsequently.

Solutions for a Specific Device Geometry

Certain specific measurements and calculations will be described subsequently for devices with the specific base-region geometry shown in Fig. 7.2a. This geometry is characteristic of alloy-junction transistors. The junctions, denoted by indices 1 and 2, respectively, are circular and have unequal radii R_1 and R_2. The base region is slab-shaped and extends well beyond the larger radius R_2.

The boundary-value problem defined by Eqs. 7.6 and 7.7 has been solved for this geometry in the approximation, commonly valid for commercial alloy-junction transistors, that

(a) Pairs can diffuse linear distances of the order of the base thickness W with but little chance of recombination.

(b) $R_1 \gg W$.

(c) $(R_2 - R_1) \gg W$.

(d) L_2, the effective diffusion length for pairs in the region $r > R_2$, is small compared to R_2. $L_2 = \left(\dfrac{1}{\tau D} + \dfrac{2S}{WD} \right)^{-\frac{1}{2}}$.

The normalized pair-density distributions

$$\frac{P_1}{\exp{(V_{1b'}/V_T)}} \quad \text{and} \quad \frac{P_2}{\exp{(V_{2b'}/V_T)}}$$

and the pair-density distribution $P_{b'}$ depend on the axial co-ordinate z

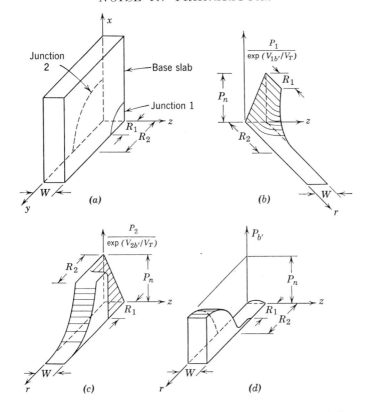

Fig. 7.2. Pair-density distributions for the alloy-junction triode.

and radial co-ordinate r approximately as shown in Figs. 7.2b, c, and d, respectively. Near the junctions, the shapes of these pair-density distributions are negligibly influenced by recombination. Thus $\dfrac{P_1}{\exp\ (V_{1b'}/V_T)}$ and $\dfrac{P_2}{\exp\ (V_{2b'}/V_T)}$ are described to first approximation, in the region inside the radius R_1, by the expressions $P_n \dfrac{z}{W}$ and $P_n \left(1 - \dfrac{z}{W}\right)$, respectively, and, in the region between the radii R_1 and R_2, by the expressions $P_n \sin \dfrac{z}{L_1} \exp\left(-\dfrac{r - R_1}{L_1}\right)$ and $P_n \left[1 - \sin \dfrac{z}{L_1} \exp\left(-\dfrac{r - R_1}{L_1}\right)\right]$, respectively, where $L_1 = \dfrac{2W}{\pi}$. Far from the junctions, the shapes of these pair-density distributions are strongly influenced by recombination. Thus $\dfrac{P_2}{\exp\,(V_{2b'}/V_T)}$ and $P_{b'}$ are

described to first approximation, in the region beyond the radius R_2,

by the expressions $P_n \exp\left(-\dfrac{r - R_2}{L_2}\right)$ and $P_n\left[1 - \exp\left(-\dfrac{r - R_2}{L_2}\right)\right]$,

respectively, where L_2 is the effective diffusion length cited in approximation d.

The transfer currents I_{jk} are obtained from these P_j's via the integrations shown in Eqs. 7.8. For the approximations cited above, these currents are given to first approximation by

$$I_{12} = \pi R_1{}^2 e D\,\frac{P_n \exp\,(V_{1b'}/V_T)}{W}$$

$$I_{1b'} = \frac{\pi R_1{}^2 W}{2}\,e\,\frac{P_n \exp\,(V_{1b'}/V_T)}{\tau}$$

$$+\; 2\pi R_1 L_1 e S P_n \exp\,(V_{1b'}/V_T) \qquad (7.15)$$

$$I_{2b'} = \left(\pi R_2{}^2 - \frac{\pi R_1{}^2}{2} + 2\pi R_2 L_2\right) W e\,\frac{P_n \exp\,(V_{2b'}/V_T)}{\tau}$$

$$+\; (\pi R_2{}^2 - \pi R_1{}^2 + 4\pi R_2 L_2) e S P_n \exp\,(V_{2b'}/V_T)$$

The three transfer currents I_{21}, $I_{b'1}$, and $I_{b'2}$ are obtained from Eqs. 7.15 by use of Eqs. 7.9.

Special Diode and Triode Operating Conditions

The triode described above can be converted into a diode by connecting its heavily doped p-type electrodes together. Such a device will be called a double diode, and its connected electrodes will be denoted by the index d.

The pair-density distributions P_j and transfer currents I_{jk} of a triode and of its corresponding double diode are closely related. The distributions $P_{b'}$ for the two devices are identical. The distribution P_d associated with the double-diode electrode d is related to the distributions P_1 and P_2 associated with the triode electrodes 1 and 2 as follows:

$$\frac{P_d}{\exp\,(V_{db'}/V_T)} = \frac{P_1}{\exp\,(V_{1b'}/V_T)} + \frac{P_2}{\exp\,(V_{2b'}/V_T)} \qquad (7.16)$$

For the particular triode geometry described on page 352, the distribution $\dfrac{P_2}{\exp\,(V_{2b'}/V_T)}$ extends over a much larger region of the base

than the distribution $\dfrac{P_1}{\exp\,(V_{1b'}/V_T)}$.

The double-diode transfer currents $I_{db'}$ and $I_{b'd}$ are related to the triode transfer currents $I_{1b'}$, $I_{2b'}$, and $I_{b'1}$, $I_{b'2}$ as follows:

$$\frac{I_{db'}}{\exp{(V_{db'}/V_T)}} = \frac{I_{1b'}}{\exp{(V_{1b'}/V_T)}} + \frac{I_{2b'}}{\exp{(V_{2b'}/V_T)}},$$

$$I_{b'd} = I_{b'1} + I_{b'2}$$

(7.17)

The diode small-signal schematic, including noise generators, is shown in Fig. 7.3a.

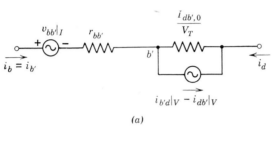

(a)

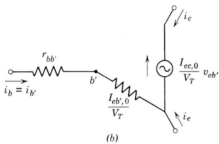

(b)

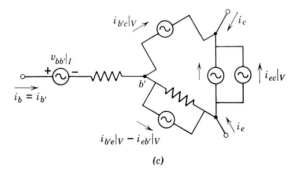

(c)

Fig. 7.3. Representations of the diode and triode.

Triodes are normally and advantageously operated with the collector biased to saturation; that is, with exp $(V_{cb'}/V_T)$ so small that I_{ce} and $I_{cb'}$, the transfer currents flowing *from* the collector *to* the emitter and base, respectively, are zero for all practical purposes. Thus the conductance $I_{cb',0}/V_T$, the transconductance $I_{ce,0}/V_T$, and the noise generators $i_{cb'}|_V$ and $i_{ce}|_V$ are all zero, and the small-signal schematics of Figs. 7.1b and c reduce to those shown in Figs. 7.3b and c, respectively.

The triode collector-to-base current-amplification factor α_{cb}, defined as the ratio i_c to i_b, is seen from Fig. 7.3b to be given by

$$\alpha_{cb} = \frac{i_c}{i_b} = \frac{I_{ec,0}}{I_{eb',0}} \qquad (7.18)$$

An alloy-junction triode operated with the small junction as emitter and the large junction as collector will be called a normal triode. Since for this operation $I_{ec,0}$ is to be identified with $I_{12,0}$ and $I_{eb',0}$ with $I_{1b',0}$, the current-amplification factor, according to Eqs. 7.15 and 7.18, is given approximately by

$$\alpha_{cb,\,norm} = \frac{I_{12,0}}{I_{1b',0}} = \frac{R_1}{W} \frac{D}{\left(\dfrac{R_1}{2\tau} + \dfrac{4S}{\pi}\right)W} \qquad (7.19)$$

An alloy-junction triode operated with the large junction as emitter and the small junction as collector will be called an inverted triode. Since for this operation $I_{ec,0}$ is to be identified with $I_{21,0}$, and $I_{eb',0}$

TABLE 7.1

Typical Values of Quantities Characterizing the Alloy-Junction Triode

Primary Quantities

R_1	0.030 cm	ρ_n	3 ohm-cm
R_2	0.060 cm	$r_{bb'}$	200 ohms at small base currents
W	0.004 cm	$I_{b',0}\dfrac{\partial r_{bb'}}{\partial I_{b'}}$	-40 ohms at large base currents
τ	40×10^{-6} sec		
$\overline{S_0}$	350 cm/sec	$\psi(f)$	$\dfrac{10^{-7}}{f}$ cm^4/sec

Secondary Quantities

P_n	10^{12} cm^{-3}	$I_{b'1,0}$	0.050×10^{-6} ampere
L_2	0.015 cm	$I_{b'2,0}$	1.36×10^{-6} ampere
$\alpha_{cb,\,norm}$	102	$I_{b'd,0}$	1.41×10^{-6} ampere
$\alpha_{cb,\,inv}$	3.7	$\dfrac{I_{12,0}}{\exp(V_{1b'}/V_T)}$	5.1×10^{-6} ampere

with $I_{2b',0}$, the current-amplification factor, according to Eqs. 7.15 and 7.18, is given approximately by

$$
\alpha_{cb,\,inv} = \frac{I_{21,0}}{I_{2b',0}}
$$

$$
= \frac{R_1}{W} \frac{R_1 D}{\left[\left(R_2{}^2 - \frac{R_1{}^2}{2} + 2R_2 L_2 \right) \frac{W}{\tau} + (R_2{}^2 - R_1{}^2 + 4R_2 L_2) S \right]} \tag{7.20}
$$

Typical values of quantities characterizing the germanium alloy-junction triode and its corresponding double diode are listed in Table 7.1. Those quantities designated as secondary were computed from the primary quantities via formulas listed above or, in the case of P_n, from the base resistivity ρ_n. The quantities $\psi(f)$ and $\partial R_{bb'}/\partial I_{b'}$ will be explained in Section 3.

2. Diffusion–Recombination Noise

Two Theorems from Tube Noise

Two theorems from the theory of fluctuations in temperature-limited tube currents will be useful here.

The first theorem concerns a current I consisting of many short current pulses of equal charge content Q distributed randomly in time.† It states: At frequencies low compared to the reciprocal of the time duration of one pulse, fluctuations in the current I are described by

$$
\overline{i^2} = 2Q^2 N_0 \, \Delta f \tag{7.21}
$$

where N_0 is the average number of pulses per unit time. If the pulses are unidirectional, QN_0 equals the direct current I_0, and Eq. 7.21 may be written

$$
\overline{i^2} = 2QI_0 \, \Delta f \tag{7.22}
$$

The second theorem concerns several currents I_{jk} flowing through a common terminal j but several circuit branches jk ($k = 1, 2, \cdots$). The total current $\sum_k I_{jk}$ is to consist of many short unidirectional current pulses of equal content Q distributed randomly in time; a given pulse is to pass through but one of the circuit branches jk; and the probability of passage through a given branch jk is to be the same for all pulses. This theorem states: At frequencies low compared to

† The charge content Q is equal to $\int I \, dt$ integrated over one pulse.

the reciprocal of the time duration of one pulse, fluctuations in the currents I_{jk} are described by

$$\overline{(i_{jk})^2} = 2QI_{jk,0}\,\Delta f \tag{7.23}$$

and are all uncorrelated.

In the sections immediately following, fluctuations in N-electrode-device transfer currents I_{jk} will be considered in the approximation that fluctuations in the creations, motions, and recombinations of pairs are completely random. It will be concluded, with the aid of the two theorems above, that the noise-current generators $i_{jk}|_V$ characterizing this diffusion–recombination noise are all uncorrelated and are described, at low frequencies, by

$$\overline{(i_{jk}|_V)^2} = 2eI_{jk,0}\,\Delta f \tag{7.24}$$

Since fluctuations are to be considered with the voltages $V_{jb'}$ held constant, it will be convenient to consider devices without base-lead resistances and with all junction electrodes a-c short-circuited to the base.

Diffusion–Recombination Noise in Diodes

Consider a pair injected into the diode base. Pair injection consists of the flow of a hole of charge $+e$, where e is the magnitude of the charge on the electron, across the junction, and the flow of an electron of charge $-e$ through the external short circuit, as shown schematically in Fig. 7.4a. The injected pair diffuses about the base. If it recombines in the base, the total process has amounted electrically to the passage of one short current pulse of content $+e$ through the circuit db'.† Since injections of pairs that recombine in the base occur randomly in time, the totality of all such processes results in a transfer current $I_{db'}$ with an a-c component $i_{db'}|_V$ described, at low frequencies, by

$$\overline{(i_{db'}|_V)^2} = 2eI_{db',0}\,\Delta f \tag{7.25}$$

If the injected pair instead diffuses back to the junction (this is usually much more probable), the pair is extracted. Extraction consists of the flow of a hole back across the junction and of an electron back through the external short circuit, as shown schematically in Fig. 7.4b. In this case, the total process has amounted electrically to the passage of one short current pulse of content $+e$ through the

---

† A current in the loop jk is considered positive if it carries, effectively, positive charge from the electrode j to the electrode k inside the device and from k to j through the external short circuit.

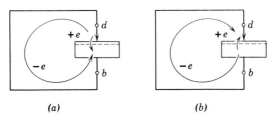

Fig. 7.4.   Pair injection and extraction at a diode junction.

circuit db' plus the passage of one short current pulse of content $-e$ through the same circuit a pair-diffusion time later.† Since a pair-diffusion time is short, the total process can be viewed as the passage of one short current pulse of content $+e + (-e) = 0$ through the circuit db'. The totality of all such processes results in a transfer current with no d-c component and with no a-c component at frequencies low compared to the reciprocal of an average pair-diffusion time.

Next, consider a pair generated thermally in the diode base. If the pair recombines in the base, no electric currents result. If the pair instead diffuses to the junction, it is extracted and recombines through the external short circuit, as shown schematically in Fig. 7.4b. In this case, the total process has amounted electrically to the passage of one short current pulse of content $+e$ through the circuit $b'd$. Since extractions of thermally generated pairs occur randomly in time, the totality of all such processes results in a transfer current $I_{b'd}$ with an a-c component $i_{b'd}|_V$ described, at low frequencies, by

$$\overline{(i_{b'd}|_V)^2} = 2eI_{b'd,0}\,\Delta f \tag{7.26}$$

Because the diode is a-c short-circuited externally, the currents associated with pair extractions do not alter the junction bias $V_{db'}$. Thus the a-c components of pair generation in the base and pair injection at the junction are uncorrelated, and the currents $i_{db'}|_V$ and $i_{b'd}|_V$ are uncorrelated.

Diffusion–Recombination Noise in N-Electrode Devices

Consider a pair injected at the junction j. Subsequent processes of pair recombination in the base and pair return to the injecting electrode are described as above for the diode.

† A diffusion time means here the time required for a pair to diffuse away from and back to the junction.

If the injected pair instead diffuses to some other junction k, it is extracted there and recombines through the external short circuit. In this case, the total process has amounted electrically to the passage of one short current pulse of content $+e$ through the circuit branch jb' plus the passage, a pair-diffusion time later,† of one short current pulse of content $+e$ through the branch $b'k$. Since a pair-diffusion time is short, the total process can be viewed, at low frequencies, as the passage of one short current pulse of content $+e$ through the circuit branch jk. Since extractions at junction k of pairs injected at junction j occur randomly in time, the totality of all such processes results in a transfer current I_{jk} with an a-c component $i_{jk}\big|_V$ described by Eq. 7.24.

Since pairs injected at junction j are injected randomly in time, and since their recombinations in the base and extractions at the various junctions follow fixed probabilities, the currents $i_{jk}\big|_V$ and $i_{jl}\big|_V$, for $k \neq l$, are uncorrelated.

The argument above is readily applied to transfer currents $I_{b'k}$ associated with pairs originating at the special electrode $j = b'$. "Pairs injected at the junction j" is merely replaced by "pairs generated thermally in the base region."

Because all junction electrodes are a-c short-circuited to the base, the currents associated with pair extractions at the junctions do not alter the junction biases $V_{kb'}$. Thus the a-c components of pair generation in the base and pair injection at the several junctions are all uncorrelated, and the currents $i_{jk}\big|_V$ are all uncorrelated.

In summary, diffusion–recombination noise in arbitrarily terminated N-electrode devices is described, at low frequencies, by Eqs. 7.13, where the mutually uncorrelated currents $i_{jk}\big|_V$ are described by Eq. 7.24. In the language of circuit theory, in addition to the conductance and transconductance elements cited in connection with the signal theory, every electrode j is connected to every other electrode k by the noise-current generator $i_{jk}\big|_V - i_{kj}\big|_V$, where this generator is described by

$$\overline{(i_{jk}\big|_V - i_{kj}\big|_V)^2} = 2e(I_{jk,0} + I_{kj,0})\,\Delta f \qquad (7.27)$$

and where these generators are all uncorrelated.

This particular representation of diffusion–recombination noise was first proposed for the diode by Anderson and van der Ziel[2] and for the triode by Giacoletto.[3] The representation of the N-electrode device was first proved and stated by North.[4] North's treatment, based on his mesh model, yielded a representation valid even at very high

† A diffusion time means here the time required for a pair to diffuse from the junction j to the junction k.

frequencies. At frequencies low compared to the reciprocal of an average pair-diffusion time, his representation reduces to that described above.

In the event that the N-electrode device is at thermal equilibrium, $I_{jk,0} = I_{kj,0}$. Thus Eqs. 7.14 describing the signal properties of the device and Eqs. 7.27 describing the noise properties reduce to

$$i_j = \sum_k \frac{I_{jk,0}}{V_T} (v_{jb'} - v_{kb'}) = \sum_k \frac{I_{jk,0}}{V_T} v_{jk} \tag{7.28}$$

and

$$\overline{(i_{jk}\big|_V - i_{kj}\big|_V)^2} = 4eI_{jk,0}\,\Delta f = 4kT\frac{I_{jk,0}}{V_T}\Delta f \tag{7.29}$$

respectively, where v_{jk} is the small-signal voltage of electrode j relative to electrode k. In the language of circuit theory, Eqs. 7.28 state that every electrode j is connected to every other electrode k by the conductance $I_{jk,0}/V_T$. The conductance $I_{jk,0}/V_T$ and the noise-current generator $(i_{jk}\big|_V - i_{kj}\big|_V)$ in parallel with it are related, according to Eqs. 7.29, by Nyquist's formula. That is, at thermal equilibrium the diffusion–recombination noise *is* the thermal noise of the intrinsic device. Since the signal and noise properties were derived from the same physical model, this result must necessarily obtain.

White Noise in the Base-Lead Element

Noise in the base-lead semiconductor is represented in circuit schematics by the noise-voltage generator $v_{bb'}\big|_I$. See Figs. 7.1c, 7.3a, and 7.3c.

Thermal noise in this element is described by Nyquist's formula:

$$\overline{(v_{bb'}\big|_I)^2} = 4kTr_{bb'}\,\Delta f \tag{7.30}$$

Fluctuations in the base-current-carrier population result in an additional noise voltage $\Delta v_{bb'}\big|_I$ that increases rapidly with direct base current $I_{b',0}$. The white and $1/f$ components of this additional noise are discussed in the preceding chapter by Professor van der Ziel and in Section 3, page 369, of the present chapter, respectively. The white component was small for the measurements described in this chapter and will be neglected here.

White Noise in Diodes

White noise in diodes is conveniently summarized with the aid of Fig. 7.3a and Eqs. 7.27 and 7.30. Its contribution to the diode equivalent noise resistance is

$$R_{eq}{}^{WH} = r_{bb'} + \frac{1}{4kT\,\Delta f}\left(\frac{V_T}{I_{db',0}}\right)^2 \frac{1}{\overline{(i_{b'd}|_V - i_{db'}|_V)^2}}$$

$$= r_{bb'} + \frac{1}{2}\left(1 + \frac{I_{b'd,0}}{I_{db',0}}\right)\frac{V_T}{I_{db',0}} \tag{7.31}$$

$V_T/I_{db',0}$ is the small-signal resistance of the junction. The quantity $\frac{1}{2}[1 + (I_{b'd,0}/I_{db',0})]$ equals unity for zero diode bias and approaches one half and infinity for heavy forward and reverse biases, respectively.

Noise in the reverse-biased diode is more conveniently described by the equivalent noise current of a fictitious noise-current generator across its terminals. For saturation biasing, that is, for exp $(V_{db'}/V_T)$ so small that $I_{db'}$ is zero for all practical purposes, $I_{eq}{}^{WH}$ is equal to the saturation current $I_{b'd,0}$. This relationship follows directly from Fig. 7.3a and Eq. 7.27 and has been well substantiated experimentally.

White Noise in Triodes

White noise in triodes, with the collector biased to saturation, is conveniently summarized with the aid of Fig. 7.3b and c and Eqs. 7.27 and 7.30. For either the common-emitter or the common-base connection, the white-noise contribution to the triode noise factor is

$$F^{WH} = 1 + \frac{r_{bb'}}{R_G} + \frac{(R_G + r_{bb'})^2}{4kTR_G\Delta f}\left[\overline{(i_{b'e}|_V - i_{eb'}|_V)^2} + a^2\overline{(i_{ec}|_V)^2}\right.$$
$$\left. + (1 + a)^2\overline{(i_{b'c}|_V)^2}\right]$$

$$= 1 + \frac{r_{bb'}}{R_G} + \frac{R_G}{2V_T}\left(1 + \frac{r_{bb'}}{R_G}\right)^2\left[(I_{b'e,0} + I_{eb',0}) + a^2I_{ec,0}\right.$$
$$\left. + (1 + a)^2I_{b'c,0}\right] \tag{7.32}$$

where R_G is the source resistance terminating the triode input and where

$$a = \left[\frac{1}{\alpha_{cb}} + \frac{V_T}{(R_G + r_{bb'})I_{ec,0}}\right] \tag{7.33}$$

For a high-α_{cb} triode, the $I_{b'e,0}$ term is negligible, and

$$I_{ec,0} \approx I_{e,0} \quad \text{and} \quad I_{eb',0} \approx \frac{I_{e,0}}{\alpha_{cb}} \tag{7.34}$$

Thus, in this case, Eq. 7.32 may be approximated by

$$F^{WH} = 1 + \frac{r_{bb'}}{R_G} + \frac{R_G}{2V_T}\left(1 + \frac{r_{bb'}}{R_G}\right)^2\left[\frac{I_{e,0}}{\alpha_{cb}} + a^2I_{e,0} + (1 + a)^2I_{b'c,0}\right] \tag{7.35}$$

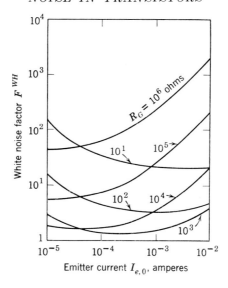

Fig. 7.5. White noise at the output of a normal triode.

Terms proportional to the quantity a would clearly be negligible if a were of the order of $1/\alpha_{cb}$. Thus, in Eq. 7.35, a may be approximated by

$$a \approx \frac{V_T}{(R_G + r_{bb'})I_{e,0}} \tag{7.36}$$

The curves in Fig. 7.5 show F^{WH} versus emitter current $I_{e,0}$ and source resistance R_G, as computed from Eqs. 7.35 and 7.36 for a normal triode with the typical characteristics listed in Table 7.1. F^{WH} has a minimum value ≈ 1.4 for $R_G \approx 10^3$ ohms and $I_{e,0} \approx 3 \times 10^{-4}$ ampere, though this minimum is broad.

In the representation used here, the white noise is ascribed to four uncorrelated noise generators. See Fig. 7.3c. All four of these generators contribute to F^{WH}. However, for R_G large, F^{WH} is dominated at low currents by noise from the $i_{b'c}|_V$ generator, and at high currents by noise from the $i_{eb'}|_V$ generator; for R_G small, F^{WH} is dominated at low currents by noise from the $i_{ec}|_V$ generator, and at high currents by noise from the $v_{bb'}|_I$ generator. In some very interesting and thorough experiments, Guggenbühl and Strutt[5]† have operated junction transis-

† The notation of these authors differs from that used here. However, their resolution of noises from different generators is, in the notation used here, the resolution described above.

tors under these four extreme conditions and have in each instance observed noise factors in agreement with Eq. 7.35. Thus all four white-noise generators have been independently verified experimentally.

Adjustment of Emitter Current and Source Resistance for Minimum Noise Factor in a High-α_{cb} Triode

In view of Eq. 7.36, the first two terms in the square bracket of Eq. 7.35 are proportional to $I_{e,0}$ and $1/I_{e,0}$, respectively. The sum of these terms is therefore minimized with respect to emitter current when these two terms are equal, that is, for $a = 1/\alpha_{cb}^{1/2}$. Since this optimum value of a is small compared to unity, the quantity $(1 + a)$ in the third term of the square bracket may be taken as unity. Substitution of $a = 1/\alpha_{cb}^{1/2}$ in Eqs. 7.36 and 7.35 yields

$$I_{e,0,\text{opt}} = \alpha_{cb}^{1/2} \frac{V_T}{R_G + r_{bb'}} \tag{7.37}$$

and

$$F^{WH} = \left(1 + \frac{1}{\alpha_{cb}^{1/2}}\right)\left(1 + \frac{r_{bb'}}{R_G}\right) + \frac{R_G I_{b'c,0}}{2V_T}\left(1 + \frac{r_{bb'}}{R_G}\right)^2 \tag{7.38}$$

respectively. These equations are valid for any value of source resistance R_G.

For minimization with respect to R_G, it is convenient to anticipate that the quantities $1/\alpha_{cb}^{1/2}$, $r_{bb'}/R_{G,\text{opt}}$, and $I_{b'c,0}R_{G,\text{opt}}/2V_T$ are all small compared to unity. If products of these small quantities are neglected, Eq. 7.38 reduces to

$$F^{WH} = 1 + \frac{1}{\alpha_{cb}^{1/2}} + \frac{r_{bb'}}{R_G} + \frac{R_G I_{b'c,0}}{2V_T} \tag{7.39}$$

The sum of the last two terms is minimized with respect to source resistance R_G when these two terms are equal; that is, for

$$R_{G,\text{opt}} = \left(\frac{2V_T r_{bb'}}{I_{b'c,0}}\right)^{1/2} \gg r_{bb'} \tag{7.40}$$

Substitution of Eq. 7.40 into Eqs. 7.37 and 7.39 yields

$$I_{e,0,\text{opt}} = \left(\frac{\alpha_{cb} V_T I_{b'c,0}}{2r_{bb'}}\right)^{1/2} \tag{7.41}$$

and

$$(F^{WH})_{\text{min}} = 1 + \frac{1}{\alpha_{cb}^{1/2}} + \left(\frac{2r_{bb'} I_{b'c,0}}{V_T}\right)^{1/2} \tag{7.42}$$

respectively. For a normal triode with the typical characteristics listed in Table 7.1, the values corresponding to Eqs. 7.40, 7.41, and 7.42 are $R_{G,\text{opt}} = 2.7 \times 10^3$ ohms, $I_{e,0,\text{opt}} = 0.93 \times 10^{-4}$ ampere, and $(F^{WH})_{\min} = 1.25 = 1.0$ db.

3. $1/f$ Surface Noise

Fluctuations in Surface Recombination Velocity

In this section the excess noise due to correlated generations and recombinations of electron-hole pairs will be considered. The details of the correlations depend on the detailed characteristics of the charge-carrier traffic across the semiconductor forbidden band. Since these characteristics are poorly known at present, no microscopic theory of the noise will be undertaken.

In Sections 1 and 2, the probabilities of pair generations and recombinations were characterized by the phenomenological constants τ and S. In an analogous phenomenological description of excess noise, correlations in pair generations and recombinations may be viewed as equivalent to slow fluctuations in the probabilities τ and S.

Fluctuations in both τ and S might occur. North[4] has proposed a detailed noise model in which thermal fluctuations in surface potentials produce, effectively, local fluctuations in surface recombination velocity S that are approximately independent of the pair density P. His model will be followed here in the phenomenological sense that S will be *assumed* to undergo local fluctuations that are approximately independent of P. The assumed fluctuations in S will be related to noise properties which are directly measurable. The magnitude and spectrum of the fluctuations in S will be arbitrarily chosen to account for the magnitude and spectrum of observed transistor $1/f$ noise. However, it will be shown that significant features of the observed noise, other than its magnitude and spectrum, are indeed explained[6] by the model, that is, by the concept of correlated generations and recombinations of pairs. A phenomenological treatment of fluctuations in volume lifetime τ would be very similar.

A recombination surface is envisioned as consisting of many small area elements a_i characterized by individual surface recombination velocities $S_i = S_{i,0} + s_i$, which fluctuate independently. The net recombination rate at a_i is $S_i(P - P_n)a_i$, where P is the local pair density. Since recombination may be regarded as the negative of generation, this may be written

$$-G_i = S_i(P - P_n)a_i \tag{7.43}$$

where $G_i = G_{i,0} + g_i$ is the net pair generation rate. The fluctuation s_i drives a fluctuation g_i described by

$$-g_i = s_i(P_0 - P_n)a_i \qquad (7.44)$$

The average surface recombination velocity $\bar{S}_0$, usually obtained empirically, is defined by summing $G_{i,0}$ over a representative small area da containing many elements a_i:

$$-\sum_i G_{i,0} = (P_0 - P_n)\sum_i S_{i,0}a_i \equiv (P_0 - P_n)\bar{S}_0\, da \qquad (7.45)$$

Similarly, the spectral density $\psi(f)$ characterizing the excess noisiness of a surface, usually obtained empirically, is defined by summing the independent fluctuations $g_i{}^2$ over the area da, transforming to the frequency domain, and averaging:

$$\sum_i \overline{g_i{}^2} = (P_0 - P_n)^2 \sum_i \overline{s_i{}^2 a_i{}^2} \equiv (P_0 - P_n)^2\, \psi(f)\, \Delta f\, da \qquad (7.46)$$

In words, fluctuations in S produce, at recombination surfaces, *apparent* noisy generations of pairs of mean-square rate $(P_0 - P_n)^2\, \psi(f)$ per unit frequency and per unit area.

For etched germanium surfaces, $\psi(f)$ in the audio region proves empirically to be $\approx 10^{-7 \pm 1}/f$ cm^4 per sec, where the frequency f is expressed in cycles per second. ψ is not quite independent of P_0; empirically it decreases somewhat with increasing P_0.

The Theory of the Point Source

Since the area da is small, and since the spectral density $\psi(f)$ is appreciable only at low frequencies, the generation described by Eq. 7.46 can be regarded as originating from a point source that is quasistationary in time. In this section, the circuit-schematic representation of a point source of pairs of quasistationary strength $G = G_0 + g$ introduced at some position $\mathbf{r}$ in the base of an N-electrode device will be derived. The representation of $1/f$ surface noise will be obtained subsequently by summing over many point sources. These sources are distributed, in accordance with Eq. 7.46, over the base recombination surface. While for surface noise the d-c component of the generation G_0 is zero and the points of generation $\mathbf{r}$ all lie on the base recombination surface, these quantities will not be so restricted here. In Section 5 the point-source theory will be applied to irradiation noise; for this case the d-c generation is not zero, and the points of generation lie throughout the base volume.

In the statement of the boundary-value problem, pair generations were described by the terms P_n/τ and SP_n of Eqs. 7.1 and 7.3, respectively. With the addition of a point source, a generation $\delta(\mathbf{r})G$ must be added to these, where $\delta(\mathbf{r})$ is Dirac's delta function. The generations P_n/τ and SP_n were the sources producing the pair-density distribution $P_{b'}$. The additional generation $\delta(\mathbf{r})G$ produces an additional analogous† pair-density distribution $\Delta P_{b'}$. $\Delta P_{b'}$ satisfies the same conditions as $P_{b'}$, except that P_n/τ and SP_n are replaced by $\delta(\mathbf{r})G$. Thus $\Delta P_{b'}$ satisfies the equation of continuity,

$$D\nabla^2 \Delta P_{b'} - \frac{\Delta P_{b'}}{\tau} + \delta(\mathbf{r})G = 0 \tag{7.47}$$

and the boundary conditions,

$$\Delta P_{b'} = 0 \qquad\qquad \text{at all junctions}$$

$$\Delta R_{b'} = -D\frac{\partial \Delta P_{b'}}{\partial n} = S\,\Delta P_{b'} - \delta(\mathbf{r})G \quad \text{at the base surface} \tag{7.48}$$

where $\Delta R_{b'}$ is the additional net pair surface recombination rate per unit area. It is to be remembered that these particular equations, together with the accompanying equations for the other P_j's, are merely one convenient way of stating a set of conditions linear in P; they are not conditioned by a particular termination for the N-electrode device.

The processes of pair generation at the point source, pair recombination in the base, pair extraction at the junctions, and pair diffusion as represented by the pair-density distribution $\Delta P_{b'}$ describe $(N - 1)$ additional transfer currents $\Delta I_{b'j}$, $(j \neq b')$ flowing *from* the b' electrode *to* the $(N - 1)$ junction electrodes. $\Delta I_{b'j}$ is given by

$$\Delta I_{b'j} = -e \int_J D\frac{\partial \Delta P_{b'}}{\partial n}\,da = e\,w_j(\mathbf{r})G \qquad (j \neq b') \tag{7.49}$$

where $w_j(\mathbf{r})$ is the probability that a pair generated at $\mathbf{r}$ will reach the junction j. $\Delta P_{b'}$ and $\Delta I_{b'j}$ are proportional to G but are independent of the junction voltages.

In the device current–voltage characteristics given by Eqs. 7.10 and 7.13, the transfer currents $I_{b'j}$ and $i_{b'j}|_V$ must be everywhere increased by increments $\Delta I_{b'j}$ and $\Delta i_{b'j}|_V$, respectively, where $\Delta I_{b'j}$ is given by Eq. 7.49, and $\Delta i_{b'j}|_V$ by the corresponding small-signal equation

$$\Delta i_{b'j}\big|_V = e\,w_j(\mathbf{r})g \qquad (j \neq b') \tag{7.50}$$

† Indeed, $P_{b'}$ could itself be regarded as produced by a distribution of point sources of pairs of strength P_n/τ per unit volume and SP_n per unit area.

In the language of circuit theory, Eq. 7.50 states that, with the addition of a point source of pairs in the device base, the b' electrode is connected to every junction electrode j by the additional current generator $\Delta i_{b'j}|_V$. While the point source augments the expressions for the electrode currents I_j and introduces new generators in the small-signal schematic, it does not change the expressions for the old elements in the schematic.

The small-signal representation of the point source is shown for the particular cases of the diode and of the triode with collector biased to saturation, in Figs. 7.6a and b, respectively. Since it is convenient to treat white noise and $1/f$ noise separately, the generators $i_{jk}|_V$ and $v_{bb'}|_I$ representing diffusion–recombination noise and thermal noise, respectively, have been omitted. The voltage generator $\Delta v_{bb'}|_I$ shown in Fig. 7.6 represents a base-conductivity-modulation effect associated with the point source itself. The origin of this generator is discussed in the following section.

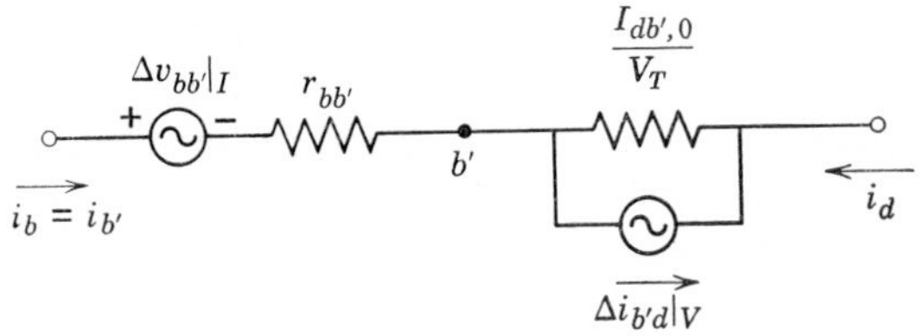

(a)

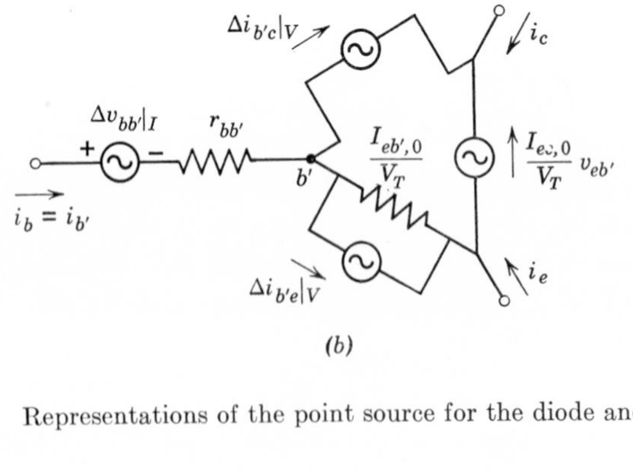

(b)

Fig. 7.6.   Representations of the point source for the diode and triode.

The Base-Conductivity Modulation Effect

Let the voltage $V_{bb'}$ of the external base electrode b relative to the b' region be related to the base current $I_{b'}$ by the factor $R_{bb'}$:

$$V_{bb'} = R_{bb'}I_{b'} \tag{7.51}$$

For a given device, the pair-density distributions P_j have fixed geometrical shapes and vary in amplitude only through the factors $\exp(V_{jb'}/V_T)$, except for $\Delta P_{b'}$, which is proportional to G. Since these distributions extend somewhat into the base-lead region, they modify the electric conductivity of this region. $R_{bb'}$ must therefore be regarded as a function of G and of the $(N-1)$ junction voltages $V_{jb'}$, and the small-signal equation corresponding to Eq. (7.51) is

$$v_{bb'} = R_{bb',0}i_{b'} + I_{b',0}\left(\frac{\partial R_{bb'}}{\partial G}\bigg|_V \, g + \sum_j \frac{\partial R_{bb'}}{\partial V_{jb'}}\bigg|_G v_{jb'}\right) \tag{7.52}$$

Thus the base-lead element must be represented, in general, by a resistance $R_{bb',0}$ and a complicated voltage generator proportional to $I_{b',0}$.

For the measurements described in this chapter, every device junction, except perhaps one emitter denoted by the index e, will be reverse-biased to saturation. That is, for $j \neq e$, $\exp(V_{jb'}/V_T)$ will be so small that $R_{bb'}$ will be independent of $V_{jb'}$, and the device transfer currents I_{jk} will be zero. Subject to this limitation, it will now be demonstrated that the base-lead element can be represented by a resistance and a voltage generator proportional to $I_{b',0}\,g$.

In this single-emitter case, $\dfrac{\partial R_{bb'}}{\partial V_{jb'}}\bigg|_G = 0$ for $j \neq e$; also, the small-signal base current $i_{b'}$, according to Eq. 7.13 with the irrelevant diffusion–recombination-noise generators (white noise) $i_{jk}|_V$ set equal to zero, is given by

$$i_{b'} = -\frac{I_{eb',0}}{V_T}v_{eb'} + \sum_j \Delta i_{b'j}\bigg|_V \qquad (j \neq b') \tag{7.53}$$

Thus, using Eqs. 7.50 and 7.53, Eq. 7.52 may be written

$$v_{bb'} = R_{bb',0}i_{b'} + I_{b',0}\left[\frac{\partial R_{bb'}}{\partial G}\bigg|_V \, g - \frac{V_T}{I_{eb',0}}\frac{\partial R_{bb'}}{\partial V_{eb'}}\bigg|_G \left(i_{b'} - eg\sum_j w_j\right)\right] \tag{7.54}$$

If $R_{bb'}$ is regarded as a function of G and $I_{b'}$ (rather than of G and $V_{eb'}$), the quantity $-\dfrac{V_T}{I_{eb',0}}\dfrac{\partial R_{bb'}}{\partial V_{eb'}}\bigg|_G$ is readily identified as $\dfrac{\partial R_{bb'}}{\partial I_{b'}}\bigg|_G$. Thus, as asserted, Eq. 7.54 may be written

$$v_{bb'} = r_{bb'}i_{b'} + \Delta v_{bb'}\big|_I \tag{7.55}$$

where

$$r_{bb'} = R_{bb',0} + I_{b',0}\frac{\partial R_{bb'}}{\partial I_{b'}}\bigg|_G \tag{7.56}$$

and

$$\Delta v_{bb'}\big|_I = I_{b',0}\left(\frac{\partial R_{bb'}}{\partial G}\bigg|_V - e\frac{\partial R_{bb'}}{\partial I_{b'}}\bigg|_G \sum_j w_j\right)g \tag{7.57}$$

Equation 7.55 is represented by the base-lead elements shown in Fig. 7.6.

The base-lead resistance in the small-signal schematic is commonly measured empirically as the ratio $v_{bb'}/i_{b'}$ when no point source is present, that is, for $G = 0$† and therefore for $\Delta v_{bb'}\big|_I = 0$. Thus, provided the d-c component G_0 of the point source is zero, $r_{bb'}$ as defined by Eq. 7.55 *is* the small-signal base-lead resistance that is measured empirically. At any operating point where $\partial R_{bb'}/\partial I_{b'}$ is appreciable, $r_{bb'}$, according to Eq. 7.56, will differ from $R_{bb',0}$, the d-c value of the total-voltage–total-current ratio $R_{bb'}$.

1/f Surface Noise in N-Electrode Devices

The uncorrelated point sources of pairs characterizing surface noise are described by Eq. 7.46. There is no d-c component of generation G_0, or, rather, the corresponding d-c pair generation at recombination surfaces has already been accounted for by the saturation currents $I_{b'j,0}$. In the small-signal schematic, each point source is separately represented by $(N - 1)$ perfectly correlated current generators $\Delta i_{b'j}\big|_V$ and one perfectly correlated voltage generator $\Delta v_{bb'}\big|_I$, as shown for the diode and the triode, with collector biased to saturation, in Fig. 7.6. The total surface noise is represented by N partially correlated generators $\sum_{da} \Delta i_{b'j}\big|_V$ and $\sum_{da} \Delta v_{bb'}\big|_I$ obtained by summing the individual generators representing the individual point sources.

---

† See, for example, L. J. Giacoletto.[7] Measurements of $r_{bb'}$ are carried out at signal levels where noise is negligible. Thus source strengths G associated with 1/f surface noise are effectively zero.

The squares and cross products of the summed current generators are described by

$$\overline{\left(\sum_{da} \Delta i_{b'j}\Big|_V\right)\left(\sum_{da'} \Delta i_{b'k}\Big|_V\right)} = e^2 \sum_{da}\sum_{da'} w_j(\mathbf{r})\, w_k(\mathbf{r}')\, \overline{g(\mathbf{r})\, g(\mathbf{r}')}$$

$$= e^2 \sum_{da} w_j(\mathbf{r})\, w_k(\mathbf{r})\, \overline{g^2(\mathbf{r})} \tag{7.58}$$

$$= e^2\, \psi(f)\, \Delta f \int_{\mathrm{BS}} w_j(\mathbf{r})\, w_k(\mathbf{r})(P_0 - P_n)^2\, da$$

If the $\dfrac{\partial R_{bb'}}{\partial G}\bigg|_V$ term of $\Delta v_{bb'}\big|_I$ is neglected, the summed voltage generator is given approximately by

$$\sum_{da} \Delta v_{bb'}\bigg|_I \approx -\, I_{b',0}\, \frac{\partial R_{bb'}}{\partial I_{b'}}\bigg|_G \sum_j \left(\sum_{da}\Delta i_{b'j}\bigg|_V\right) \tag{7.59}$$

This expression is especially simple because the quantity $I_{b',0}\dfrac{\partial R_{bb'}}{\partial I_{b'}}$ can be obtained experimentally.†   For high-α_{cb} triodes, the neglected $\dfrac{\partial R_{bb'}}{\partial G}\bigg|_V$ term, which measures the effect of $\Delta P_{b'}$ on $R_{bb'}$ for a unit change in G, is small. This follows because the pair-density distribution $\Delta P_{b'}$ is small owing to the fact that noise pairs are generated close to the junctions where $(P_0 - P_n)$ is large. However, even for diodes and low-α_{cb} triodes where noise pairs are generated further from the junctions, Eq. 7.59, with the aid of experimentally determined values of $I_{b',0}\dfrac{\partial R_{bb'}}{\partial I_{b'}}$, accounts to first approximation for observed values of the summed voltage generator.

The square of the voltage generator $\sum_{da}\Delta v_{bb'}\big|_I$, as approximated by Eq. 7.59, and its cross products with the current generators $\sum_{da}\Delta i_{b'j}\big|_V$

---

† For $G = 0$, Eq. 7.56 can be written

$$R_{bb'} = \frac{1}{I_{b'}}\int_0^{I_{b'}} r_{bb'}\, dI_{b'}$$

Thus $R_{bb'}(I_{b'})$ can be obtained by numerical integration of measured values of the small-signal base-lead resistance $r_{bb'}(I_{b'})$, and $I_{b'0}\dfrac{\partial R_{bb'}}{\partial I_{b'}}$ can then be computed from Eq. 7.56.

are described by sums of integrals of the type occurring in Eq. 7.58. The applicability of these formulas to the particular cases of the diode and triode will be discussed in detail below.

The summed current and voltage generators are only partially correlated because the probabilities $w_j(\mathbf{r})$ depend differently on $\mathbf{r}$. Their spectra and cross-spectra are all proportional to the quantity $\psi(f)$ characterizing the excess noisiness of the surface and, because of the factors $(P_0 - P_n)^2$, increase rapidly with d-c injection.

1/f Surface Noise in Diodes

Surface noise in diodes is conveniently summarized with the aid of Fig. 7.6a and Eqs. 7.59 and 7.58. Its contribution to the diode equivalent noise resistance is

$$
\begin{aligned}
R^s{}_{\mathrm{eq}} &= \frac{1}{4kT\,\Delta f} \overline{\left(\frac{V_T}{I_{db',0}} \sum_{da} \Delta i_{b'd} \bigg|_V - \sum_{da} \Delta v_{bb'} \bigg|_I \right)^2} \\
&= \frac{1}{4kT\,\Delta f} \left(\frac{V_T}{I_{db',0}} + I_{b',0}\,\frac{\partial R_{bb'}}{\partial I_{b'}} \right)^2 \overline{\left(\sum_{da} \Delta i_{b'd} \bigg|_V \right)^2} \qquad (7.60)\\
&= \frac{e\,\psi(f)}{4V_T} \left(\frac{V_T}{I_{db',0}} + I_{b',0}\,\frac{\partial R_{bb'}}{\partial I_{b'}} \right)^2 \int_{\mathrm{BS}} w_d{}^2(\mathbf{r})(P_0 - P_n)^2\,da
\end{aligned}
$$

Now, since $(P_0 - P_n)$ equals $P_n[\exp(V_{db',0}/V_T) - 1]$ times a factor dependent upon τ, $\bar{S}_0$ and the diode geometry, and since the diode direct current $I_{d,0}$ is given by the sum of the net recombination currents in the volume and at the surface,

$$
I_{d,0} = e \int_{\mathrm{BV}} \frac{(P_0 - P_n)}{\tau}\,dv + e \int_{\mathrm{BS}} \bar{S}_0(P_0 - P_n)\,da \qquad (7.61)
$$

the integral in Eq. 7.60 evidently equals $I_{d,0}{}^2$ times a factor dependent upon τ, $\bar{S}_0$ and the diode geometry. For a triode with the particular geometry described in Section 1, page 352, operated as a double diode,

$$
\begin{aligned}
w_d(\mathbf{r}) &\approx 1 && \text{for} \quad r < R_2 \\
&\approx \exp[-(r - R_2)/L_2] && \text{for} \quad r > R_2 \qquad (7.62)
\end{aligned}
$$

The unity value of the probability $w_d(\mathbf{r})$ inside the radius R_2 follows from the original assumption that pairs can diffuse linear distances of the order of the base thickness W with but little chance of recombination. The length L_2 characterizing the decrease of $w_d(\mathbf{r})$ beyond R_2

is the same effective diffusion length L_2 that characterizes the shapes of the pair-density distributions P_d and $P_{b'}$ beyond this radius. See Section 1, page 352. Computation of the integrals in Eqs. 7.60 and 7.61 for this particular geometry yields

$$\int_{BS} w_d{}^2(\mathbf{r})(P_0 - P_n)^2 \, da$$

$$\approx \frac{I_{d,0}{}^2(R_2{}^2 - R_1{}^2 + R_2 L_2)}{\pi e^2[(R_2{}^2 + 2R_2 L_2)(W/\tau) + (R_2{}^2 - R_1{}^2 + 4R_2 L_2)\bar{S}_0]^2} \quad (7.63)$$

The solid curve in Fig. 7.7 shows $R^S{}_{eq}$ at 1 *cycle per second* versus $I_{d,0}$, as computed from Eqs. 7.60 and 7.63 for a forward-biased double diode with the typical characteristics listed in Table 7.1.† The principal features of $R^S{}_{eq}$ are a rapid rise at low currents, a broad maximum, a sharp minimum at $I_{d,0} \approx 10^{-3}$ ampere where contributions from the $\sum\limits_{da} \Delta i_{b'd} \big|_V$ and $\sum\limits_{da} \Delta v_{bb'} \big|_I$ generators interfere destructively, and a final rapid rise at high currents. The contribution from the $\sum\limits_{da} \Delta i_{b'd} \big|_V$ generator dominates $R^S{}_{eq}$ below the interference minimum; that from $\sum\limits_{da} \Delta v_{bb'} \big|_I$ dominates $R^S{}_{eq}$ above the interference minimum.

If the $\dfrac{\partial R_{bb'}}{\partial G} \bigg|_V$ term of $\sum\limits_{da} \Delta v_{bb'} \big|_I$ had been retained in the computation, the $\sum\limits_{da} \Delta i_{b'd} \big|_V$ and $\sum\limits_{da} \Delta v_{bb'} \big|_I$ generators would not be perfectly correlated, and $R^S{}_{eq}$ would be modified at high currents approximately as shown by the dotted curve in Fig. 7.7. The derivation of this result is tedious and will be omitted.

The dot–dash curve in Fig. 7.7 shows $R^{WH}{}_{eq}$ vs. $I_{d,0}$, as computed from Eq. 7.31 for a forward-biased double diode with the typical characteristics listed in Table 7.1. Except for $I_{d,0} \lesssim 10^{-6}$ ampere, $R^S{}_{eq}$ at 1 cycle per second is much larger than $R^{WH}{}_{eq}$. Since $R^S{}_{eq}$, proportional to ψ, varies approximately as $1/f$, this dominance passes away at higher frequencies.

One example of double-diode noise versus forward current is shown

---

† For devices with the particular geometry described in Section 1, page 352, the quantity $I_{b',0} \dfrac{\partial R_{bb'}}{\partial I_{b'}}$ proves to be approximately independent of the base current $I_{d,0}$ at large base currents. See Fig. 6 and Appendix II of reference 6. Thus a constant value -40 ohms is listed for this quantity in Table 7.1.

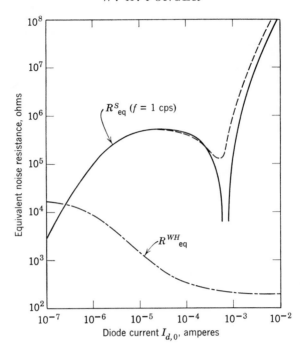

Fig. 7.7.   Noise in forward-biased diodes.

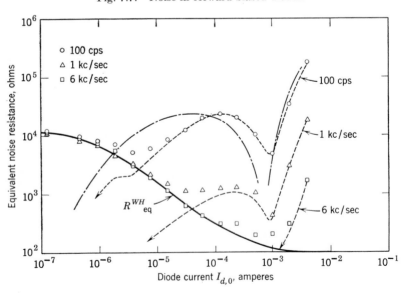

Fig. 7.8.   Noise observed across a forward-biased diode.

in Fig. 7.8. Equivalent noise resistances observed at 100, 1000, and 6000 cycles per sec are shown by circles, triangles, and squares, respectively. These measured values can be resolved into components characteristic of R^{WH}_{eq} and R^{S}_{eq}, respectively. The solid curve shows R^{WH}_{eq} computed, according to Eq. 7.31, from observed values of $I_{b'd,0}$ and $r_{bb'}$ for this unit. The dotted curves show, at each frequency, the excess noise $R^{W}_{eq} - R^{WH}_{eq}$. The excess noise varies approximately as $1/f$ and exhibits the dependence on current characteristic of surface noise. The dot–dash curve, adjusted vertically arbitrarily, shows the expected current dependence of surface noise computed, according to Eqs. 7.60 and 7.63, from observed values of $I_{b'd,0}$ and $I_{b',0}$ ($\partial R_{bb'}/\partial I_{b'}$) for this unit. The observed and expected current dependences agree very well.

Noise in Reverse-Biased Diodes

Noise in the reverse-biased diode is more conveniently described by the equivalent noise current of a fictitious noise-current generator across its terminals. For saturation biasing, that is, for exp $(V_{db'}/V_T)$ so small that the transfer current $I_{db'}$ is zero for all practical purposes, I^{S}_{eq} is given by

$$I^{S}_{eq} = \frac{1}{2e\,\Delta f}\overline{\left(\sum_{da}\Delta i_{b'd}\,\bigg|_V\right)^2} = \frac{e\,\psi(f)}{2}\int_{BS} w_d^2(\mathbf{r})(P_0 - P_n)^2\,da \qquad (7.64)$$

This relationship follows directly from Fig. 7.6a and Eq. 7.58. For a triode with the particular geometry described in Section 1, page 352, operated as a double diode, the integral in Eq. 7.64 is given by Eq. 7.63, where, for saturation biasing, $I_{d,0}^2$ equals $I_{b'd,0}^2$. For a double diode with the typical characteristics listed in Table 7.1, I^{S}_{eq} is comparable to $I_{b'd,0}$, the equivalent noise current characterizing the white noise for saturation biasing, for $f \approx 70$ cycles per second. Since I^{S}_{eq}, proportional to ψ, varies approximately as $1/f$, surface noise in reverse-biased diodes is small compared to the white noise over most of the audio region. This small, voltage-independent $1/f$ noise in diodes reverse-biased to saturation has only recently been identified experimentally.[8]

For sufficiently large reverse biases, p-n junctions pass noisy leakage currents in excess of their saturation currents. Leakage noise is represented in the diode schematic by a voltage-dependent $1/f$ noise-current generator in parallel with the white-noise $i_{b'd}\,|_V$ generator; that is, across the diode junction. In the early days of junction fabrication, leakage noise was large, even for reverse biases as small as 1 volt. Nowadays, for well-made junctions at reverse biases

$\lesssim 10$ volts, leakage noise is smaller than $1/f$ surface noise and small compared to the white noise in the audio region. In poorer diodes and in well-made diodes at larger reverse biases, leakage noise is large compared to $1/f$ surface noise and large compared to the white noise in the audio region.

1/f Surface Noise in Triodes

Surface noise in triodes, with the collector biased to saturation, is conveniently summarized with the aid of Fig. 7.6b and Eqs. 7.58 and 7.59. Surface noise appearing at the triode output will be expressed by its contribution ΔF^S to the triode noise factor. For either the common-emitter or the common-base connection, ΔF^S is given by

$$\Delta F^S = \cfrac{\cfrac{1}{4kTR_G\Delta f}}{\left[(R_G + r_{bb'}) \sum_{da} \Delta i_{b'e} \Big|_V + (R_G + r_{bb'})(1 + a) \sum_{da} \Delta i_{b'c} \Big|_V + \sum_{da} \Delta v_{bb'} \Big|_I \right]^2}$$

$$(7.65)$$

where R_G is the source resistance terminating the triode input and a is the quantity defined by Eq. 7.33. The terms involving a are small except at low emitter currents. If these terms and the $\dfrac{\partial R_{bb'}}{\partial G}\Big|_V$ term of $\displaystyle\sum_{da} \Delta v_{bb'} \Big|_I$ are neglected, Eq. 7.65 reduces to

$$\Delta F^S = \frac{e\,\psi(f)}{4V_T} R_G \left(1 + \frac{R_{bb',0}}{R_G}\right)^2 \int_{BS} (w_e + w_c)^2 (P_0 - P_n)^2 \, da \quad (7.66)$$

The summed probability $[w_e(\mathbf{r}) + w_c(\mathbf{r})]$ for a triode is identical with the probability $w_d(\mathbf{r})$ for its corresponding double diode.

To the approximation of the schematic shown in Fig. 7.3b, the triode output resistance is infinite for either the common-emitter or the common-base connection. Noise appearing at the input terminals is therefore independent of the load across the output terminals. For this reason, in considering noise appearing at the input terminals, it will be convenient to consider the triode as a two-terminal device whose two terminals *are* the input terminals.

Surface noise appearing at the triode input will be expressed by its contribution R^S_{eq} to the equivalent noise resistance of the input terminals considered as a two-terminal device. R^S_{eq} is given, for the common-emitter connection, by

$$R^S_{eq} = \frac{1}{4kT\,\Delta f}\overline{\left(\frac{V_T}{I_{eb',0}}\sum_{da}\Delta i_{b'e}\Big|_V + \frac{V_T}{I_{eb',0}}\sum_{da}\Delta i_{b'c}\Big|_V - \sum_{da}\Delta v_{bb'}\Big|_I\right)^2}$$

$$= \frac{e\,\psi(f)}{4V_T}\left(\frac{V_T}{I_{eb',0}} + I_{b',0}\frac{\partial R_{bb'}}{\partial I_{b'}}\right)^2\int_{\mathrm{BS}}w_d{}^2(\mathbf{r})(P_0 - P_n)^2\,da \quad (7.67)$$

and, for the common-base connection, by

$$R^S_{eq} = \frac{1}{4kT\,\Delta f}$$

$$\overline{\left[(r_{\mathrm{in}} - r_{bb'})\sum_{da}\Delta i_{b'e}\Big|_V - r_{bb'}\sum_{da}\Delta i_{b'c}\Big|_V - \sum_{da}\Delta v_{bb'}\Big|_I\right]^2}$$

$$= \frac{1}{4kT\,\Delta f}\overline{\left[r_{\mathrm{in}}\sum_{da}\Delta i_{b'e}\Big|_V - R_{bb',0}\sum_{da}\left(\Delta i_{b'e}\Big|_V + \Delta i_{b'c}\Big|_{V'}\right)\right]^2} \quad (7.68)$$

where r_{in}, the input resistance of the common-base triode, is given by

$$r_{\mathrm{in}} = \frac{V_T}{I_{e,0} + I_{b'e,0}} + \frac{r_{bb'}}{1 + \alpha_{cb}} \quad (7.69)$$

The integrals occurring in Eqs. 7.66, 7.67, and 7.68 extend over recombination surfaces on both sides of the base slab and are rather complicated. However, $(P_0 - P_n)$ differs from the pair-density distribution $P_{e,0}$ associated with the emitter electrode by, at most, an amount P_n. For pair injections at the emitter large compared to P_n, $(P_0 - P_n)$ may therefore be replaced by $P_{e,0}$. Approximated in this way, the integrals are readily carried out and equal $[P_n\exp(V_{eb',0}/V_T)]^2$ times factors dependent, in general, upon τ, $\bar{S}_0$, and the triode geometry. $P_n\exp(V_{eb',0}/V_T)$ will be expressed here in terms of the more easily measured transfer current $I_{ec,0} = |I_{c,0} + I_{b'c,0}|$.

For a triode with the particular geometry described in Section 1, page 352, the integral occurring in Eqs. 7.66 and 7.67 is given, for the normal connection, by

$$\int_{\mathrm{BS}}w_d{}^2(\mathbf{r})(P_0 - P_n)^2\,da \approx \int_{\mathrm{BS}}w_d{}^2(\mathbf{r})P_{1,0}{}^2\,da \approx \frac{2W^3 I_{ec,0}{}^2}{\pi^2 e^2 D^2 R_1{}^3} \quad (7.70)\dagger$$

† For meaning of subscripts 1,0 and 2,0, see Section 1, p. 352.

and, for the inverted connection, by

$$\int_{BS} w_d{}^2(\mathbf{r})(P_0 - P_n)^2 \, da \approx \int_{BS} w_d{}^2(\mathbf{r}) P_{2,0}{}^2 \, da$$

$$\approx \frac{W^2 I_{ec,0}{}^2}{\pi e^2 D^2 R_1{}^4} \left(R_2{}^2 - R_1{}^2 + R_2 L_2\right) \quad (7.71)$$

In the normal-triode case, the approximation of $(P_0 - P_n)$ by $P_{1,0}$ in Eq. 7.70 reduces the locations of the noise sources to a narrow ring-shaped surface around the periphery of the emitter junction. In the inverted-triode case, the locations of the noise sources still extend over large recombination surfaces on both sides of the base slab, just as in the diode case.

The three integrals occurring in Eq. 7.68 may be approximated similarly. For a triode with this same type of geometry operated normally in the common-base connection, surface noise appearing at the input proves to be described approximately by

$$R^s{}_{eq} \approx \frac{e \, \psi(f)}{4 V_T} \left(\frac{1}{2} r_{in}{}^2 - \frac{4}{3} r_{in} R_{bb',0} + R_{bb',0}{}^2 \right) \int_{BS} w_d{}^2(\mathbf{r}) P_{1,0}{}^2 \, da$$

$$(7.72)$$

The particular form of Eq. 7.72 arises from an arbitrary expression of two of the three integrals in Eq. 7.68 as multiples of the third.

Noise at the Triode Output

The solid curves in Fig. 7.9 shows ΔF^S at *1 cycle per second* versus emitter current $I_{e,0}$ and source resistance R_G, as computed from Eqs. 7.66 and 7.70 for a normal triode with the typical characteristics listed in Table 7.1.[†] For this high-current approximation, ΔF^S, for a given emitter current, is minimal for $R_G = R_{bb',0}$, though the minimum is broad.

If the a terms of Eq. 7.65 had been retained in the computation and if $(P_0 - P_n)$ had not been approximated by the pair-density distribution $P_{1,0}$, ΔF^S would be modified at low currents approximately as shown by the dotted curves in Fig. 7.9. The high-current approximation given by the solid curves is seen to be accurate down to

---

† For this computation, $R_{bb',0}$ was taken equal to 200 ohms, independent of base current.

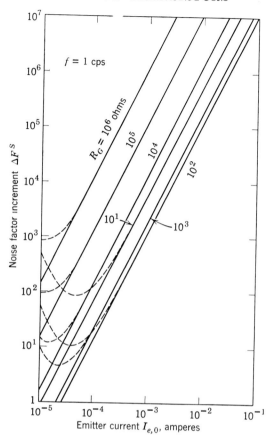

Fig. 7.9.   Surface noise at the output of a normal triode.

$I_{e,0} \approx 10^{-4}$ ampere.   Reduction of $I_{e,0}$ below this level would reduce ΔF^S very little.

Noise contributions from the generators $\sum_{da} \Delta i_{b'e} \big|_{V'}$, $\sum_{da} \Delta i_{b'c} \big|_{V'}$, and $\sum_{da} \Delta v_{bb'} \big|_{I}$ all interfere *constructively* at the triode output.   The contribution from $\sum_{da} \Delta v_{bb'} \big|_{I}$ is small, however.   Neglect of this generator would change the factor $[1 + (R_{bb',0}/R_G)]^2$ in Eq. 7.66 to $[1 + (r_{bb'}/R_G)]^2$. Since the difference between $r_{bb'}$ and $R_{bb',0}$ is ordinarily small compared

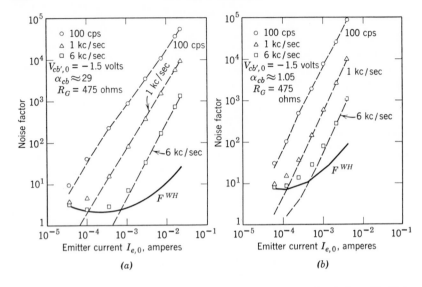

Fig. 7.10. (a) Noise observed at the output of a normal triode. (b) Noise observed at the output of an inverted triode.

to R_G, the base-conductivity-modulation effect, as represented by $\sum_{da} \Delta v_{bb'}\big|_{I'}$, ordinarily contributes negligibly to triode output noise.

ΔF^S at 1 cycle per sec is much larger than F^{WH} for nearly all $I_{e,0}$ and R_G. Compare Figs. 7.5 and 7.9. Since ΔF^S, proportional to ψ, varies approximately as $1/f$, this dominance passes away at higher frequencies. At $f = 10^3$ cycles per sec and $R_G = 10^3$ ohms, ΔF^S and F^{WH} are comparable for $I_{e,0} \approx 10^{-3}$ ampere.

Examples of normal- and inverted-triode output noise versus emitter current are shown in Fig. 7.10. Noise factors observed at 100, 1000, and 6000 cycles per sec are shown by circles, triangles, and squares, respectively. These measured values can be resolved into components characteristic of F^{WH} and ΔF^S, respectively. The solid curves show F^{WH} computed, according to Eq. 7.32, from observed values of $I_{b'c,0}$ and $r_{bb'}$ for these units. The dotted curves show, at each frequency, the excess noise $F - F^{WH}$. The excess noise varies approximately† as $1/f$ and as $I_{ec,0}^{3/2}$.

---

† For real germanium surfaces, ψ evidently decreases somewhat with increasing pair injection, since ΔF^S typically increases somewhat more slowly than the second power of the current.

Noise at the Triode Input

The curves in Fig. 7.11 show R^S_{eq} *at 1 cycle per sec* versus emitter current, as computed from Eqs. 7.67, 7.70, and 7.72 for a normal triode with the typical characteristics listed in Table 7.1.†

For the common-emitter connection, surface noise appearing at the triode input depends on current in a manner very similar to that of surface noise across the forward-biased diode. The minimum in triode R^S_{eq} at $I_{e,0} \approx 10^{-1}$ ampere results from interference between the $\sum_{da} \Delta v_{bb'}\Big|_I$ and $\left(\sum_{da} \Delta i_{b'e}\Big|_V + \sum_{da} \Delta i_{b'c}\Big|_V\right)$ generators. Because the normal-triode and double-diode quantities $I_{b',0} \dfrac{\partial R_{bb'}}{\partial I_{b'}}$ are approximately equal, this minimum and the corresponding minimum in double-diode noise occur at approximately equal device *base* currents. Compare Eqs. 7.60 and 7.67. If the $\dfrac{\partial R_{bb'}}{\partial G}\Big|_V$ term of $\sum_{da} \Delta v_{bb'}\Big|_I$ had been retained in the computation, the modification of R^S_{eq} for the high-α_{cb} triode

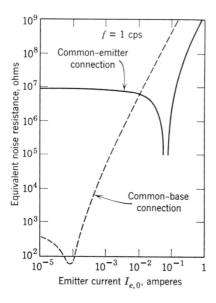

Fig. 7.11. Surface noise at the input of a normal triode.

† For this computation, $R_{bb',0}$ was taken equal to 200 ohms, independent of base current.

considered here would be much smaller than the corresponding modification for the double diode shown by the dotted curve in Fig. 7.7.

For the common-base connection, surface noise appearing at the triode input depends very differently on current. The base-conductivity-modulation effect, as represented by $\sum\limits_{da} \Delta v_{bb'}\big|_I$, contributes negligibly to this noise. The minimum in R^S_{eq} at $I_{e,0} \approx 10^{-4}$ ampere results from a feedback effect of $r_{bb'}$ and would be characteristic of any noise source represented by a generator in the same position as $\Delta i_{b'e}\big|_V$.

For either the common-emitter or the common-base connection, the correlation between triode output and input surface noise, as a function of emitter current, reverses sign when the input noise, as described by R^S_{eq}, passes through a minimum.

Examples of normal- and inverted-triode input noise versus emitter current, for the common-emitter connection, are shown in Fig. 7.12. Observed equivalent noise resistances, minus the small contributions expected from white noise, are shown by circles, triangles, and squares at 100, 1000, and 6000 cycles per sec, respectively. This excess noise varies approximately as $1/f$ and exhibits the dependence on current characteristic of surface noise: that is, the dependence exhibited by the solid curve in Fig. 7.11.

The inverted-triode example shows the rapid rise of R^S_{eq} just beyond the interference minimum. It is difficult to operate high-α_{cb} triodes at the large currents ($|I_{b',0}| \approx 10^{-3}$ ampere) required to observe this rise. An especially low-α_{cb} unit was chosen for the normal-triode example shown in Fig. 7.12 so that at least the beginnings of this rise could be detected.

Surface noise appearing at the triode output and surface noise appearing at the triode input should bear a definite relationship that is characteristic of the generator representation of the noise but independent of the empirical quantity ψ. For a triode in the common-emitter connection, this relationship is, according to Eqs. 7.66 and 7.67,

$$R^S_{eq} = \left(\frac{V_T}{I_{eb',0}} + I_{b',0}\frac{\partial R_{bb'}}{\partial I_{b'}}\right)^2 \frac{\Delta F^S}{R_G[1 + (R_{bb',0}/R_G)]^2} \tag{7.73}$$

The curves in Fig. 7.12 show expected values of R^S_{eq} computed, according to Eq. 7.33, from observed values of $V_T/I_{eb',0}$, $I_{b',0}(\partial R_{bb'}/\partial I_{b'})$, R_G, $R_{bb',0}$, and ΔF^S for these units. The observed and expected input noises agree very well.

The triode noise measurements just cited were made in the common-emitter connection. Montgomery[9] has measured noises at the triode

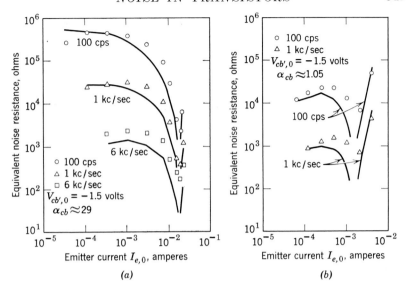

Fig. 7.12. (a) Excess noise observed at the input of a common-emitter normal triode. (b) Excess noise observed at the input of a common-emitter inverted triode.

output and input and their correlation for the common-base connection. It has been shown[6] that his observations are in harmony with expectations based on the noise models discussed above.

Properties of the Noisy-Pair-Generations Model

It has been shown that the surface-noise model is consistent with the $1/f$ noise observed in forward-biased diodes, with the $1/f$ noise observed at the triode output, with the $1/f$ noise observed at the triode input, and with the ratio of the last two. The extent that $1/f$ noise might be explained by alternative models will be considered in detail in this section and in the following one.

Now except at low pair-injection levels where $1/f$ noise is small and therefore difficult to measure, the surface noises expected in the diode and at the triode output and input depend upon the strength and locations of the assumed noisy point sources of pairs only through the factor $\sum_j w_d{}^2(\mathbf{r}_j)\overline{g^2(\mathbf{r}_j)}$, where $\overline{g^2(\mathbf{r}_j)}$, given by Eq. 7.46, describes the strength of the source at $\mathbf{r}_j$. See Eqs. 7.60, 7.66, 7.67, and 7.72. Thus, if some other distribution of noisy point sources were assumed, expected diode and triode output and input noises would depend upon the new source-strength distribution $\overline{g^2(\mathbf{r}_j)}$ only through the same factor. The *ratio* of noises expected at the triode output and input cancels this factor and is therefore independent of the assumed distribution $\overline{g^2(\mathbf{r}_j)}$.

It has been shown that no one-generator representation can account for the ratio of $1/f$ noises observed at the triode output and input.[6] The suitability for this purpose of the multigenerator representation of noisy-pair generations, even though the source-strength distribution $\overline{g^2(\mathbf{r}_j)}$ need not be known in detail, is actually quite striking, as evidenced in Fig. 7.12. The correctness of the noisy-pair-generations concept for $1/f$ noise is therefore probable.

If this concept is accepted, the observed dependences of diode and triode $1/f$ noise on current requires that $\overline{g^2(\mathbf{r}_j)}$ be proportional, except at very low currents, to $P_0{}^n(\mathbf{r}_j)$, where the exponent n is approximately 2 or somewhat smaller. That is,

$$\sum_j w_d{}^2(\mathbf{r}_j) \, \overline{g^2(\mathbf{r}_j)} \approx \frac{\Delta f}{f} \int_{\text{base}} w_d{}^2(\mathbf{r}) \, A(\mathbf{r}) \, P_0{}^2(\mathbf{r}) \, d\sigma \qquad (7.74)$$

where $A(\mathbf{r})(\Delta f/f) \, d\sigma$, for a unit d-c pair density $P_0(\mathbf{r})$, is the source strength in the element of integration $d\sigma$. The region of integration extends over all parts of the base, in general both surfaces and volumes, where $A(\mathbf{r})$ is not zero.

Since observed noises depend on $A(\mathbf{r})$ only through the integral in Eq. 7.74, one can gain experimental information on $A(\mathbf{r})$ only by varying the spatial distribution of the weighting factor $P_0{}^2(\mathbf{r})$ in this integral. The ratio of triode output and input $1/f$ noises failed to yield information on $A(\mathbf{r})$ because output and input noises responded to the same pair-density distribution $P_0(\mathbf{r})$. Three different pair-density distributions $P_0(\mathbf{r})$ have been discussed in this chapter: the distribution $P_{d,0}(\mathbf{r})$ which approximates $P_0(\mathbf{r})$ in the double-diode connection, and the distributions $P_{1,0}(\mathbf{r})$ and $P_{2,0}(\mathbf{r})$ which approximate $P_0(\mathbf{r})$ in the normal- and inverted-triode connections, respectively. The ratios of noises associated with these three distributions provide two conditions on the sought source distribution $A(\mathbf{r})$.

The Locations of Noisy-Pair Generations

The relationship between noises appearing at the input of the common-emitter inverted triode and across its corresponding forward-biased double diode is given by

$$\frac{R_{\text{eq,inv}}}{R_{\text{eq,di}}} \approx \frac{\left(\dfrac{V_T}{I_{eb',0}} + I_{b',0}\dfrac{\partial R_{bb'}}{\partial I_{b'}}\right)^2 \displaystyle\int_{\text{base}} w_d{}^2(\mathbf{r})\, A(\mathbf{r})\, P_{2,0}{}^2(\mathbf{r})\, d\sigma}{\left(\dfrac{V_T}{I_{db',0}} + I_{b',0}\dfrac{\partial R_{bb'}}{\partial I_{b'}}\right)^2 \displaystyle\int_{\text{base}} w_d{}^2(\mathbf{r})\, A(\mathbf{r})\, P_{d,0}{}^2(\mathbf{r})\, d\sigma} \tag{7.75}$$

See Eqs. 7.60, 7.67, and 7.71, where $A(\mathbf{r})$, for the special surface-noise model, was constant over the base recombination surface and zero elsewhere. For a device with the particular geometry described in Section 1, page 352, the pair-density distributions $P_{2,0}$ and $P_{d,0}$, the transfer currents $I_{eb',0}$ and $I_{db',0}$, and the inverted-triode and double-diode quantities $\partial R_{bb'}/\partial I_{b'}$ are practically identical, respectively, for equal device base currents.† Thus, for equal device base currents, $R_{\text{eq,inv}} \approx R_{\text{eq,di}}$ for any reasonable source-strength distribution $A(\mathbf{r})$.

While the relationship between noises associated with the pair-density distributions $P_{2,0}$ and $P_{d,0}$ therefore proves to yield no information on $A(\mathbf{r})$, it does provide a condition that must be satisfied by any noisy-pair-generations model. It has been found that the $1/f$ noise in transistors satisfies this condition. Figure 7.13 shows examples of double-diode and inverted-triode R_{eq} versus device base current. The $1/f$ components of R_{eq} observed at 100 cycles per sec are shown by dotted and solid curves for five units in the double-diode and common-emitter inverted-triode connections, respectively. For equal device base currents, the noises of corresponding units agree within experimental error.

The relationship between noises appearing at the output of the inverted triode and at the output of its corresponding normal triode is given, for equal device input terminations R_G, by

$$\frac{\Delta F_{\text{inv}}}{\Delta F_{\text{norm}}} \approx \frac{\displaystyle\int_{\text{base}} w_d{}^2(\mathbf{r})\, A(\mathbf{r})\, P_{2,0}{}^2(\mathbf{r})\, d\sigma}{\displaystyle\int_{\text{base}} w_d{}^2(\mathbf{r})\, A(\mathbf{r})\, P_{1,0}{}^2(\mathbf{r})\, d\sigma} \tag{7.76}$$

provided the difference between the inverted- and normal-triode quantities $R_{bb',0}$ is small compared to R_G. See Eqs. 7.66, 7.70, and 7.71. Since for a device with the particular geometry described in Section 1, page 352, the distributions $P_{2,0}$ and $P_{1,0}$ are very different, the ratio $\Delta F_{\text{inv}}/\Delta F_{\text{norm}}$ depends somewhat critically on the source-strength distribution $A(\mathbf{r})$.

---

† Since $\partial R_{bb'}/\partial I_{b'}$ measures the effect of the pair-density distribution $P_0(\mathbf{r})$ on $R_{bb'}$ for a unit change in base current, equality of the inverted-triode and double-diode pair-density distributions $P_{2,0}$ and $P_{d,0}$ implies equality of the inverted-triode and double-diode quantities $\partial R_{bb'}/\partial I_{b'}$.

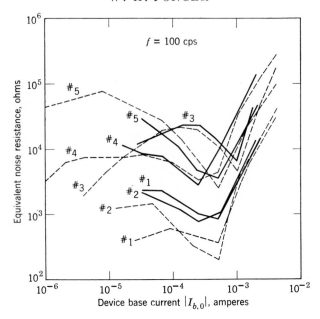

Fig. 7.13. Comparison of double-diode (dotted lines) and inverted-triode (solid lines) $1/f$ noise.

According to North's surface-noise model, $A(\mathbf{r})$ is approximately constant over base recombination surfaces and zero elsewhere. Thus, for this model and for equal device *collector* currents,

$$\frac{\Delta F^{S}_{\text{inv}}}{\Delta F^{S}_{\text{norm}}} \approx \frac{\int_{\text{BS}} w_d^{2}(\mathbf{r})\, P_{2,0}^{2}(\mathbf{r})\, da}{\int_{\text{BS}} w_d^{2}(\mathbf{r})\, P_{1,0}^{2}(\mathbf{r})\, da} \approx \frac{\pi}{2} \frac{R_2^{2} - R_1^{2} + R_2 L_2}{R_1 W} \quad (7.77)$$

If fluctuations in volume lifetime τ, as opposed to fluctuations in surface recombination velocity S, were responsible for $1/f$ noise, $A(\mathbf{r})$ would instead be approximately constant over the base volume. For this model and for equal device *collector* currents,

$$\frac{\Delta F^{V}_{\text{inv}}}{\Delta F^{V}_{\text{norm}}} \approx \frac{\int_{\text{BV}} w_d^{2}(\mathbf{r})\, P_{2,0}^{2}(\mathbf{r})\, dv}{\int_{\text{BV}} w_d^{2}(\mathbf{r})\, P_{1,0}^{2}(\mathbf{r})\, dv} \approx 3 \frac{R_2^{2} + R_2 L_2/2}{R_1^{2}} \quad (7.78)$$

If pairs were generated noisily at the junctions themselves, $A(\mathbf{r})$ would be approximately constant over the areas of the junctions and zero elsewhere. For this model and for equal device *collector* currents,

$$\frac{\Delta F^J_{\text{inv}}}{\Delta F^J_{\text{norm}}} \approx \frac{\int_J w_d^2(\mathbf{r}) \, P_{2,0}^2(\mathbf{r}) \, da}{\int_J w_d^2(\mathbf{r}) \, P_{1,0}^2(\mathbf{r}) \, da} \approx \frac{R_2^2}{R_1^2} \qquad (7.79)$$

where J is the area of both junctions.

Figure 7.14 shows examples of normal- and inverted-triode output noise versus collector current. The $1/f$ components of noise factors observed at 100 cycles per sec are shown by dotted and solid curves for five units in the inverted and normal connections, respectively.

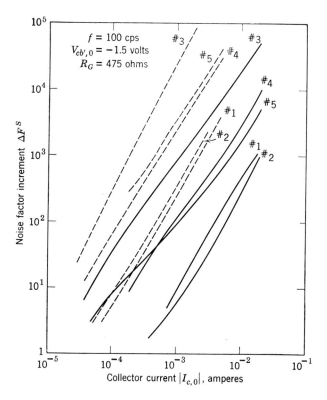

Fig. 7.14. Comparison of normal-triode (solid lines) and inverted-triode (dotted lines) $1/f$ noise.

For equal collector currents, the noises of corresponding inverted and normal triodes stand in the ratio ≈ 30. For these particular units, the ratios computed according to Eqs. 7.77, 7.78, and 7.79 are ≈ 60, ≈ 20, and ≈ 6, respectively. Noisy-pair generations at recombination surfaces or in the bulk are consistent with the observed ratio within expected errors.† Noisy-pair generations at the junctions themselves are not.

4. Design of Low-Noise Transistor Amplifiers

Noise Factor of a Multistage Amplifier

The noise factor of a multistage amplifier has been formulated quite generally.[10] A slight modification of this formulation will be followed here.

Let $F_j(R_G)$ be the noise factor of the jth stage when operated *singly* from a source resistance R_G. It is assumed that the relations $F_j(R_G)$ are known. Let the stages be a-c coupled as shown in Fig. 7.15a.

To the approximation of the schematic shown in Fig. 7.3b, the triode output resistance is infinite for either the common-emitter or the common-base connection. Thus R_j terminates the input of the jth stage, and the noise at the amplifier output may be considered as arising from uncorrelated noise-current generators i_j, as shown in Fig. 7.15b, described by

$$\overline{(i_j)^2} = F_j(R_j) \left(4kT \frac{1}{R_j} \Delta f \right) \tag{7.80}$$

Let A_j be the short-circuit current gain of the jth stage. Then currents at the amplifier output driven by the generator i_{j+1} could be equally well accounted for by a generator of strength i_{j+1}/A_j in parallel with the generator i_j. By repetitions of this idea, output currents driven by the generator i_{j+1} could be equally well accounted for by a generator of strength $i_{j+1}/(A_1 A_2 \cdots A_j)$ in parallel with the generator i_1. Thus the noise factor of the total amplifier is seen to be given by

---

† The ratio ≈ 60 computed according to the surface-noise model is known to be an upper limit. For the units studied here, the extent of the pair-density distribution $P_{1,0}$ along the base recombination surface just beyond R_1 was somewhat greater than would be computed from the idealized parallel-plane model described in Section 1, page 352. Thus, observed normal-triode base currents and surface noises should both be somewhat larger than computed ideally. This was indeed verified for the base currents.

$$F = \cfrac{1}{4kT\,\cfrac{1}{R_1}\,\Delta f}\left[\overline{(i_1)^2} + \frac{\overline{(i_2)^2}}{(A_1)^2} + \frac{\overline{(i_3)^2}}{(A_1 A_2)^2}\right.$$

$$\left. + \frac{\overline{(i_4)^2}}{(A_1 A_2 A_3)^2} + \cdots\right]$$

$$= F_1(R_1) + R_1\left[\frac{1}{(A_1)^2}\frac{F_2(R_2)}{R_2} + \frac{1}{(A_1 A_2)^2}\frac{F_3(R_3)}{R_3}\right.$$

$$\left. + \frac{1}{(A_1 A_2 A_3)^2}\frac{F_4(R_4)}{R_4} + \cdots\right]$$

(7.81)

If the stages are instead coupled by ideal transformers, as shown in Fig. 7.15c, R_j must be taken, in Eq. 7.81, as the output resistance of transformer T_j. The current gain A_j must be increased by the current gain of transformer T_{j+1}, the element added between the generators i_j and i_{j+1}.

Equation 7.81 is not intended as a basis for elaborate analytical minimization of F. Arrangements can be made such that noise contributions from the second and higher stages are all very small. The first-stage operating point that minimizes $F_1(R_1)$ therefore accurately approximates the first-stage operating point that minimizes F. The requirement that the $F_2(R_2)/R_2$ term of F be small then establishes limits for the operating point of the second stage, the requirement that the $F_3(R_3)/R_3$ term be small then establishes limits for the operating point of the third stage, and so on. These limits prove to be severe for, at most, the second stage.

Specialization of the Operating Conditions

Hereafter it will be assumed that transistors of all stages have the typical characteristics listed in Table 7.1 and, in order that the gains A_j be maximized, are operated as normal triodes in the common-emitter connection. $F_j(R_j)$ is given by

$$F_j(R_j) = F^{WH}(R_j) + \Delta F^S(R_j) \tag{7.82}$$

where F^{WH} and ΔF^S versus source resistance and emitter current are given by the curves shown in Figs. 7.5 and 7.9, respectively. The gain A_j, as computed with the aid of the schematic shown in Fig. 7.3b, is given by

$$A_j = -\alpha_{cb}\cfrac{R_j}{R_j + r_{bb'} + \cfrac{V_T}{I_{eb',0}}}\left(\frac{N_1}{N_2}\right)_{j+1} \tag{7.83}$$

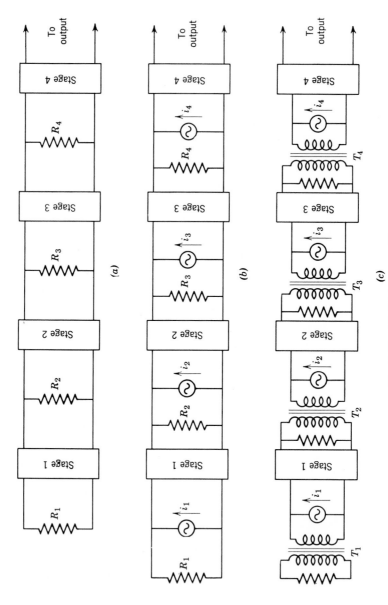

Fig. 7.15. Multistage amplifiers.

where $(N_1/N_2)_{j+1}$ is the current step-up ratio of transformer T_{j+1}. In order that these gains be large, the emitter currents should be large. ΔF^S, according to the high-current approximation of Eqs. 7.66 and 7.70, is given by[†]

$$\Delta F^S = \frac{\psi(f)}{2\pi^2 e V_T D^2} \left(\frac{W}{R_1}\right)^3 R_G \left(1 + \frac{R_{bb'0}}{R_G}\right)^2 I_{ec,0}{}^2 \qquad (7.84)$$

Thus, for large currents, the quantity $\Delta F^S(R_j)/R_j$ is minimized with respect to R_j if R_j is large compared to $R_{bb',0}$.

Suppose the amplifier noise is dominated by the white noise. $F_1{}^{WH}$ is minimized for a source resistance R_1 of $\approx 3 \times 10^3$ ohms and an emitter current $I_{e,0}$ of $\approx 10^{-4}$ ampere. For this first-stage operating point, it is readily verified that the gain A_1, even if no coupling transformer T_2 is used, is so large that noise from higher stages may be neglected. Thus $F_{\min} \approx (F_1{}^{WH})_{\min} \approx 1.25$. For the first-stage operating point just cited, $\Delta F_1{}^S \approx 50/f$; see Fig. 7.9. Thus the initial assumption that the white noise is somewhat larger than the $1/f$ noise is valid only if the frequency f is somewhat larger than ≈ 50 cycles per sec.

Suppose the amplifier noise is dominated by the $1/f$ noise. $\Delta F_1{}^S$, according to the curves shown in Fig. 7.9, is minimized for a source resistance R_1 of $\approx 10^3$ ohms and an emitter current $I_{e,0}$ of $\approx 3 \times 10^{-5}$ ampere. This first-stage operating point will therefore be assumed. The second stage can be operated at currents such that Eq. 7.84 accurately describes its $1/f$ noise. For R_2 large compared to $R_{bb',0}$, the noise contributed by this stage is smaller than that contributed by the first stage, according to Eq. 7.81 and the curves shown in Fig. 7.9, if the second-stage emitter current is smaller than $\approx 7(N_1/N_2)_2 \times 10^{-5}$ ampere, where $(N_1/N_2)_2$ is the current step-up ratio of coupling transformer T_2. It is readily verified that the gain A_1A_2, even if no coupling transformers T_2 and T_3 are used, can be arranged so large that noise from the third stage is negligible. Thus $F_{\min} \approx (\Delta F_1{}^S)_{\min} \approx 5/f$; see Fig. 7.9. For the first-stage operating point cited in this case, $F_1{}^{WH} \approx 2$; see Fig. 7.5. Thus the initial assumption that the $1/f$ noise is somewhat larger than the white noise is valid only if the frequency f is somewhat smaller than ≈ 2 cycles per sec.

In the region from ≈ 1 to ≈ 100 cycles per sec, both white and $1/f$ noises contribute importantly to the noise factor, and the problem is more complicated. However, since the operating points that mini-

[†] In Eq. 7.84, R_1 is the radius of the smaller transistor junction. See Fig. 7.2. The source resistance R_G corresponds to the resistance R_j occurring in Fig. 7.15.

mize the noises individually are not very different, the operating point that minimizes them together is restricted to a narrow range. Indeed, the optimum choice of the first-stage source resistance and emitter current varies only from $\approx 3 \times 10^3$ ohms and $\approx 10^{-4}$ ampere for complete domination of the noise by the white noise to $\approx 10^3$ ohms and $\approx 3 \times 10^{-5}$ ampere for complete domination by the $1/f$ noise.

Broad-Band Amplifiers

One is often interested in the noise performance of a broad-band amplifier. For this case, an average noise factor $\bar{F}$, as opposed to spot-noise factor F considered previously, has been introduced. $\bar{F}$ and F are related by

$$\bar{F} = \frac{\int_0^\infty F(f)\, G(f)\, df}{\int_0^\infty G(f)\, df} \tag{7.85}$$

where $G(f)$ describes the frequency dependence of amplifier power gain.

Noise performances of representative broad-band amplifiers are shown in Fig. 7.16. The two solid curves show $\bar{F}$ versus first-stage

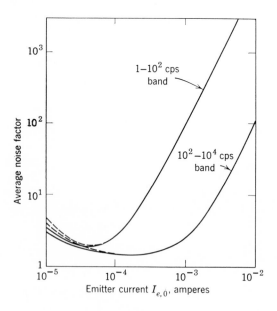

Fig. 7.16.   Noise performances of broad-band amplifiers.

emitter current, as computed from Eq. 7.85 and the curves shown in Figs. 7.5 and 7.9, for two amplifiers with rectangular pass bands extending from 1 to 10^2 and from 10^2 to 10^4 cycles per sec, respectively. For the computation of these curves, R_1 was taken as 10^3 ohms, and noise from stages higher than the first was neglected. If noise from higher stages is included, $\bar{F}$ increases at low currents to the values shown by the dotted curves in Fig. 7.16, for the choices $R_2 = 10^4$ ohms, $(N_1/N_2)_2 = 1$, and $I_{(e,0)2\text{nd stage}} = 2(I_{e,0})_{1\text{st stage}}$. For both amplifiers, the rapid increase of $\bar{F}$ with current at high emitter currents is due to $1/f$ noise.

Dependence of Triode Noise on Collector Voltage

As a junction j of an N-electrode device is reverse-biased to saturation, its associated pair-density distribution P_j becomes negligibly small, and, within the approximations followed in this chapter, the device performance becomes independent of the particular value of $V_{jb'}$. In particular, for a triode with collector reverse-biased to saturation, F^{WH} and ΔF^S are independent of the collector bias $V_{cb'}$.

For real triodes, the base thickness W depends slightly on collector voltage.[11] Since F^{WH} and ΔF^S depend on W, this effect promotes a slight dependence of F^{WH} and ΔF^S on d-c collector bias $V_{cb',0}$. For ΔF^S, this slight dependence has been verified quantitatively.[6] However, to a useful first approximation, F^{WH} and ΔF^S *are* independent of the collector voltage.

For sufficiently large reverse biases, triode collector junctions are potent sources of $1/f$ leakage noise. Leakage noise is represented in the schematic by a voltage-dependent $1/f$ noise-current generator in parallel with the white-noise $i_{b'c}|_V$ generator. This noise was responsible for the large noise factors of early junction transistors. Nowadays, for well-made transistors at collector biases $\lesssim 10$ volts, leakage noise is small compared to the white noise in the audio region.

If leakage noise *is* present in a particular unit, it can be reduced by lowering $V_{cb',0}$. Such a reduction forms the basis of Volkers and Pedersen's "hushed" transistor amplifier.[12] However, since transistors with negligible leakage noise are available in quantity, it is preferable to use these for low-noise applications.

Design of Transistors

Transistor design parameters that affect surface noise are placed in evidence by Eq. 7.84. ΔF^S is proportional to $\psi(f)(W/R_1)^3 1/D^2$. Thus the quantity ψ and base thickness W should be small, while the emitter radius R_1 and pair-diffusion constant D should be large. Detailed prescriptions for lowering ψ are presently unknown.

For present-day transistors, the white noise is already small. The slight improvements in F^{WH} that could be effected by lowering $r_{bb'}$ and $I_{b'c,0}$ and by raising α_{cb} (see Eq. 7.42) are unimportant. However, at frequencies comparable to the reciprocal of the time required for pairs to diffuse across the base, diffusion–recombination noise increases above the white-noise level characteristic of this noise at lower frequencies. Thus, for high-frequency applications, this diffusion time should be short.

5. Irradiation Noise

Bursts of Pairs Generated by Single Fast Particles

A fast particle passing through, or stopping in, a semiconductor commonly generates, either directly or through its fast secondaries, one or many electron-hole pairs. The average energy $\bar{w}$ lost by the particle for each pair produced is of the order of a few times the band-gap energy. $\bar{w}$ is ≈ 3.0 and ≈ 3.6 electron volts for germanium[13] and silicon,[14] respectively. The total energy ΔE lost by the particle is less than or equal to its initial energy if it passes through or stops in the semiconductor, respectively. For a relativistic particle of charge Ze traversing a path length λ in a material of density ρ,[15]

$$\Delta E \approx 2Z^2\rho\lambda \times 10^6 \text{ electron volts} \tag{7.86}$$

If the particle speed is somewhat smaller than the speed of light, ΔE is larger than this "minimum" energy loss.

The bursts of pairs generated by fast particles near p-n junctions may be detected by pulse techniques. Suppose a fast particle generates M_g pairs within a diffusion length L of an unbiased junction diode. Most of these pairs diffuse to the junction. Let the diode and its associated load be characterized by a small-signal resistance R and a small-signal capacitance C. The capacitance C is charged to a voltage $V \approx M_ge/C$ in a time $\approx L^2/D$ and is discharged in a time $\approx RC$. If M_ge is larger than $(kTC)^{1/2}$, the root-mean-square charge of thermal origin on a capacitance C in a network at thermal equilibrium, the pulse due to the fast particle is larger than the root-mean-square noise.

If this pulse is amplified by a broad-band amplifier with frequency limits f_1 and f_2, its observed rise time is the longer of L^2/D and $1/2\pi f_2$, and its observed decay time is the shorter of RC and $1/2\pi f_1$. The decay time should be somewhat longer than the rise time so that the full pulse amplitude is developed. The amplifier pass band should be no wider than necessary so that noise of amplifier origin is minimized.

Example of Detection of Single Fast Particles

For example, for a relativistic particle of charge number $Z = 1$ traversing a silicon diode with p-type base with a thickness W of 0.020 cm, ΔE is $\approx 2\rho W \times 10^6 \approx 2 \times 2.5 \times 0.02 \times 10^6 \approx 1.0 \times 10^5$ electron volts, and M_g is $\Delta E/\bar{w} \approx 2.8 \times 10^4$ pairs. Most of these pairs reach the junction, in a time of $\approx W^2/D \approx 13 \times 10^{-6}$ sec. If the resistance R and capacitance C of the diode and its associated load are 0.9×10^6 ohms and 44×10^{-12} farad, RC is $\approx 40 \times 10^{-6}$ sec. The root-mean-square number of electronic charges of thermal origin on the capacitance C is $(kTC)^{1/2}/e \approx 2.6 \times 10^3$, a factor of 10 smaller than M_g. If the amplifier pass band extends from 2 to 50 kc per sec, the pulse rise and decay times are not limited by the amplifier. For a tube amplifier with an average equivalent noise resistance of 3×10^3 ohms over this band, the equivalent root-mean-square noise voltage of amplifier origin, referred to the amplifier input, is $(4kTR_{eq}\,\Delta f)^{1/2} \approx 1.6 \times 10^{-6}$ volt. This is small compared to the pulse amplitude V of $M_g e/C \approx 100 \times 10^{-6}$ volt.

In summary, under these conditions, pulses due to fast particles have a rise time of $\approx 13 \times 10^{-6}$ sec, a decay time of 40×10^{-6} sec, and, if produced by relativistic particles of charge number $Z = 1$, should be ≈ 10 times larger than the root-mean-square noise. Figure 7.17 shows linear oscilloscope traces taken during a study[16] of pulses produced under these conditions by approximately relativistic β particles from a Y^{90} source ($E_{max} \approx 2.2$ Mev). Figure 7.17a shows several pulses along a slow time base; Fig. 7.17b shows the background noise in the absence of β-particle irradiation; Fig. 7.17c shows two pulses along a faster time base. The expected and observed pulse height, rise, and decay characteristics agree within expected errors.

Under these conditions and for fast particles arriving randomly in time at intensities greater than $\approx 10^4$ particles per sec,† the pulses caused by individual fast-particle incidences overlap, and a large Gaussian noise results. This "irradiation noise" is the principal subject of this section. A large flux of fast particles can produce a large irradiation noise even if the pulses due to single fast-particle incidences are small compared to the background noise.

McKay[13] has studied pair irradiation-generation in the barrier layers of semiconductor diodes. In this case, the generated charge is

---

† A flux of 10^4 relativistic particles per cm² sec of charge number $Z = 1$ corresponds to an ionization intensity of ≈ 1.5 roentgens per hour.

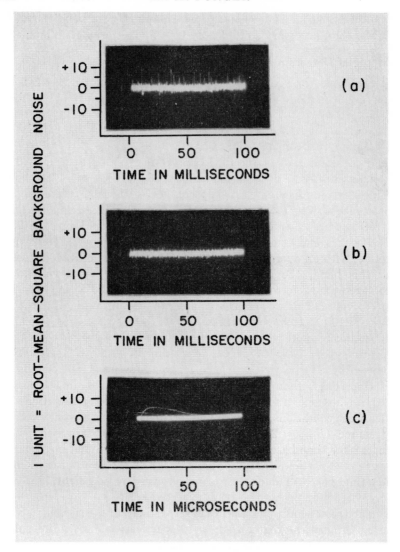

Fig. 7.17. Pulses due to beta particles.

swept onto the diode capacitance by the barrier field in a time shorter than 10^{-8} sec, and the rise time of the observed pulse is usually limited by the amplifier. If the amplifier is very fast, the pulse decay time, as determined by the diode RC time constant, may be chosen as short as 10^{-7} sec. Such pulses are much shorter than pulses whose rise times are limited by pair diffusion and, therefore, would be preferred in high-speed-counting applications. However, it is the slower diffusion-limited pulses discussed above that form the building blocks of irradiation noise in transistors and that therefore are of primary interest here.

Irradiation Noise in N-Electrode Devices

Irradiation noise in N-electrode devices has interesting conceptual similarities to both diffusion–recombination noise and $1/f$ surface noise. Like surface noise, it has to do with correlated generations of pairs in the device base, the correlated generations in this case being those resulting from single fast-particle incidences. Thus irradiation noise can be treated by a generalization of the point-source method used in Section 3 for surface noise and can be represented in the device schematic by a collection of noise-current generators $\Delta i_{b'j}|_V$ lying in the same positions as the $1/f$ surface noise generators. Like diffusion–recombination noise, irradiation noise has to do with random and uncorrelated creations, motions, and recombinations of pairs, though in this case the unit of uncorrelated creation is the burst of pairs resulting from a single fast-particle incidence, not the single pair itself. Thus the irradiation-noise generators $\Delta i_{b'j}|_V$ cited above are characterized by white frequency spectra. The uncorrelated noises associated with pair creations, motions, and recombinations will not be calculated separately. Instead, as in Section 2 for diffusion–recombination noise, well-known tube-noise theorems will be utilized.

Suppose fast particles are incident upon the base of an N-electrode device. Let t_n be the time of incidence of the nth particle and let $G(n, \mathbf{r})$ be the density distribution of *pair generation* resulting from this incidence. The point-source generation rate $\delta(\mathbf{r})\,G(t)$ of Section 3, page 367, must be replaced by the extended-source generation rate $\sum_n G(n, \mathbf{r})\,\delta(t_n)$. The transfer-current increments $\Delta I_{b'j}$ corresponding to this generation are described by

$$\Delta I_{b'j} = e \sum_n M_j(n)\,\delta(t_n) \tag{7.87}$$

where $M_j(n)$ is the number of pairs excited by the nth incidence *that reach the jth junction*. These $M_j(n)$ pairs arrive at junction j over an interval comparable to a pair-diffusion time. However, at frequencies low compared to the reciprocal of an average pair-diffusion time, the delta function $\delta(t_n)$ in Eq. 7.87 accurately describes their arrival in time.

In the device current–voltage characteristics given by Eqs. 7.10 and 7.13, the transfer currents $I_{b'j}$ and $i_{b'j}|_V$ must be everywhere increased by increments $\Delta I_{b'j}$ and $\Delta i_{b'j}|_V$ respectively,† where $\Delta i_{b'j}|_V$ is the small-signal variation in $\Delta I_{b'j}$ when the voltage $V_{jb'}$ is held constant. While irradiation of the device augments the expressions for the electrode currents I_j and introduces new current generators $\Delta i_{b'j}|_V$ in the small-signal schematic, it does not change the expressions for the old elements in the schematic, provided these are stated in terms of the transfer currents I_{jk}.

The currents $\Delta i_{b'j}|_V$ contain two components: first, signal components induced by the signal content of the incident radiation, and, second, noise components caused by fluctuations in the incidence rate of fast particles and by fluctuations associated with the probabilistic nature of both pair irradiation–generation and pair diffusion to the junctions.

Associated with the irradiation–generation of pairs in the base, there is a base-conductivity-modulation effect that is represented in the small-signal schematic by a voltage generator in the base-lead arm. See Section 3, page 369. However, since the direct base current $I_{b',0}$ is usually small in irradiated devices, this generator is usually small and will be neglected here.

The Noise-Current Generators $\Delta i_{b'j}|_V$

If the fast particles arrive randomly in time, the description of the resulting noise is effected by the familiar analysis of noise in multiplication by secondary emission.[20]

Let N_0 be the average number of fast-particle incidences per unit time. Let $w(M_j, M_k)$ be the probability that an incidence results in exactly M_j pairs reaching the junction j, and exactly M_k pairs reaching the junction k. Let $\Delta I_{b'j}(M_j, M_k)$ and $\Delta I_{b'k}(M_k, M_j)$ be the contributions of this particular class of incidences to the transfer currents $\Delta I_{b'j}$ and $\Delta I_{b'k}$, respectively.

Since incidences of this class occur randomly in time, fluctuations in the currents $\Delta I_{b'j}(M_j, M_k)$ are described, according to the first theorem of Section 2, page 357, by

$$\overline{[\Delta i_{b'j}(M_j, M_k)|_V]^2} = 2(M_j e)^2 [w(M_j, M_k)N_0] \, \Delta f \qquad (7.88)$$

Now since

---

† For such a representation of the p-n junction photocell, see J. N. Shive.[17] For such a representation of the p-n junction β-ray cell, see W. G. Pfann and W. van Roosbroeck,[18] also, P. Rappaport, J. J. Loferski, and E. G. Linder.[19]

$$\frac{\Delta I_{b'j}(M_j, M_k)}{M_j} = \frac{\Delta I_{b'k}(M_k, M_j)}{M_k} \qquad (7.89)$$

fluctuations in the cross product of the currents $\Delta I_{b'j}(M_j, M_k)$ and $\Delta I_{b'k}(M_k, M_j)$ are described by

$$\overline{\Delta i_{b'j}(M_j, M_k)\big|_V \, \Delta i_{b'k}(M_k, M_j)\big|_V} = M_j M_k \, w(M_j, M_k) 2e^2 N_0 \, \Delta f \qquad (7.90)$$

The total transfer current $\Delta I_{b'j}$ is given by

$$\Delta I_{b'j} = \sum_{M_j} \sum_{M_k} \Delta I_{b'j}(M_j, M_k) \qquad (7.91)$$

Since the individual currents making up this sum are all uncorrelated, fluctuations in the product of the transfer currents $\Delta I_{b'j}$ and $\Delta I_{b'k}$ are described by

$$\overline{\Delta i_{b'j}\big|_V \Delta i_{b'k}\big|_V} = \sum_{M_j} \sum_{M_k} \overline{\Delta i_{b'j}(M_j, M_k)\big|_V \, \Delta i_{b'k}(M_k, M_j)\big|_V}$$

$$= \overline{M_j M_k} 2e^2 N_0 \, \Delta f \qquad (7.92)$$

where $\overline{M_j M_k}$ is the mean value of $M_j M_k$ obtained by averaging this product over all incidences. N_0 is related to the d-c components of the transfer currents by the formulas

$$\Delta I_{b'j,0} = \overline{M_j} e N_0 \qquad (7.93)$$

where $\overline{M_j}$ is the mean value of M_j.

If electron-hole pairs are generated singly (this occurs, for example, if the incident particles are photons whose energies lie in the fundamental absorption band of the base semiconductor), the transfer currents $\Delta I_{b'j}$ meet the requirements of the second theorem of Section 2, page 357. Thus their fluctuations are described by

$$\overline{(\Delta i_{b'j}\big|_V)^2} = 2e \, \Delta I_{b'j,0} \, \Delta f \qquad (7.94)$$

and are all uncorrelated. Equations 7.94 may also be recognized as a special case of Eqs. 7.92. For, if pairs are generated singly, $M_j = 1$ or 0. Thus $M_j^2 = M_j$, and $\overline{M_j^2}/\overline{M_j} = 1$. Moreover, if $M_j = 1$, $M_k = 0$ for $k \neq j$. Thus $\overline{M_j M_k} = 0$.

The quantities $\overline{M_j M_k}$ are more easily conceived than computed, as is also true of multiplication by secondary emission. Simple approximate values will prove useful, however, in the particular examples considered below.

Irradiation Noise in Diodes

Irradiation noise in diodes is conveniently summarized with the aid of Fig. 7.6a and Eq. 7.92. The base-conductivity-modulation effect, as represented by the generator $\Delta v_{bb'}|_I$ in Fig. 7.6, will be neglected here. The irradiation-noise contribution to the diode equivalent noise resistance is

$$R^R_{\text{eq}} = \frac{1}{4kT\,\Delta f}\left(\frac{V_T}{I_{db',0}}\right)^2 \overline{(\Delta i_{b'd}|_V)^2} \tag{7.95}$$

where

$$\overline{(\Delta i_{b'd}|_V)^2} = \overline{M_d^2}\,2e^2 N_0\,\Delta f = \left(\frac{\overline{M_d^2}}{\overline{M_d}}\right)2e\,\Delta I_{b'd,0}\,\Delta f \tag{7.96}$$

Noise in the reverse-biased diode is more conveniently described by the equivalent noise current of a fictitious noise-current generator across its terminals. For saturation biasing, I^R_{eq} is equal to $(\overline{M_d^2}/\overline{M_d})\,\Delta I_{b'd,0}$. This relationship follows directly from Fig. 7.6a and Eq. 7.92.

If the pairs are generated close to the junction, most of them reach the junction. Thus $(\overline{M_d^2})^{1/2} \approx \overline{M_d} \approx \overline{M_g}$, where $\overline{M_g}$ is the average number of pairs generated for each fast-particle incidence. Since $\overline{M_g}$ may be large, irradiation noise may be large. This large noise has been observed by Loferski et al.[16] for germanium and silicon diodes bombarded by high-energy electrons. In their experiments, the value of $\overline{M_g}$, estimated via Eq. 7.96 from the observed noise and direct current $\Delta I_{b'd,0}$, was $\approx 10^5$ and was consistent with the value estimated previously[21] via Eq. 7.93 from the observed incident electron current eN_0.

For diodes irradiated by photons in the fundamental absorption band of the base semiconductor, $\overline{M_d^2}/\overline{M_d} = 1$. In particular, for reverse saturation biasing, I^R_{eq} is then equal to $\Delta I_{b'd,0}$, and the equivalent noise current characterizing both the diffusion–recombination noise and the photo noise is equal to the diode direct reverse current $|I_{d,0}| = I_{b'd,0} + \Delta I_{b'd,0}$. This relationship has been verified for low-leakage photodiodes by Slocum and Shive,[22] and, extremely accurately, by Pearson et al.[23]

Irradiation noise in forward-biased diodes is complicated by the simultaneous presence of $1/f$ surface noise.

The noise-current generators characterizing surface noise are proportional to integrals of the form $\psi(f)\int_{\text{BS}} w_j^2(\mathbf{r})(P_0 - P_n)^2\,da$, where $(P_0 - P_n)$ is the excess d-c pair density. Large excess pair densities can be excited by irradiation–generation as well as by injection at emitters. Such an excitation of surface noise would account for the principal features of the large $1/f$ noise observed by Gianola[24] in forward-biased photodiodes, provided the quantity ψ decreased somewhat with increasing pair density for his units. The corresponding $1/f$ noise excited in forward-biased diodes by high-energy electrons is less important, since in this case the white noise excited by the irradiation is very large.

Irradiation Noise in Triodes

Irradiation noise in triodes, with the collector biased to saturation, is conveniently summarized with the aid of Fig. 7.6b and Eq. 7.92. The generator $\Delta v_{bb'}|_I$ will again be neglected. For either the common-emitter or the common-base connection, the irradiation–noise contribution to the triode noise factor is

$$\Delta F^R = \frac{1}{4kT \, \Delta f} R_G \left(1 + \frac{r_{bb'}}{R_G}\right)^2 \overline{[\Delta i_{b'e}|_V + (1 + a) \, \Delta i_{b'c}|_V]^2} \quad (7.97)$$

where a is the quantity defined by Eq. 7.33. The terms involving a are small except at low emitter currents. If these terms are neglected, Eq. 7.97 reduces, according to Eq. 7.92, to

$$\Delta F^R = \overline{(M_e + M_c)^2} \, \frac{eN_0 R_G}{2V_T} \left(1 + \frac{r_{bb'}}{R_G}\right)^2 \quad (7.98)$$

Triodes are not commonly used to detect radiations, and irradiation noise in such devices will be viewed as a nuisance here. Since a good triode can be operated such that its noise factor is approximately unity for $R_G \approx 10^3$ ohms, it is desirable that ΔF^R, for a triode exposed to radiations, be less than unity for this value of source resistance.

If the pairs are generated close to the junctions, most of them reach the junctions. Thus $\overline{(M_e + M_c)^2} \approx (\overline{M_g})^2$, where $\overline{M_g}$ is the average number of pairs generated for each fast-particle incidence. Low-energy photons incident upon a triode are normally excluded from the base by the casing of the triode. For incident relativistic particles of charge Ze, $\overline{M_g} = \Delta E / \bar{w}$, where the energy loss ΔE is given by Eq. 7.86. In general, the path length λ occurring in Eq. 7.86 is of the order of the triode base thickness W. Thus, for a germanium triode with the typical characteristics listed in Table 7.1, $\overline{M_g}$ is $\approx (2Z^2 \times 5.4 \times 0.004 \times 10^6)/3.0 \approx 1.4Z^2 \times 10^4$ pairs, and, for $R_G = 10^3$ ohms, ΔF^R is $\approx Z^4 N_0 \times 10^{-6}$. Thus irradiation noise is negligible if $Z^4 N_0$, the weighted number of fast particles per unit time traversing the triode base within a diffusion length of the junctions, is less than $\approx 10^6$ sec^{-1}. For a triode with the particular geometry described in Section 1, page 352, N_0 would be of the order of the fast-particle intensity per unit area multiplied by the area $\pi R_2{}^2$ of the larger junction. Cosmic-ray intensities are of the order of 10^{-1} particles per cm^2 sec. However, particle intensities greater than 10^6 particles per cm^2 sec can be realized near man-made sources.

Single Pulses in Triodes

It is possible that pulses associated with single fast-particle incidences are large even if the averaged noise is small. However, it will now be shown that such pulses are small compared to other triode noises unless the charge number Z of the fast particles is large.

The characteristics of individual pulses depend upon capacitive effects which would not be explainable from the simplified triode schematic shown in Fig. 7.3b. Emitter-base and collector-base capacitances $C_{b'e}$ and $C_{b'c}$, respectively, must be added to this schematic.[25] The generated charge resulting from single fast-particle incidences collects on these two capacitances in a time $\approx W^2/D$, where W is the triode base thickness, and D is the pair-diffusion constant. The initial division of charge between the two capacitances depends on the particular trajectory of the fast particle. However, a normal-co-ordinate solution of the charge-decay problem shows that, regardless of the initial charge division, the collected charge is rapidly reapportioned between the two capacitances, and then both capacitances discharge with the same time constant $\tau = (1/a)[(C_{b'e}V_T/I_{ec,0}) + R_L C_{b'c}]$, where R_L is the triode load resistance, and a is the dimensionless quantity defined in Eq. 7.33.

The average number of fast-particle incidences near the junctions during an interval τ is $N_0\tau$, and the root-mean-square fluctuation from this number is $(N_0\tau)$. If N^*_0 is defined as the value of N_0 for which $\Delta F^R = 1$, then the number of superimposed single pulses required to match the amplitude of the root-mean-square background noise is approximately $(N^*_0\tau)^{1/2}$. Thus the amplitude of a single pulse relative to the background noise is $\approx (N^*_0\tau)^{-1/2}$.

For a germanium triode with the typical characteristics listed in Table 7.1, N^*_0, for irradiation by relativistic particles of charge number $Z = 1$, is $\approx 10^6$ particles per sec. A typical value of the decay constant τ is $\approx 10^{-5}$ sec. Thus pulses due to single incidences of singly charged particles are of the order of one third the amplitude of the root-mean-square background noise.

Pulses due to fast particles of large charge number Z or due to high-energy nuclear disintegrations in the base semiconductor are much larger than pulses due to singly charged fast particles and are therefore larger than the background noise. However, such pulses occur infrequently. Even at the top of the atmosphere, the cosmic-ray intensity of particles with charge number $Z > 2$ is $\approx 10^{-2}$ particle per cm^2 sec,[26] and the rate of cosmic-ray-induced nuclear disintegrations in germanium is $\approx 10^{-2}$ disintegration per cm^3 per sec.†[27] Thus these larger pulses occur at rates of the order of a few per day.

Summary

The basic N-electrode device consists of $(N - 1)$ heavily doped p-type electrodes attached to an approximately equipotential N-type base region b'. $(N^2 - N)$ transfer currents I_{jk} $(k \neq j)$ are introduced, where I_{jk} is the electric current associated with the diffusion of pairs *from* the electrode j *to* the electrode k. Under conditions approximated in real devices, the small-signal representation of the device consists of the following elements:

1. *Signal elements.* Every junction electrode j is connected to the b' electrode by the conductance $\dfrac{I_{jb',0}}{V_T}$ and to every other junction electrode k by the transconductance generator $\left(\dfrac{I_{jk,0}}{V_T} v_{jb'} - \dfrac{I_{kj,0}}{V_T} v_{kb'} \right)$, where $v_{jb'}$ is the small-signal voltage of the jth junction electrode relative to the b' region.

2. *Diffusion–recombination-noise elements.* Every electrode j is connected to every other electrode k by the noise-current generator $i_{jk}|_V - i_{kj}|_V$. Fluctuations in the currents $i_{jk}|_V$ are described by

$$\overline{(i_{jk}|_V)^2} = 2eI_{jk,0}\,\Delta f$$

and are all uncorrelated.

3. *$1/f$-surface-noise elements.* The b' electrode is connected to every junction electrode j by the noise-current generator $\displaystyle\sum_{da} \Delta i_{b'j}\Big|_V$. Fluctuations in the currents $\displaystyle\sum_{da} \Delta i_{b'j}\Big|_V$ are described by

$$\left(\sum_{da} \Delta i_{b'j}\Big|_V \right)\left(\sum_{da} \Delta i_{b'k}\Big|_V \right) = e^2\,\psi(f)\,\Delta f \int_{BS} w_j(\mathbf{r})\,w_k(\mathbf{r})(P_0 - P_n)^2\,da$$

† The disintegration rate quoted above was measured for silver bromide emulsion. However, the disintegration rate in germanium would be approximately the same.

where BS is the base recombination surface, $(P_0 - P_n)$ the local excess d-c pair density, $w_j(\mathbf{r})$ the probability that a pair generated at $\mathbf{r}$ will reach the junction j, and $\psi(f)$ a quantity characterizing the excess noisiness of the surface. For etched germanium surfaces, $\psi(f) \approx 10^{-7 \pm 1}/f$ cm^4 per sec.

4. *Irradiation-noise elements.* The b' electrode is connected to every junction electrode j by the noise-current generator $\Delta i_{b'j}|_V$. Fluctuations in the currents $\Delta i_{b'j}|_V$ are described by

$$\overline{\Delta i_{b'j}|_V \, \Delta i_{b'k}|_V} = \overline{M_j M_k} 2e^2 N_0 \, \Delta f$$

where N_0 is the average number of fast-particle incidences per unit time and $\overline{M_j M_k}$ is the average product of the numbers of pairs reaching the junctions j and k for each fast-particle incidence. In the case that electron-hole pairs are generated singly (for example, in photo-irradiation), fluctuations in the currents $\Delta i_{b'j}|_V$ are described by the simpler formulas

$$\overline{(\Delta i_{b'j}|_V)^2} = 2e \, \Delta I_{b'j,0} \, \Delta f$$

and are all uncorrelated.

The generators representing $1/f$ surface noise, irradiation noise, and that portion of diffusion–recombination noise associated with the transfer currents $I_{b'j}$ share common positions in the device schematic. This situation obtains because these noises are all associated with noisy-pair generations in the device base.

Thermal noise in the base-lead resistance $r_{bb'}$ is represented by a noise-voltage generator $v_{bb'}|_I$ in series with $r_{bb'}$. Fluctuations in the voltage $v_{bb'}|_I$ are described by

$$\overline{(v_{bb'}|_I)^2} = 4kTr_{bb'} \, \Delta f$$

Base-conductivity-modulation noise is represented by an additional noise-voltage generator $\Delta v_{bb'}|_I$ in series with $r_{bb'}$ that increases rapidly with direct base current. The noise contributed by this generator at the triode output is negligibly small, though the corresponding contributions at the triode input and across the diode are large for large device base currents.

The triode noise factor F is given by

$$F = F^{WH} + \Delta F^S + \Delta F^R$$

where F^{WH}, ΔF^S, and ΔF^R, the contributions of thermal and diffusion–recombination noises, of $1/f$ surface noise, and of irradiation noise, respectively, to F are given by Eqs. 7.32, 7.84, and 7.98, respectively.

For transistors with the typical characteristics listed in Table 7.1, the white-noise contribution F^{WH} is minimized at ≈ 1.25 for a source resistance R_G of $\approx 3 \times 10^3$ ohms and for an emitter current $I_{e,0}$ of $\approx 10^{-4}$ ampere. The $1/f$-surface-noise contribution ΔF^S is minimized for a source resistance R_G of $\approx 10^3$ ohms and for an emitter current $I_{e,0}$ of $\approx 3 \times 10^{-5}$ ampere. In general, ΔF^S can be made smaller than unity in the audio region. The irradiation-noise contribution ΔF^R is negligibly small unless the triode is exposed to strong artificial sources of radiations.

Triodes with poor collectors exhibit an additional large $1/f$ leakage noise. Such units should not be used as low-noise amplifiers.

The contents of this chapter have been strongly influenced by the writer's association in noise researches with Drs. D. O. North and L. J. Giacoletto. Drs. Giacoletto and L. S. Nergaard read the manuscript and made numerous useful suggestions which have been incorporated into the text.

REFERENCES

1. W. Shockley, (a) *Bell System Tech. J.*, **28**, 435 (1949); (b) *Electrons and Holes in Semiconductors*, D. Van Nostrand Co., New York, (1950).
2. R. L. Anderson and A. van der Ziel, *Trans. IRE PGED*-1, 20 (1952).
3. L. J. Giacoletto, IRE–AIEE Conf. on Semiconductor Device Research, University of Minnesota (1954); also *Transistors I*, p. 296, RCA Laboratories, Princeton, N. J. (1956).
4. D. O. North, IRE–AIEE Conf. on Semiconductor Device Research, University of Pennsylvania (1955); also *Phys. Rev.*, to be published.
5. W. Guggenbühl and M. J. O. Strutt, *Arch. Elektr. Übertragung* **9**, 259 (1955).
6. W. H. Fonger, IRE–AIEE Conf. on Semiconductor Device Research, University of Pennsylvania (1955); also *Transistors I.*, p. 239, RCA Laboratories, Princeton, N. J. (1956).
7. L. J. Giacoletto, *RCA Rev.*, **14**, 269 (1953).
8. W. H. Fonger, to be published.
9. H. C. Montgomery, *Proc. IRE*, **40**, 1461 (1952).
10. IRE Standards on Electron Devices, "Methods of Measuring Noise," *Proc. IRE*, **41**, 890 (1953).
11. J. M. Early, *Proc. IRE*, **40**, 1401 (1952).
12. W. K. Volkers and N. E. Pedersen, *Tele-Tech*, **14**, 82 (1955).
13. K. G. McKay, *Phys. Rev.*, **84**, 829 (1951).
14. K. G. McKay and K. B. McAfee, *Phys. Rev.*, **91**, 1079 (1953).
15. B. Rossi and K. Greisen, *Revs. Mod. Phys.*, **13**, 240, Fig. 2 (1941).
16. W. H. Fonger, J. J. Loferski, and P. Rappaport, *J. Appl. Phys.* **29**, 588 (1958).
17. J. N. Shive, *Proc. IRE*, **40**, 1410 (1952).
18. W. G. Pfann and W. van Roosbroeck, *J. Appl. Phys.*, **25**, 1422 (1954).
19. P. Rappaport, J. J. Loferski, and E. G. Linder, *RCA Rev.*, **17**, 100, (1956).

20. A. van der Ziel, *Noise*, p. 113, Prentice-Hall, New York (1954).

21. P. Rappaport, *Phys. Rev.*, **93**, 246 (1954).

22. A. Slocum and J. N. Shive, *J. Appl. Phys.*, **25**, 406 (1954).

23. G. L. Pearson, H. C. Montgomery, and W. L. Feldmann, *J. Appl. Phys.*, **27**, 91 (1956).

24. U. F. Gianola, *J. Appl. Phys.*, **27**, 51 (1956).

25. L. J. Giacoletto, *RCA Rev.*, **15**, 506, Fig. 24 (1954).

26. M. F. Kaplon, B. Peters, H. L. Reynolds, and D. M. Ritson, *Phys. Rev.*, **85**, 295 (1952).

27. U. Camerini, T. Coor, J. H. Davies, P. H. Fowler, W. O. Lock, H. Muirhead, and N. Tobin, *Phil. Mag. London*, **40**, 1073 (1949).

Index